Advanced Calculus

HARPERCOLLINS COLLEGE OUTLINE

Advanced Calculus

Paul Duchateau, Ph.D.
Colorado State University

HarperPerennial
A Division of HarperCollins*Publishers*

An American BookWorks Corporation Production

Project Manager: William R. Hamill
Editor: Robert A. Weinstein

Library of Congress Catalog Card Number: 91-58266
ISBN: 0-06-467139-9

92 93 94 95 96 ABW/RRD 10 9 8 7 6 5 4 3 2 1

Contents

Preface

A course in advanced calculus has been a part of the undergraduate mathematics curriculum for many years, and over a period of time at least two separate courses of advanced calculus courses have evolved. One of these courses view elementary calculus from an advanced standpoint and thus serves as an introduction to more advanced real analysis. These courses serve as an introduction to reading and writing proofs and frequently emphasize single variable calculus. Another common approach to advanced calculus concentrates on those topics which are generally slighted or omitted entirely from beginning courses in calculus. Often these courses emphasize topics from multivariable calculus.

This book contains sufficient material that it could serve as a text for either type of course. It may be used as a supplementary text for a course in elementary real analysis, or it may serve as a reference illustrating some of the more advanced applications of calculus. The numerous solved problems make it particularly suitable to a program of self study.

The book has been divided into three parts:

I. Real Valued Functions of One Real Variable
II. Real Valued Functions of Several Real Variables
III. Calculus of Vector Valued Functions

Part I contains seven chapters providing a standard treatment of advanced calculus of functions of one real variable, the material which forms the core of any introduction to real analysis. This part of the text emphasizes proofs and abstract concepts.

Multivariable calculus, the differentiation and integration of real valued functions of several real variables is the subject of Part II of the book. Chapter 8 presents the theory and various applications of partial differentiation including implicit and inverse function theorems and the several variable Taylor's theorem. Multiple integrals, including discussions of change of variables and evaluation in terms of iterated integrals, is the content of chapter 9.

The calculus of vector valued functions of several real variables is covered in Part III. Definitions and identities for vector differential operators, including maximum-minimum principles for the Laplacian, are found in chapter 10, while chapter 11 is devoted to the vector integral calculus; topics here include line and surface integrals, the divergence theorem and the theorems of Green and Stokes. Parts II and III of this book include the advanced material usually not covered in introductory courses in calculus. The solved problems include a mixture of proofs and advanced applications.

By emphasizing Part I of this book, one obtains an excellent introduction to real analysis while a course focusing on advanced topics in calculus would be the result of stressing parts II and III. All three parts could be covered in a two-semester treatment of single and multi-variable advanced calculus.

The solved problems include numerous proofs and illustrative examples. Chapters 1 through 7 include several supplementary problems that can be used for purposes of review and to test the student's comprehension of the material.

Finally, I would like to express my appreciation to my family for their continued support and I would like to thank Bill Hamill for his hard work in helping to produce this text.

1

The Real Numbers

In a beginning course in calculus the emphasis is on introducing the techniques of the subject. An advanced calculus course builds on this foundation, extending the techniques and reexamining the fundamentals in a more rigorous light. A rigorous treatment of calculus requires more than just a heuristic understanding of the real numbers. On the other hand, a systematic construction of the real numbers is perhaps more than is required at this level. Therefore, in this chapter, we shall present a set of axioms which are sufficient to imply all relevent properties of the real numbers. These axioms describe a set of abstract objects whose properties include those of the real numbers of our experience. However, the axioms allow us to deduce a much more detailed portrait of the real numbers than is available as the result of computational experience.

For computational purposes we represent each real number as a decimal expansion and for puposes of visualization, we often find it convenient to think of the real numbers as being in one to one correspondence with the points of a line. We will consider special subsets of the real numbers including the natural numbers, integers and rational numbers.

The set of real numbers consists of rational and irrational numbers and we will use the axioms to show that each of these sets of numbers is everywhere dense in the reals. A set of numbers is said to be everywhere dense in the reals if a number from the set can be found between any two reals. Additional properties of the real numbers developed from the axioms in this chapter include:

- *PRINCIPLE OF MATHEMATICAL INDUCTION,*
- *MONOTONE SEQUENCE THEOREM (Theorem 1.7)*
- *NESTED INTERVAL THEOREM (Theorem 1.8)*
- *BOLZANO-WEIERSTRASS THEOREM (Theorem 1.10)*

Each of these results plays an important role in subsequent chapters in providing a rigorous treatment of the calculus of functions of one variable.

Terminology

SETS

We shall speak frequently in this chapter about sets of real numbers. A set of real numbers is just a collection of real numbers. The numbers in the set are referred to as the elements of the set. A set of numbers may be defined by listing the elements of the set or by specifying conditions for membership in the set. If M is a set, all of whose elements are also contained in another set, O, we say that M is a subset of O. A set containing no elements is said to be empty. For example, if the conditions for membership in the set are self contradictory, then the set will be empty.

THE REAL NUMBERS

The real number system contains several important subsets of numbers:

NATURAL NUMBERS sometimes called the "counting numbers" these are the numbers 1, 2, 3, ... The set of natural numbers is denoted by $\mathbb{N}$

INTEGERS the integers consist of the natural numbers together with the numbers 0, –1, –2, –3, ... The set of integers will be denoted by $\mathbb{Z}$. Note that $\mathbb{N}$ is a subset of $\mathbb{Z}$.

RATIONAL NUMBERS denoted by $\mathbb{Q}$, any number that can be expressed as the quotient of two integers belongs to the set of rational numbers. Clearly $\mathbb{Z}$ is a subset of $\mathbb{Q}$.

IRRATIONAL NUMBERS any number that is not rational is said to be irrational.

The *real numbers* are composed of the set of all rational numbers together with the set of all irrational numbers. We denote the set of real numbers by $\mathbb{R}$. An even larger number system, the *complex numbers* will not be discussed in this book.

REPRESENTATIONS FOR REAL NUMBERS

For computational purposes it is often convenient to express real numbers in terms of a *decimal representation*. e.g. $5/4 = 1.250$, $1/3 = .333\ldots$, $\sqrt{2} = 1.414\ldots$ and $\pi = 3.141592\ldots$. For rational numbers, the decimal representation either terminates, as in the case of 5/4, or it repeats, as in the case of $1/3 = .333\ldots$ or $1/11 = .090909\ldots$ For an irrational number, the decimal representation does not terminate nor is there any repetition of any block of numbers of finite length.

THE REAL LINE

For purposes of visualization it is often helpful to think of the *real line* in which we consider the points on a line as corresponding to real numbers. In particular, if we label one point as corresponding to the real number zero, then the points to the right of this point correspond to the positive real numbers and the points to the left of zero correspond to the negative real numbers. Zero is neither positive nor negative.

AN AXIOMATIC DEFINITION OF THE REAL NUMBERS

While this rather primitive description of the real numbers is sufficient for a discussion of calculus on an elementary level, a more precise knowledge is needed for a more rigorous treatment of the subject. We shall now present a set of axioms from which it will be possible to deduce all the essential properties of the real numbers. These axioms are grouped into three classes: field axioms, order axioms, and the completeness axiom.

Field Axioms

For all real numbers x, y, z we have

1. $x + y = y + x$
2. $(x + y) + z = x + (y + z)$
3. There exists a real number 0 such that $x + 0 = x$ for all x
4. For each x in $\mathbb{R}$ there exists w in $\mathbb{R}$ such that $x + w = 0$
5. $xy = yx$
6. $(xy)z = x(yz)$
7. There exists a real number $1 \neq 0$ such that $x1 = x$ for all x
8. For each $x \neq 0$ there exists w in $\mathbb{R}$ such that $xw = 1$
9. $x(y + z) = xy + xz$

FIELDS

Any set of objects that satisfies these nine axioms is called a *field* for the operations of addition and multiplication. The real numbers form a field but are not the only example of a field.

SUBTRACTION AND DIVISION

The element w in Axiom 4 is usually denoted $(-x)$. Then subtraction $x - y$ is defined to mean, $x + (-y)$. The element w in Axiom 8 is usually denoted by x^{-1}. Then division y/x for x different from zero is defined to mean yx^{-1}.

Example 1.1: Division by Zero

We have defined $z = y/x$ to mean $z = yx^{-1}$ which is to say $zx = yx^{-1}x = y$. That is, z is the unique number such that $zx = y$. In the case of division by zero, we have $z = y/0$ which requires that z be the unique number such that $z0 = y$. If $y = 0$, then this is satisfied by *every* number z, and if y is different from zero then there is *no* value z for which this holds. As a result we are forced to consider division by zero as an *undefined operation.* No consistent meaning can be associated with the expression $y/0$.

In addition to the field axioms, the real numbers satisfy the *order axioms.*

Order Axioms

The set P of positive real numbers satisfies

10. if $x, y \in P$ then $x + y \in P$
11. if $x, y \in P$ then $xy \in P$
12. if $x \in P$ then $-x$ is not in P
13. if x is real then exactly one of the following must hold $x = 0, x \in P$, or $-x \in P$.

ORDERED FIELDS

Any field which satisfies Axioms 10 through 13 is said to be an *ordered field.*

Mathematical Induction

The real numbers are an ordered field but this is not the only example of an ordered field. For example, the natural numbers form an ordered field. In fact, N is the smallest subset of $\mathbb{R}$ with the following properties.

INDUCTION PROPERTIES

(a) 1 is in the set
(b) if x is in the set then $x + 1$ must also belong to the set

Since $\mathbb{N}$ is the smallest subset of $\mathbb{R}$ with these properties, it follows that any subset of R having the properties must contain $\mathbb{N}$. This is the basis of the principle of mathematical induction.

Example 1.2: Mathematical Induction

In order to prove the following formula for the sum of the first N integers,

$$1 + 2 + 3 + \cdots + N = N(N + 1)/2, \tag{1.1}$$

we note first that (1.1) holds in the case that $N = 1$; i.e., 1 belongs to the set of integers N for which (1.1) is valid. Next, suppose that this set contains the integer m; i.e., suppose (1.1) holds with $N = m$

$$1 + 2 + \cdots + m = m(m + 1)/2.$$

If we add $m + 1$ to each side of this expression we obtain

$$1 + 2 + \cdots + m + m + 1 = m(m + 1)/2 + m + 1 = (m + 1)(m + 2)/2.$$

But this is just (1.1) with $N = m + 1$ and it follows that if (1.1) holds for m it must also hold for $m + 1$. Then the set of integers for which (1.1) is valid has both of the properties (a) and (b). This proves that (1.1) holds for all natural numbers N.

Inequalities

In an ordered field, and for the real numbers in particular, we can define $x < y$ to mean $y - x \in P$, and $x > y$ to mean $x - y \in P$. In addition, $x \leq y$ means $x < y$ or

$x = y$, with a similar definition for $x \geq y$. The order axioms imply the following properties for $<$:

Theorem 1.1

Theorem 1.1 For real numbers a, b, and c

i) if $a \neq 0$ then $a^2 > 0$
ii) if $a < b$ and $b < c$ then $a < c$
iii) if $a < b$, then for any c, $a + c < b + c$
iv) if $a < b$ and $0 < c$ then $ac < bc$

ABSOLUTE VALUE

If we define $|x|$, the *absolute value* of a real number x by,

$$|x| = \begin{matrix} x & \text{if} & x \geq 0 \\ -x & \text{if} & x < 0, \end{matrix}$$

then we can prove

Theorem 1.2

Theorem 1.2 For all x, y in $\mathbb{R}$,

i) $|xy| = |x|\,|y|$
ii) $|x + y| < |x| + |y|$ (TRIANGLE INEQUALITY)

INTERVALS

Using the notation of absolute value and inequalities, we can describe certain subsets of real numbers and visualize them as subsets of the real line. For given values $a < b$, the set of points

$$I = \{x \in \mathbb{R} : a \leq x \leq b\} = [a, b]$$

is a *closed interval* on the real axis, denoted by [a, b]. The set of points

$$I = \{x \in \mathbb{R} : a < x < b\} = (a, b)$$

is an *open interval* on the real axis, denoted by (a, b). The intervals

$$[a, b) = \{x \in \mathbb{R} : a \leq x < b\} \quad \text{and} \quad (a, b] = \{x \in \mathbb{R} : a < x \leq b\}$$

are neither open, nor closed, but are sometimes called *half-open* or *half-closed.* The intervals mentioned above are examples of *bounded intervals*. The intervals

$$\{x \in \mathbb{R}: x > a\} = (a, \infty) \quad \{x \in \mathbb{R} : x \geq a\} = [a, \infty)$$

$$\{x \in \mathbb{R} : x < a\} = (-\infty, a) \quad \{x \in \mathbb{R} : x \leq a\} = (-\infty, a]$$

$$(x \in \mathbb{R}\} = (-\infty, \infty)$$

are all examples of *unbounded intervals.*

The Symbol Infinity

The symbols $-\infty$ and ∞ are not real numbers, they are symbols having the following order properties, $-\infty < x$, and $x < \infty$ for all real numbers x.

GLB AND LUB OF A SET

A real number b is said to be an *upper bound* for a set S of real numbers if $b \geq x$ for all x in S. We say that β is a *least upper bound* for S if β is an upper bound for S and if $\beta \leq b$ for all upper bounds b for S. *Lower bound* and *greatest lower bound* are defined in an analogous way. The least upper bound is also referred to as the *supremum* or *sup* and the greatest lower bound is often called the *infimum* or *inf* of the set.

Completeness Axiom

14. Every nonempty set of real numbers having an upper bound has a least upper bound.

COMPLETE ORDERED FIELD

An ordered field which satisfies the completeness axiom is called a complete, ordered field. The real numbers are, essentially, the only example of a complete ordered field. All properties of the real numbers may now be deduced from the axioms 1 through 14. These properties include:

Theorem 1.3

Theorem 1.3 Every nonempty set of real numbers having a lower bound has a greatest lower bound.

Theorem 1.4 Archimedean Law

Theorem 1.4 If $x \in \mathbb{R}$ then there exists an $n \in N$ such that $x < n$.

COROLLARY 1.5

Between any two real numbers there is a rational number.

COROLLARY 1.6

Between any two real numbers there is an irrational number.

Sequences

We introduce here the notion of a sequence. Sequences will be discussed in greater detail in Chapter 2.

Definition A *sequence* is an ordered set of real numbers that are in one to one correspondence with the natural numbers. The correspondence is indicated by labelling the terms in the sequence in order: $a_1, a_2, a_3, \ldots$ Alternatively, a sequence is a real valued function with domain equal to the natural numbers.

Example 1.3 Sequences

Each of the following is an example of a sequence,

2, 4, 6, …	$a_n = 2n \quad n \in N$
1/2, 1/3, …	$a_n = 1/(n+1) \quad n \in N$
2, 4, 8, 16, …	$a_n = 2^n \quad n \in N$

CONVERGENCE

Definition The sequence $\{a_n\}$ is *convergent* with limit equal to A if, for each $\in > 0$, there is a positive integer N depending on $\in$ such that $|a - A| < \in$ n for all $n > N$.

MONOTONE SEQUENCES

A sequence $\{a_n\}$ is said to be *increasing* if $a_n \leq a_{n+1}$ for $n \in \mathbb{N}$; the sequence is said to be *decreasing* if $a_n \geq a_{n+1}$ for $n \in \mathbb{N}$. A sequence is said to be *monotone* if it is either increasing or decreasing. It is a consequence of the completeness axiom that monotone sequences of real numbers converge if they are bounded.

Theorem 1.7 Monotone Sequence Theorem

Theorem 1.7 Let the increasing sequence of numbers $\{a_n\}$ be bounded above. Then the sequence has a least upper bound α, and the sequence converges to α. If the decreasing sequence $\{b_n\}$ is bounded below then the sequence has a greatest lower bound, β, and the sequence converges to β. We can summarize by saying a monotone sequence is convergent if it is bounded.

We have also the so-called "nested interval theorem",

Theorem 1.8 The Nested Interval Theorem

Theorem 1.8 Let $I_1, I_2, \ldots$ denote a family of closed, bounded intervals with I_{n+1} contained together with its endpoints in I_n for $n \in \mathbb{N}$. Then there exists $x \in \mathbb{R}$ such that $x \in I_n$ for every n. Suppose further that the length of I_n tends to zero with increasing n. Then these "nested intervals" have exactly one point in common.

Example 1.4 Nested Intervals

It is important to realize that the intervals must be both closed and bounded in order for this theorem to hold.

1.4 (a) Consider the nested intervals $I_n = \{x \in \mathbb{R} : 0 < x < 1/n\}$ for $n \in \mathbb{N}$. If $x \in \mathbb{R}$ belongs to I_n for every n, then $x > 0$ and, since $x < 1/n$ for every n, x must also satisfy $x \leq 0$. It is evident that no such x can exist. This example does not contradict Theorem 1.8 since the intervals are bounded but not closed.

1.4 (b) The nested intervals $\{I_n = x \in \mathbb{R} : n \leq x < \infty\}$ $n \in \mathbb{N}$ have no point that is common to all the I_n since such a number x would have to satisfy $x > n$ for all $n \in \mathbb{R}$. This would contradict Theorem 1.4 and hence no such point can exist. This does not violate the nested interval theorem since these intervals are not bounded.

POINTS SETS IN R

We have been discussing sets of real numbers, or equivalently, sets of points on the real line. We refer to these sets as *point sets*. A point set that contains only finitely many points is called a *finite set*. Otherwise the set is said to be an *infinite set*. We now introduce some additional terminology we shall find convenient in the chapters to come .

OPEN AND CLOSED SETS

Definition For a in $\mathbb{R}$ and for $\epsilon > 0$, the set $N_\epsilon(a) = \{x \text{ in } \mathbb{R}: a - \epsilon < x < a + \epsilon\}$ is called an *ϵ-neighborhood* of the point a.

Definition A point set U in $\mathbb{R}$ is said to be open if, for each x in U there is an ϵ-neighborhood of x that is contained in U.

Definition If S denotes a point set in $\mathbb{R}$, then the *complement* of S is the point set consisting of all points in $\mathbb{R}$ that are not in S.

Definition A point set F in $\mathbb{R}$ is said to be *closed* if its complement is an open set.

COMPACT AND CONNECTED SETS

Definition A point x in $\mathbb{R}$ is an *accumulation point* or *cluster point* for a point set S in $\mathbb{R}$ if every ϵ-neighborhood of x contains at least one point of S other than x. The point x may or may not belong to S.

Definition A subset of the reals is said to be *compact* if it is closed and bounded.

Definition Two sets are said to be *disjoint* if the sets have no points in common.

Definition A subset of the reals is said to be *connected* if and only if it cannot be represented as the union of two nonempty disjoint sets neither of which contains a limit point of the other.

Example 1.5 Point Sets

1.5 (a) The interval $(a, b) = \{x : a < x < b\}$ is open. To see this note that if x_0 is in (a, b) and we choose $\epsilon > 0$ but smaller than either of the numbers $x_0 - a$ or $b_0 - x$, then $N_\epsilon(x_0)$ is contained in (a, b).

1.5 (b) The interval $[a, b] = \{x : a \leq x \leq b\}$ is closed since its complement consists of the point set $\{x : x < a\}$ together with the point set $\{x : x > b\}$; i.e., we say the complement consists of the *union* of these two sets. Then we can show as in the previous example that the complement is open.

1.5 (c) Note that in the sense defined here, open and closed are not opposites. For example, the point set $(a, b] = \{x : a < x \leq b\}$ is *neither* open nor closed. If we define the empty set to be an open set, then the entire real line is *both* open and closed; i.e. it is open by definition and closed since its complement, the empty set, is open.

1.5 (d) Each point of the interval [a, b] is an accumulation point for the point set I = (a, b). Note that the points a and b do not belong to I. The natural numbers, ℕ is a set of numbers that has no accumulation points.

1.5 (e) Every closed, bounded interval is compact and connected. Every finite point set is compact. To see this note that the complement of a finite point set is open. Then the finite point set is closed and since it is finite, it must also be bounded.

1.5 (f) The set $\{x : x \neq 0\}$ is not connected since it can be represented as the union of the disjoint sets $\{x : x < 0\}$ and $\{x : x > 0\}$. The point $x = 0$ is an accumulation point of both sets but belongs to neither of them. A set that consists of a single point is connected as is any interval, regardless of whether it is open, closed or otherwise. Point sets on the real line have the following important properties,

Properties of Point Sets

Theorem 1.9 Characterization of Closed Sets

Theorem 1.9 A point set F in ℝ is closed if and only if it contains all its accumulation points.

Theorem 1.10 Bolzano-Weierstrass

Theorem 1.10 If A is a bounded, infinite set in ℝ then A must have at least one accumulation point.

Theorem 1.11

Theorem 1.11 Let A denote a subset of ℝ and let p denote the least upper bound for A (or the greatest lower bound) if p does not belong to A then p must be an accumulation point for A.

Theorem 1.12

Theorem 1.12 Each nonempty compact set in ℝ contains its least upper bound and its greatest lower bound.

Theorem 1.13 Characterization of Compact Sets

Theorem 1.13 A is a compact set in ℝ if and only if every sequence in A contains a subsequence that converges to a point of A. Here a subsequence of a sequence a_n is obtained by eliminating some of the a_n's and keeping the remaining terms in their original order.

Theorem 1.14 Characterization of Connected Sets

Theorem 1.14 A subset of R is connected if and only if it is an interval or a single point.

SOLVED PROBLEMS

Representations of Real Numbers

PROBLEM 1.1

Convert each of the following rational numbers to decimal representation:
(a) 1/8, 11/80 (b) 1/7, 1/27, 1/271
(c) 1/6, 17/66

SOLUTION 1.1

By dividing, or by using a calculator, we find

1.1 (a) $1/8 = .125000 \ldots$; $11/80 = .1375000 \ldots$

1.1 (b) $1/7 = .142857\underline{142857}$; $1/27 = .037037\underline{037}$; $1/271 = .00369\underline{00369}$

1.1 (c) $1/6 = .166\underline{6}$; $17/66 = .25757\underline{57}$

Each of the rational numbers in group (a) converts to a *terminating* decimal. In group (b), each of the decimals is *periodic* of length six, three and five, respectively. The decimals in group (c) are examples of *eventually periodic* representations. In the periodic and eventually periodic decimals, the repeating portion of the decimal has been underlined.

PROBLEM 1.2

Convert each of the following periodic decimals to its equivalent quotient of integers:

(a) $.00271\underline{00271}$
(b) $.1555\underline{5}$, and $.4999\underline{9}$
(c) $.500\underline{0}$

SOLUTION 1.2

The theory of geometric series is required in order to *prove* that each of these periodic decimals converges to a unique real number. This will be discussed in Chapter 6. However, if we assume that this has already been proved then the following procedure can be used to find the equivalent quotient of integers for each decimal.

1.2 (a) Let $\quad x = .0027100271 \ldots$

Then $\quad 10^5 x = 271.0027100271 \ldots$

By subtracting the first equation from the second, it follows that

$$99,999\, x = 271.$$

Thus $\quad x = 1/369\,.$

1.2 (b) Let $\quad x = .1555 \ldots \qquad y = .4999 \ldots$

Then $\quad 10x = 1.555 \ldots \qquad 10y = 4.999 \ldots$

$\quad 100x = 15.555 \ldots \qquad 100y = 49.999 \ldots$

and it follows by subtracting line 2 from line 3 that

$$90x = 14; \qquad 90y = 45$$
$$x = 14/90 \qquad y = 1/2.$$

1.2 (c) Let $x = .500\ldots$

Then $10x = 5.$ and $x = 1/2.$

Note that $y = .4999\ldots$ and $x = .500\ldots$ are each equivalent to the quotient 1/2. We simply interpret this to mean that a decimal representation that is eventually periodic, terminating in an infinite string of nines, is by definition equivalent to a terminating decimal. The equivalent terminating decimal is obtained by dropping the string of nines and increasing the last remaining digit by one.

PROBLEM 1.3

Prove that the square of an odd integer must be odd.

SOLUTION 1.3

N is an odd integer if it is of the form, $N = 2m + 1$ for $m \in \mathbb{Z}$. Then $N^2 = 4m^2 + 4m + 1$ is of the form $2k + 1$ for k equal to the integer $2m^2 + 2m$. That is, N^2 is odd.

PROBLEM 1.4

Prove that $\sqrt{2}$ is an irrational number.

SOLUTION 1.4

Suppose $\sqrt{2} = p/q$ where p/q has been reduced to lowest terms; i.e. the integers p and q have no factors in common except ±1. Then $p^2 = 2q^2$ and p^2 is an even integer. Then p must also be an even integer or, by the previous problem, its square would be an odd integer. Say that $p = 2k$ for some integer k. Then $q^2 = 2k^2$ and hence q^2 is even which implies that q must also be even. But then p and q have the factor 2 in common, which is in contradiction to our original assumption that p/q has been reduced to lowest terms. Then there exists no quotient p/q of integers that is equal to $\sqrt{2}$.

Mathematical Induction

PROBLEM 1.5

Prove by mathematical induction that for each positive integer M,

$$\sum_{k=1}^{M} k^2 = M(M + 1)(2M + 1)/6 \qquad (1)$$

SOLUTION 1.5

Let the set of integers M for which (1) is valid be denoted by A. Then 1 is in A since $1^2 = 2 \cdot 3/6$. Now suppose that A contains the positive integer m. Then it follows from (1) that

$$\sum_{k=1}^{M} k^2 + (m+1)^2 = m(m+1)(2m+1)/6 + (m+1)^2$$
$$= \frac{2m^3 + 9m^2 + 13m + 6}{6} \qquad (2)$$

But it is easy to check that $2m^3 + 9m^2 + 13m + 6 = (m+1)(m+2)(2m+3)$ and if we substitute this into (2) we obtain

$$\sum_{k=1}^{m+1} k^2 = (m+1)(m+2)(2m+3)/6$$

This is the formula (1) with m replaced by m + 1, and it follows that m + 1 is in A if m is in A. Then by the induction principle A contains all positive integers.

PROBLEM 1.6

Prove by induction that if $x > -1$, then

$$(1+x)^n \geq 1 + nx \qquad \text{for all n in } \mathbb{N}. \qquad (1)$$

SOLUTION 1.6

Let A denote the set of positive integers n for which (1) holds. Then clearly 1 belongs to A. Next suppose A contains the positive integer m; i.e., (1) holds with n = m. Multiplying both sides of the expression by the positive number (1 + x) leads to,

$$(1+x)^{m+1} \geq (1+mx)(1+x) = 1 + (m+1)x + mx^2 \geq 1 + (m+1)x$$

i.e.,
$$(1+x)^{m+1} \geq 1 + (m+1)x\,.$$

But this is just (1) with n replaced by m + 1. It follows that m + 1 is in A if m is in A. Then by the induction principle, A contains all positive integers.

The Order Axioms

PROBLEM 1.7

Show that if a, b are real numbers with $a \cdot b > 0$, then either $a > 0$ and $b > 0$, or else $a < 0$ and $b < 0$.

SOLUTION 1.7

If $a \cdot b > 0$ then neither a nor b is equal to zero, otherwise the product would equal zero. Since a is not zero, it follows from Axiom 13 that either $a > 0$ or else $a < 0$. Suppose $a > 0$. Then $a^{-1} > 0$, otherwise we would have $1 = a \cdot a^{-1} < 0$. Then

$$b = 1 \cdot b = (a^{-1} \cdot a) \cdot b = a^{-1} \cdot (a \cdot b) > 0,$$

where the last inequality follows by Axiom 11. Similarly, if we suppose $a < 0$, then $a^{-1} < 0$ and

$$-b = -1 \cdot b = (-a^{-1} \cdot a) \cdot b = -a^{-1} \cdot (a \cdot b) > 0.$$

Then by Axiom 12, $b < 0$.

PROBLEM 1.8

If a and b are real numbers, then prove that $0 < a < b$ implies :

$$\text{(a) } a^2 < b^2 \qquad \text{and} \qquad \text{(b) } \sqrt{a} < \sqrt{b}.$$

SOLUTION 1.8

If $0 < a < b$ then $b - a > 0$ and $b + a > 0$, and Axiom 11 implies $b^2 - a^2 > 0$. This proves (a). Now we have the identity,

$$(\sqrt{b} - \sqrt{a})(\sqrt{b} + \sqrt{a}) = b - a.$$

But $b - a > 0$ and $\sqrt{b} + \sqrt{a} > 0$ and hence by the result of the previous problem, it follows that $\sqrt{b} - \sqrt{a} > 0$.

PROBLEM 1.9

Show that for real numbers $a \geq 0$ and $b \geq 0$,

$$\sqrt{a \cdot b} \leq (a + b)/2. \qquad (1)$$

SOLUTION 1.9

In the case that $a = b$, (1) reduces to an equality. For $a > 0$, $b > 0$ and a not equal to b, $\sqrt{a} > 0$, $\sqrt{b} > 0$, and $\sqrt{a}$ is not equal to $\sqrt{b}$ by the result (b) in the previous problem. Then Theorem 1.1, part i, implies

$$(\sqrt{a} - \sqrt{b})^2 > 0;$$

i.e.

$$a - 2\sqrt{a \cdot b} + b > 0.$$

But this implies (1) by theorem 1.1, part iii.

The Completeness Axiom

PROBLEM 1.10

Prove Theorem 1.3

SOLUTION 1.10

Let S denote a nonempty set of real numbers having lower bound a. Let T denote the set of all numbers $t = -s$ for s in S. Now $a \le s$ for all s in S by definition of lower bound and hence $-s \le -a$ for every s in S. Then $-a$ is an upper bound for T and by Axiom 14, T has a *least* upper bound, $-\alpha$. Then α is a lower bound for S. In fact, α is the greatest lower bound for S since if there were a lower bound α' greater than α, it would follow that $-\alpha' < -\alpha$ is an upper bound for T. But $-\alpha$ is the *least* upper bound for T and so no such α' can exist.

PROBLEM 1.11

Prove Theorem 1.4, the Archimedean law.

SOLUTION 1.11

Let $x \in \mathbb{R}$ and let A denote the set of all natural numbers that are less than or equal to x. If A contains no elements then the theorem must hold. On the other hand, if A is not empty then A is bounded above by x and hence by Axiom 14, there exists a least upper bound, a, for A. Now $a - 1 < a$ so $a - 1$ is not an upper bound for A. It follows that there exists an $m \in A$ such that $a - 1 < m$. Then $a < m + 1$ and hence $m + 1$ is an element of $\mathbb{N}$ that is not in A; i.e., $x < n = m + 1$. This proves the theorem.

PROBLEM 1.12

Use Theorem 1.4 to prove Corollaries 1.5 and 1.6

SOLUTION 1.12

Suppose $x, y \in \mathbb{R}$ with $x < y$. If $x \ge 0$ then $y - x > 0$, $(y - x)^{-1} \in \mathbb{R}$ and then by Theorem 1.4, it follows that there exists a number m in $\mathbb{N}$ such that

$$(y - x)^{-1} < m; \quad \text{i.e. } y - x > 1/m > 0.$$

Theorem 1.4 also implies that the set of positive integers, k, such that $y \le k/m$ is not empty. Then by Theorem 1.3, the set has a smallest element, n, and hence

$$\frac{n-1}{m} < y \le \frac{n}{m}.$$

In addition,

$$x = y - (y - x) < \frac{n}{m} - \frac{1}{m} = \frac{n-1}{m}$$

and it is evident that $x < r < y$ for $r = (n - 1)/m \in \mathbb{Q}$.

If $x < 0$, then by Theorem 1.4, there exists a positive integer $k > -x$. Then $k + x > 0$ and there exists $r \in \mathbb{Q}$ such that $k + x < r < k + y$. Then $r - k \in \mathbb{Q}$ lies between x and y. This proves Corollary 1.5.

To prove Corollary 1.6, we note that by Corollary 1.5, there exists a rational number r between $x/\sqrt{2}$ and $y/\sqrt{2}$. Then the irrational number $r\sqrt{2}$ lies between x and y. This proves Corollary 1.6.

PROBLEM 1.13

Prove Theorem 1.7, the monotone sequence theorem.

SOLUTION 1.13

Let x be an *increasing* sequence that is bounded above, and let the least n upper bound be equal to α; i.e.,

$$x_1 \le x_2 \le \dots \le x_n \le x_{n+1} \le \dots \le \alpha$$

Then $x_n \le \alpha$ for all n, and in addition, since α is the *least* upper bound, for each $\in > 0$ there exists an integer N, such that $x_n > \alpha - \in$ for $n \ge N$. Now the sequence n is increasing and hence $x_N \le x_n$ for $n \ge N$. It follows that

$$\alpha - \in < x_n < \alpha \quad \text{for } n \ge N,$$

and, by definition, the sequence converges to α. The proof for a decreasing sequence is similar.

PROBLEM 1.14

Prove Theorem 1.8, the nested interval theorem.

SOLUTION 1.14

Let $I_1 = [a_1, b_1]$, $I_2 = [a_2, b_2]$, ... denote a sequence of nested intervals; i.e., I_{n+1} is contained, together with its endpoints, in I_n.Suppose also the length of I_n decreases to zero with increasing n. Then

$$a_n \le a_{n+1} \le b_{n+1} \le b_n \quad \text{for } n = 1, 2, \dots$$

and

$$b_n - a_n \to 0 \text{ as } n \to \infty.$$

Thus the sequence $\{a_n\}$ is increasing and is bounded above by b_1. Similarly, the sequence $\{b_n\}$ is decreasing and is bounded below by a_1. Then by the monotone sequence theorem, each sequence has a limit. Let

$$\lim_{n\to\infty} a_n = a \quad \text{and} \quad \lim_{n\to\infty} b_n = b\,.$$

Since $b_n - a_n$ tends to zero, it follows that a = b. Then a = b is, simultaneously, the least upper bound for the sequence $\{a_n\}$ and the greatest lower bound for the sequence $\{b_n\}$. It follows that a belongs to every one of the interval I_n. It is evident that there cannot be two such points, separated by a positive distance, say $h > 0$. If there were, then for n sufficiently large that $b_n - a_n < h$, we could not have both points in the interval I_n. Thus a is the unique point common to all the intervals. In the case that the length of I_n is not assumed to tend to zero with increasing n, we can conclude only that the interval [a, b] is contained in I_n for every n. Since $a_n < b_n$ for every n, it follows that $a_n \le b$ for every n and this, in turn, implies that $a \le b$.

POINT SETS IN $\mathbb{R}$

PROBLEM 1.15

Prove Theorem 1.9.

SOLUTION 1.15

Suppose the set F in $\mathbb{R}$ is closed, and let x be an accumulation point of F. If x is not in F, then x is in the complement of F, an open set, G. But if x belongs to this open set, G, then there is an ϵ-neighborhood of x that is contained in G. Since this neighborhood is contained in G, it contains no points of F. Then x cannot be an accumulation point for F. This proves that a closed set contains all its accumulation points.

Suppose F is a set in $\mathbb{R}$ that contains all its accumulation points. Then the complement, G, of F must be open. To see this let y denote a point of G. Since F contains all its accumulation points, y cannot be an accumulation point for F. Then there must be an ϵ-neighborhood of y that contains no points of F. But then this neighborhood must lie in G and G is thus open. This proves that a set that contains all its accumulation points is closed.

It follows that F is closed if and only if F contains all its accumulation points.

PROBLEM 1.16

Prove Theorem 1.10, the Bolzano-Weierstrass theorem.

SOLUTION 1.16

Let A denote a bounded, infinite subset of $\mathbb{R}$; that is, A is contained in a closed, bounded interval $I_1 = [a, b]$ and A contains infinitely many points. We now bisect I_1, dividing it into two equal parts. Then at least one of the two intervals $[a, (a + b)/2]$, $[(a + b)/2, b]$ contains infinitely many points of A. If this were not the case, then the union of these two intervals would contain only finitely many points but since their union also contains A, this is a contradiction. Let I_2 denote one of these two intervals containing infinitely many points. Now bisect I_2 and note that at least one of the two half-intervals of I_2 must contain infinitely many points. Choose one of these half-intervals containing infinitely many points, and denote it by I_3. Proceeding in this way, we generate a sequence of nested, closed, bounded intervals

$$I_1 \supset I_2 \supset \cdots \supset I_n \supset \cdots$$

For each n, I_n contains I_{n+1} (together with its endpoints) and since the length of $I_{n+1} = (b - a)/2^n$, the length of the intervals tends to zero with increasing n. The nested interval theorem then implies the existence of a unique point, p, common to all of the intervals.

To see that p is an accumulation point of A, let $N_\in(p) = \{x : p - \in < x < p + \in\}$ for $\in > 0$ and choose n in N sufficiently large that $(b - a)/2^n < \in$. Then p lies in I_{n+1} and the length of I_{n+1} is less than $\in$. It follows that I_{n+1} is contained in $N_\in(p)$. Since I_{n+1} contains infinitely many points of A, $N_\in(p)$ must contain points of A other than p. Then, by definition, p is an accumulation point of A. This proves the theorem.

Our knowledge of the real numbers, based on experience, includes the information that the set of real numbers contains the following subsets: the natural numbers, the integers, the rational numbers and the irrational numbers. In addition, we know that each real number has a unique decimal expansion and that the real numbers are in one to one correspondence with the points of a line.

A rigorous development of calculus requires more detailed information about the real numbers than is available from our intuition gained from computational experience. This information can be deduced from a set of axioms that completely characterize the real numbers. This axiom system contains axioms of three types: field axioms, order axioms and a completeness axiom.

The field axioms specify how the real numbers behave with respect to the operations of addition and multiplication. They imply, among other things, that division by zero is not a valid operation within the framework of the real numbers.

The order axioms imply the principle of mathematical induction and provide a structure in which we can consider such things as inequalities.

The completeness axiom leads to the discovery that the rationals and irrationals are everywhere dense in the reals; it is also the basis for the monotone sequence theorem, the nested interval theorem and the Bolzano-Weierstrass theorem. All of this information about the real numbers will be used extensively in our discussion of such topics as continuity, differentiability and integrability.

SUPPLEMENTARY PROBLEMS

1.1 Describe two properties of the real numbers that do not require the axiomatic developement for their discovery.

1.2 Describe three properties of the real numbers that require the axiomatic development for their discovery; i.e., properties that could not be known if it were not for the axiomatic development.

1.3 Show that the set of even positive integers is not bounded.

1.4 Show that if a is rational and b is irrational, then ab is irrational.

1.5 Show that for all real numbers a, b, $||a| - |b|| < |a - b|$.

1.6 Use mathematical induction to show that

$$\sum_{j=1}^{n} (2j - 1) = n^2$$

1.7 Use mathematical induction to show that for real numbers

$$\sum_{j=1}^{n} ar^j = a(1 - r^n)/(1 - r)$$

Consider the following sets of real numbers:

$A_1 = \{x_n = 1 - 1/n : n \in \mathbb{N}\}$

$A_2 = \{x_n = n - 1/n : n \in \mathbb{N}\}$

$A_3 = (2 - 1/2, 2 + 1/2) \cup (3 - 1/3, 3 + 1/3) \cup \cdots \cup (n - 1/n, n + 1/n) \cup \cdots$

$A_4 = (0, 1) \cup (1, 3/2) \cup (3/2, 7/4) \cup (7/4, 15/8) \cup \cdots$

$A_5 = \{1/2, -1/2, 1, -1, 3/2, -3/2, 2, -2, \ldots\}$

$A_6 = \{x \in (0, 1) : x \text{ is irrational}\}$

For each set, answer the following questions:

1.8 Is the set bounded above? If so, what is the LUB?

1.9 Is the set bounded below? If so, what is the GLB ?

1.10 Is the set open? Is it closed? Is it compact?

1.11 What points are interior points of the set?

1.12 What points are accumulation points of the set?

1.13 Which axiomatic property of the real numbers is central to the proof of the monotone sequence theorem?

1.14 What theorem about the real numbers is the crucial ingredient in the proof of the nested interval theorem?

2

Sequences

In the next few chapters we shall present several concepts from calculus, all of which are based on the notion of a limit. Sequences provide a convenient setting in which to discuss the notion of a limit, and so, in this chapter we develop the properties of sequences. In particular, we define convergence for sequences and present a number of theorems which can be used to tell whether a sequence converges. Finally we introduce the important Cauchy criterion for convergence.

FUNCTIONS AND SEQUENCES

A *function* can be defined as a rule which assigns to each element x in a subset, D, of the real numbers a corresponding element y in a subset, Rng, of the real numbers. We refer to the set, D, as the *domain* of the function, and to the set, Rng, as the *range* of the function. We may denote the function by writing $y = f(x)$. The function $f(x)$ is said to be *single valued* or *well defined* as long as there are not two distinct y-values in Rng that correspond to the same x in D.

In the special case that $D = \mathbb{N}$, the function is called a *sequence*. We denote sequences by writing $\{a_n\}$, $n \in \mathbb{N}$, instead of $a(n)$. A sequence $\{a_n\}$ is said to be *bounded* if there is a real number $M > 0$ such that $|a_n| \leq M$ for all n in $\mathbb{N}$. We say that the sequence $\{a_n\}$ is *convergent with limit* L, if for every $\in > 0$, there exists ν in $\mathbb{N}$ such that $|a_n - L| < \in$ for all $n > \nu$; i.e., if a_n belongs to $N_\in(L)$ for all $n > \nu$. If a sequence in not convergent, it is said to be *divergent*. Finally, we say $\{a_n\}$ tends to $+\infty$ if, for every $B > 0$ there exists ν in $\mathbb{N}$ such that $a_n > B$ for $n > \nu$, and we say a_n tends to $-\infty$ if $\{-a_n\}$ tends to $+\infty$.

Example 2.1 Sequences

2.1 (a) The sequence $\{1/n\}$ is convergent with limit equal to zero. To see this, let $\in > 0$ be given and let ν denote an integer greater than $1/\in$. Then

$$|1/n - 0| < \in \qquad \text{for every } n > \nu.$$

2.1 (b) Let $r < 1$ be given. Then the sequence $\{r^n\}$ is convergent with limit equal to zero. In this case we can show that for any $\in > 0$,

$$|r^n - 0| < \in \qquad \text{for n in N such that } n > \frac{\text{Log } \in}{\text{Log } r}$$

Note that if $r > 1$, then the sequence r^n tends to $+\infty$.

2.1 (c) The sequence $\{(-1)^n\}$ is divergent. To see this, note that $a_n = 1$ if n is even and $a_n = -1$ if n is odd. Then the only possible limits for this sequence are the values 1 and -1. But for $\in < 1/2$ there is no integer ν for which

$$|(-1)^n - 1| < \in \text{ for all } n > \nu$$

nor is it true that

$$|(-1)^n + 1| < \in \text{ for all } n > \nu$$

We list now several important facts about sequences.

Properties of Sequences Theorem 2.1

Theorem 2.1 If $\{a_n\}$ is a convergent sequence then it is bounded.

Corollary 2.2 A monotone sequence is convergent if and only if it is bounded.

Theorem 2.3 Uniqueness of Limits

Theorem 2.3 The limit of a convergent sequence is unique; i.e. a convergent sequence cannot have two distinct limits L and M.

Theorem 2.4 Squeeze Play Theorem

Theorem 2.4 Let $\{a_n\}$ be a convergent sequence with limit equal to L. Suppose that $L \leq bn \leq a_n$ for all n in N. Then $\{b_n\}$ is a convergent sequence with limit equal to L.

Theorem 2.5 Arithmetic With Sequences

Theorem 2.5 Let $\{a_n\}$, $\{b_n\}$ be convergent sequences with limits L and K, respectively. Then

a) For real numbers α and β, $\{\alpha a_n + \beta b_n\}$ is convergent with limit equal to $\alpha L + \beta K$

b) $\{a_n b_n\}$ is convergent with limit equal to KL

c) $\{a_n / b_n\}$ is convergent to L/K provided $K \neq 0$ and $bn \neq 0$ for all n

Subsequences

If f(x) is a function having domain D and range denoted by Rng, then a function g(x) whose domain is a subset, D′, of D and for which g(x) = f(x) for all x in D′, is said to be a *restriction* to D′ of the function f(x). Now suppose $\{a_n\}$ is a sequence and that $\mathbb{N}'$ denotes a subset of $\mathbb{N}$. Then the sequence $\{b_n\}$, n in $\mathbb{N}'$, is said to be a *subsequence* of $\{a_n\}$ if the domain of $\{b_n\}$ is N′ and if $a_n = b_n$ for all n in $\mathbb{N}'$. Alternatively, if we think of the sequence $\{a_n\}$ as an ordered set of real numbers that is in one to one correspondence with the natural numbers, then a subsequence of $\{a_n\}$ is an infinite ordered subset of this ordered set. We can think of the elements of the subsequence as being in one to one correspondence with a subset $\mathbb{N}'$ of $\mathbb{N}$. The notion of a subsequence is useful in discussing sequences.

Theorem 2.6

Theorem 2.6 If the sequence $\{a_n\}$ converges to limit L, then every subsequence $\{a_n\}$ must converge to L.

LIMIT POINTS

A point P is said to be a limit point for the sequence $\{a_n\}$ if $\{a_n\}$ contains a subsequence that converges to P. Note that if $\{a_n\}$ is convergent to the limit L, then L is a limit point for $\{a_n\}$. On the other hand, $\{a_n\}$ may have more than one limit point, but in this case the sequence cannot be convergent.

Theorem 2.7

Theorem 2.7 If the sequence $\{a_n\}$ has more than one limit point, then $\{a_n\}$ is divergent.

The following is a statement of the Bolzano-Weierstrass theorem in terms of subsequences.

Theorem 2.8 Bolzano-Weierstrass Theorem

Theorem 2.8 Every bounded sequence contains a convergent subsequence.

Example 2.2 Subsequences and Limit Points

2.2 (a) The sequence $\{1/n\}$ has zero as the only limit point, and the sequence is convergent to zero.

2.2 (b) The sequence, $\{1, 0, 2, 0, 3, 0, 4, 0, 5, 0, \ldots\}$ contains the subsequence $\{0, 0, \ldots\}$. Then zero is the only limit point for this sequence. However, the sequence does not converge to zero. In fact the sequence is divergent since it is not bounded.

2.2 (c) The sequence $\{(-1)^n\}$ is clearly bounded. Then by Theorem 2.8 it contains a convergent subsequence. In fact it contains two convergent subsequences: $\{1, 1, 1, \ldots\}$ and $\{-1, -1, -1, \ldots\}$. It follows that $\{(-1)^n\}$ has two limit points: + 1 and −1. Then Theorem 2.6 implies that the sequence is divergent, a fact we already obtained by other means.

We have an alternative characterization of convergence for sequences, one that does not refer explicitly to the limit of the sequence.

The Cauchy Criterion

A sequence $\{a_n\}$ is said to be a *Cauchy sequence* if, for every $\epsilon > 0$, there exists $N > 0$ such that $|a_m - a_n| < \epsilon$ for all $m, n > N$. Then we have the *Cauchy criterion* for convergence.

Theorem 2.9

Theorem 2.9 A sequence is convergent if and only if it is a Cauchy sequence.

SOLVED PROBLEMS

Limits of Sequences

PROBLEM 2.1 A CONSTANT SEQUENCE

Find the limit of the sequence $\{1, 1, 1, \dots\}$.

SOLUTION 2.1

For this sequence $a_n = 1$ for all n and it is obvious that $L = 1$. To show rigorously that $L = 1$, note that $|a_n - 1| = 0$ for all n. Then for every $\in > 0$, it is certainly the case that $|a_n - 1| < \in$, not just for large n but for all n. Then by the definition of convergence, this sequence converges to the limit 1. This is an example of a *constant sequence.*

PROBLEM 2.2 AN EVENTUALLY CONSTANT SEQUENCE

Find the limit for the sequence $\{1, 2, 3, 4, 4, 4, \dots\}$.

SOLUTION 2.2

For this sequence the limit is obviously $L = 4$, since we have $|a_n - 4| = 0$ for $n > 3$. This is an example of an eventually constant sequence.

PROBLEM 2.3

Find the limit of the sequence $\{1/2, 3/4, 7/8, \cdots, 1 - 2^{-n}, \cdots\}$

SOLUTION 2.3

Here the sequence is apparently tending to the limit 1. To prove this we write

$$|a_n - 1| = |(1 - 2^{-n}) - 1| = 2^{-n} \text{ for all } n \in \mathbb{N}. \tag{1}$$

We can prove by induction that $2^{-n} < 1/n$ for all $n \in \mathbb{N}$. Then it is evident from (1) that for all $\in > 0$, $|a_n - 1| < \in$ for all n such that, $n > 1/\in$.

PROBLEM 2.4 TENDING TO INFINITY

If $r > 1$ then show that $\{r^n\}$ tends to $+\infty$.

SOLUTION 2.4

We must show that for any $B > 0$, there exists ν in $\mathbb{N}$ such that $r^n > B$ for $n > \nu$. If we choose $\nu > \text{Log } B/\text{Log } r$ then $n > \nu$ implies $n\text{Log } r > \text{Log } B$ and this implies $r^n > B$.

PROBLEM 2.5 A DIVERGENT SEQUENCE

Prove that the sequence $\{0, 2, 0, 4, 0, 8, 0, 16, \dots\}$ is divergent.

SOLUTION 2.5

Since the terms of this sequence satisfy, $a_{2n} = 2^n$, it is clear that this sequence is not bounded. Then by Theorem 2.1, the sequence is not convergent.

PROBLEM 2.6

Find the limit of the sequence $\{1/\sqrt{n^2 + 1}\}$.

SOLUTION 2.6

For each $n \in \mathbb{N}$, we have

$$0 < 1/\sqrt{n^2 + 1} < 1/n.$$

Since $\{1/n\}$ is a convergent sequence with limit equal to zero, it follows from Theorem 2.4 that $\{1/\sqrt{n^2 + 1}\}$ is convergent with limit zero.

PROBLEM 2.7

Find the limit of the sequence $\{n/(3n + 1)\}$

SOLUTION 2.7

For each n,

$$\frac{n}{3n + 1} = \frac{1}{3 + 1/n} = \frac{a_n}{b_n}$$

Then $\{a_n\}$ and $\{b_n\}$ each can be shown to converge with limits equal to 1 and 3, respectively. Since b_n is different from zero for all n, it follows from Theorem 2.5 part (c) that $\{n/(3n + 1)\}$ converges to 1/3.

PROBLEM 2.8

Find the limit of the sequence $\{a_n\}$ where

$$a_n = \frac{3n^2 + 4n - 2}{n^2 + 3n + 2} \qquad n \in \mathbb{N}$$

SOLUTION 2.8

Note that a_n can be written as,

$$a_n = \frac{3n^2 + 4n - 2}{n^2 + 3n + 2} = \frac{3n}{n + 1} - \frac{2}{n + 2} = 3b_n - 2c_n.$$

Then we can show that $\{b_n\}$ and $\{c_n\}$ each converge with limits equal to 1 and 0 respectively. Then by part (a) of Theorem 2.5 we have that the limit of $\{a_n\}$ is equal to 3. This solution is intended to illustrate part (a) of Theorem 2.5. The same result follows more easily if we divide numerator and denominator of $\{a_n\}$ by n^2.

Properties of Sequences

PROBLEM 2.9

Prove Theorem 2.1, that convergence implies boundedness.

SOLUTION 2.9

Suppose $\{a_n\}$ is convergent with limit equal to L. Then there exists ν in $\mathbb{N}$ such that $|a_n - L| < 1$ for $n > \nu$. That is, $|a_n| < |L| + 1$ for all $n > \nu$. Let M denote the largest of the $\nu + 1$ numbers, $|a_1|, |a_2|, \ldots, |a_n|, |L| + 1$. Then $|a_n| < M$ for all n.

PROBLEM 2.10

Prove Theorem 2.3 on the uniqueness of limits.

SOLUTION 2.10

Suppose $\{a_n\}$ converges to limits L and M. Choose $\in > 0$ sufficiently small that $N_\in(L)$ and $N_\in(M)$ are disjoint. Then there exists positive integer ν such that a_n belongs to $N_\in(L)$ for all $n > \nu$ and a_n belongs to $N_\in(M)$ for all $n > \nu$. This is clearly impossible and hence distinct limits L, M cannot exist.

PROBLEM 2.11

Prove Theorem 2.4, the "squeeze play" theorem.

SOLUTION 2.11

Suppose that $\{a_n\}$ is a convergent sequence with limit equal to L, and that the b_n are "squeezed" between the a_n and the limit L; i.e.,

$$L \le b_n \le a_n \qquad \text{for all n.} \tag{1}$$

For each $\in > 0$ there exists a ν in $\mathbb{N}$ such that $|a_n - L| < \in$ for all $n > \nu$. But (1) implies that $|b_n - L| < |a_n - L|$ for all n and hence $|b_n - L| < \in$ for all $n > \nu$.

PROBLEM 2.12

Prove that the sequence $\{2^{1/n}\}$ converges to the limit 1.

Additional Convergence Proofs

SOLUTION 2.12

For each positive integer n, we have

$$2^{1/n} > 1. \tag{1}$$

Then for each n in $\mathbb{N}$, there exists a unique number, $a_n > 0$, such that

$$2^{1/n} = 1 + a_n. \tag{2}$$

By the result known as Bernoulli's inequality, proved in Problem 1.6, we have

$$2 = (1 + a_n)^n \ge 1 + na_n \qquad \text{for all n in } \mathbb{N}. \tag{3}$$

It follows from (3) that $a_n \le 1/n$ and this, together with (1) and (2) leads to

$$1 < 2^{1/n} = 1 + a_n \le 1 + 1/n \qquad \text{for all n in } \mathbb{N}.$$

Since $1 + 1/n$ converges to the limit 1, we can apply the squeeze play theorem to conclude that $2^{1/n}$ must also converge to 1. Note that introducing the auxiliary sequence $\{a_n\}$ made it easier to apply the squeeze play theorem.

PROBLEM 2.13

Prove that the sequence $\{n^{1/n}\}$ converges to the limit 1.

SOLUTION 2.13

Note that for every n in $\mathbb{N}$,

$$n^{1/n} \ge 1. \tag{1}$$

Then there exist real numbers a_n with $a_1 = 0$ and $a_n > 0$ for $n > 1$, such that

$$n^{1/n} = 1 + a_n \qquad \text{for all n in } \mathbb{N}. \tag{2}$$

Then

$$n = (1 + a_n)^n = 1 + na_n + \frac{n(n-1)}{2}a_n^2 + \cdots$$

i.e.
$$n \ge 1 + \frac{1}{2}n(n-1)a_n^2 \qquad \text{for all n in } \mathbb{N}. \tag{3}$$

We have used the binomial theorem in deriving (3). Now (3) implies

$$n - 1 \ge \frac{1}{2}n(n-1)a_n^2$$

i.e.
$$a_n^2 \le 2/n \text{ for } n > 1 \tag{4}$$

Then (4), together with (1)and (2) leads to the result,

$$1 \le n^{1/n} = 1 + a_n \le 1 + \sqrt{2/n} \qquad \text{for n in } \mathbb{N}.$$

Since $1 + \sqrt{2/n}$ converges to the limit 1, the result follows by the squeeze play theorem.

PROBLEM 2.14

Show that the sequence $\{e_n\}$ with $e_n = (1 + 1/n)^n$ for $n = 1, 2, \ldots$ converges to a limit, e, whose value is between 2 and 3.

SOLUTION 2.14

We can compute the values $e_1 = 2$, $e_2 = 2.25$, $e_3 = 2.37$, $e_4 = 2.441, \ldots$ which suggests that this is an increasing sequence. If we can prove it is increasing and prove that the sequence is bounded above by 3, then the result will follow from Corollary 2.2.

We can apply the binomial theorem to obtain

$$e_n = 1 + \frac{n}{1}\frac{1}{n} + \frac{n(n-1)}{2!}\frac{1}{n^2} + \frac{n(n-1)(n-2)}{3!}\frac{1}{n^3} + \cdots + \frac{n(n-1)\cdots 2\cdot 1}{n!}\frac{1}{n^n}$$

i.e.

$$e_n = 1 + 1 + \frac{1}{2!}(1 - 1/n) + \frac{1}{3!}(1 - 1/n)(1 - 2/n) + \cdots + \frac{1}{n!}(1 - 1/n)(1 - 2/n)\cdots\left(1 - \frac{n-1}{n}\right). \tag{1}$$

Then e_n consists of a sum of $n + 1$ terms. In the same way, we find that e_{n+1} consists of a sum of $n + 2$ terms,

$$e_{n+1} = 1 + 1 + \frac{1}{2!}\left(1 - \frac{1}{n+1}\right) + \frac{1}{3!}\left(1 - \frac{1}{n+1}\right)\left(1 - \frac{2}{n+1}\right) + \cdots + \frac{1}{(n+1)!}\left(1 - \frac{1}{n+1}\right)\left(1 - \frac{2}{n+1}\right)\cdots\left(1 - \frac{n}{n+1}\right).$$

Note that the terms in the sum for e_n are less than or equal to the corresponding terms in the sum for e_{n+1}. Thus it follows that,

$$2 \le e_1 \le e_2 \le \cdots \le e_n \le e_{n+1} \le \cdots; \tag{2}$$

i.e., $\{e_n\}$ is an increasing sequence. To see that the sequence is bounded, note that

$$(1 - k/n) \le 1 \qquad \text{for } k = 1, 2, \ldots, n. \tag{3}$$

In addition, we can show by mathematical induction that,

$$k! \ge 2^{k-1} \qquad \text{for } k = 1, 2, \ldots \tag{4}$$

Using (3) and (4) in (1), we obtain,

$$2 \le e_n \le 1 + 1 + 2^{-1} + 2^{-2} + \cdots + 2^{-n+1} \tag{5}$$

But for $n = 2, 3, \ldots$ we can show that $S_n = 2^{-1} + 2^{-2} + \cdots + 2^{-n+1} = 1 - 2^{-n+1}$. Then,

$$2 \le e_n \le 3 - 2^{-n+1} < 3 \qquad \text{for all } n \in \mathbb{N}. \tag{6}$$

It follows from (2) and (6) together that $\{e_n\}$ is a bounded increasing sequence. Then the sequence is convergent by Corollary 2.2 and since $2 \le e_n < 3$ for each n, it follows that the limit, e, must also satisfy $2 \le e < 3$.

Cauchy Sequences

PROBLEM 2.15

Prove that every convergent sequence is a Cauchy sequence.

SOLUTION 2.15

Suppose $\{a_n\}$ is convergent with limit L. We must show that for each $\in > 0$ there is a $\nu \in \mathbb{N}$ such that $|a_n - a_m| < \in$ for all m, $n > \nu$. However, since $\{a_n\}$ is convergent to L it follows that for each $\in$ we can choose ν in $\mathbb{N}$ such that $|a_n - L|$ $\in/2$ for every $n > \nu$. Then

$$|a_n - a_m| = |a_n - L + L - a_m| < |a_n - L| + |L - a_m| < \in/2 + \in/2 \text{ for m, n > n.}$$

Thus if $\{a_n\}$ is convergent, it is a Cauchy sequence.

PROBLEM 2.16

Prove that if $\{a_n\}$ is Cauchy, then it is bounded.

SOLUTION 2.16

If $\{a_n\}$ is a Cauchy sequence, then there is a ν in $\mathbb{N}$ such that for all integers m, $n > \nu$, $|a_n - a_m| < 1$. That is, $|a_n| < |a_\nu| + 1$ for $n > \nu$. Now let M denote the largest of the $\nu + 1$ numbers, $|a_1|, |a_2|, \ldots, |a_\nu|, |a_\nu| + 1$. It follows that $|a_n| < M$ for all n. Thus every Cauchy sequence is bounded.

PROBLEM 2.17

Prove that if $\{a_n\}$ is a Cauchy sequence then it is convergent.

SOLUTION 2.17

Suppose $\{a_n\}$ is a Cauchy sequence. Then, by the result of the previous problem, it is bounded and by Theorem 2.8, the Bolzano-Weierstrass theorem, it contains a convergent subsequence, $\{a_{n'}\}$, n′ in $\mathbb{N}'$, where $\mathbb{N}'$ denotes a subset of $\mathbb{N}$. Let L denote the limit of the convergent subsequence. Then we must show that $\{a_n\}$ is also convergent to L.

Since $\{a_n\}$ is Cauchy, for each $\in > 0$ there exists an integer ν such that

$$|a_n - a_m| < \in/2 \qquad \text{for all m, n} > \nu. \tag{1}$$

Also, since the subsequence is convergent to L there is a μ in the subset $\mathbb{N}'$ such that $\mu > \nu$ and

$$|a_m - L| < \in/2 \text{ for all m in } \mathbb{N}' \text{ with } m > \mu. \tag{2}$$

In particular, since $\mu > \nu$,

$$|a_n - a_m| < \in/2 \text{ for all m and } n > \mu. \tag{3}$$

Then for all $n > m$,

$$|a_n - L| = |a_n - a_m + a_m - L| < |a_n - a_m| + |a_m - L|$$

and the right side of this expression can be made less than $\in/2 + \in/2 = \in$ by choosing m in $\mathbb{N}'$ such that $m > \mu$. It follows that $\{a_n\}$ converges to L; i.e. every Cauchy sequence is convergent.

PROBLEM 2.18

Suppose that $\{a_n\}$ is such that for some constant C, $0 < C < 1$, we have

$$|a_{n+2} - a_{n+1}| \le C\,|a_{n+1} - a_n| \qquad \text{for all n in } \mathbb{N}. \tag{1}$$

Then the sequence $\{a_n\}$ is said to be a contraction. Prove that every sequence that is a contraction is a Cauchy sequence and hence converges.

SOLUTION 2.18

Suppose $\{a_n\}$ is a contraction. Then

$$\begin{aligned} |a_{n+2} - a_{n+1}| &\le C\,|a_{n+1} - a_n| \qquad \text{for all n in } \mathbb{N}. \\ &\le C^2\,|a_n - a_{n-1}| \\ &\le C^3\,|a_{n-1} - a_{n-2}| \le \cdots \le C^n\,|a_2 - a_1|. \end{aligned} \tag{1}$$

Now for integers m and n, the triangle inequality (Theorem 1.2) implies,

$$|a_m - a_n| \le |a_m - a_{m-1}| + |a_{m-1} - a_{m-2}| + \cdots + |a_{n+1} - a_n| \tag{2}$$

We apply (1) in (2) to get

$$\begin{aligned} |a_m - a_n| &\le [C^{m-2} + C^{m-3} + \cdots + C^{n-1}]\,|a_2 - a_1| \\ &\le C^{n-1} \sum_{k=0}^{m-n-1} C^k\,|a_2 - a_1| \end{aligned} \tag{3}$$

Finally, we can use mathematical induction to prove that for $C \ne 1$,

$$1 + C + C^2 + C^3 + \cdots + C^N = \sum_{k=0}^{N} C^k = \frac{1 - C^{N+1}}{1 - C} \tag{4}$$

Substituting (4) into (3) leads to

$$\begin{aligned} |a_m - a_n| &\le C^{n-1}\,\frac{1 - C^{m-n}}{1 - C}\,|a_2 - a_1| \\ |a_m - a_n| &\le \frac{C^{n-1}}{1 - C}\,|a_2 - a_1| \end{aligned} \tag{5}$$

Since $0 < C < 1$, C^{n-1} tends to zero. Then for any $\in > 0$, the right side of (5) is less than $\in$ for n sufficiently large, which is to say, $\{a_n\}$ is a Cauchy sequence.

Since a contraction is a Cauchy sequence, it is convergent. Letting m tend to infinity on the left side of (5), a_m tends to the limit, L, of the sequence. Then we have

$$|L - a_n| \le \frac{C^{n-1}}{1 - C}\,|a_2 - a_1| \qquad \text{for all n.} \tag{6}$$

PROBLEM 2.19

Let A denote a set of real numbers. Then prove the following statements are equivalent:

(a) A is compact
(b) Every sequence in A contains a subsequence that converges to a point of A

SOLUTION 2.19

(a) implies (b) If A is compact then A is closed and bounded. Then any sequence in A is bounded, and by the Bolzano-Weierstrass theorem it contains a convergent subsequence. The limit point of this subsequence is an accumulation point for A and since A is closed, this point belongs to A.

(b) implies (a) We will prove this by showing that if (a) is false then (b) must also be false. If (a) is false then A is not closed or A is not bounded or both.

Suppose A is not closed. Then there exists a point p not in A that is an accumulation point for A. Since p is an accumulation point for A, there exists a sequence of points of A that converge to p. Then by Theorem 2.6 every subsequence of this sequence must also converge to p. But Theorem 2.7 then implies that none of these subsequences can converge to a point of A. Thus (b) is false if A is not closed.

Suppose A is not bounded. For purposes of discussion, suppose A has no upper bound. Then there exists a sequence $\{p_n\}$ of points in A such that $p_1 \in A$ and $p_{n+1} > p_n + 1$. Then for all m, n with m > n, we have $p_m - p_n > 1$. Then no subsequence of $\{p_n\}$ is a Cauchy sequence. Thus (b) is false if A_n is not bounded.

We have proved that if (a) is false then (b) must be false, which is equivalent to showing that (b) implies (a). This is Theorem 1.13, providing a characterization of compactness that is equivalent to the definition.

In this chapter we introduced the notion of a sequence *as a special type of function, one whose domain of definition is the natural numbers. A sequence is said to be* convergent *to the limit L if, for each $\in > 0$ there is a ν in $\mathbb{N}$ such that a_n belongs to $N_\in(L)$ for all $n > \nu$. A sequence that is not* convergent *is divergent. The sequence $\{a_n\}$ is said to be:*

bounded	if there is an M > 0 such that $\lvert a_n \rvert < M$ for all n
increasing	if $a_{n+1} \geq a_n$ for all n
decreasing	if $a_{n+1} \leq a_n$ for all n
monotone	if it is either increasing or decreasing

In addition to the definition for convergence, there are many theorems that may be used to determine whether a sequence is convergent or divergent.

For example: Every sequence that is convergent is bounded thus every sequence that is not bounded is divergent.

A point P is a limit point of the sequence a_n if there is a subsequence of a_n that converges to P. If a sequence has more than a single limit point, it cannot be convergent; the limit of a convergent sequence is the unique limit point for that sequence.

While there are bounded sequences that do not converge, (e.g., $a_n = (-1)^n$), every bounded sequence contains a convergent subsequence. This is a version of the Bolzano-Weierstrass theorem that was stated in Chapter 1 in the form "every bounded infinite set of real numbers contains an accumulation point."

A monotone sequence is convergent if and only if it is bounded. For a sequence that is not monotone we have the Cauchy criterion which states that the sequence $\{a_n\}$ converges if and only if for every $\in > 0$ there is a ν in $\mathbb{N}$ such that

$$|a_n - a_m| < \in \qquad \textit{for all } m, n > \nu.$$

The Cauchy criterion is a characterization of convergence that is equivalent to the definition. In certain situations this characterization is more convenient to use than the definition. For example, contraction sequences are Cauchy sequences, hence they converge.

SUPPLEMENTARY PROBLEMS

Use the definition of convergence to establish the convergence or divergence of the following sequences:

2.1 $a_n = 2/\sqrt{n}$

2.2 $a_n = \dfrac{(-1)^n}{n^2+1}$

2.3 $a_n = \dfrac{3}{2+n}$

2.4 Given that $a_n = 1/n$ converges to zero, use the squeeze play theorem to prove that

$$b_n = \frac{1}{\sqrt{n^2+1}}$$

The convergence or divergence of each of the following sequences can be established by simply stating one of the theorems on sequences in the chapter. For each sequence, state the appropriate theorem.

2.5 $a_n = \sqrt{n}$

2.6 $\{a_n\} = \{1, 1/2, 1, 1/3, 1, 1/4, 1, 1/5, \ldots\}$

2.7 $a_n = 1 - 10^{-n}$

2.8 $a_n = \sum_{m=1}^{n} 1/m$

(Hint: Show that $a_{2n} - a_n$ does not tend to zero with increasing n. Then use the Cauchy criterion for convergence.)

3

Continuity

In this chapter we begin by defining the fundamental notion of continuity *for real valued functions of a single real variable. We also introduce the notion of* function limit *and use this concept to give equivalent characterizations of continuity. When trying to decide whether a given function is or is not continuous, it is often helpful to have more than one way of characterizing continuity.*

A function that is continuous on an interval has a number of special properties. Some of these important consequences of continuity include:

- *The Bounded Range Theorem (Corollary 3.7)*
- *The Extreme Value Theorem (Corollary 3.8)*
- *The Intermediate Value Theorem (Corollary 3.10)*

The notion of uniform continuity *is also introduced. This notion will be of particular importance in connection with the discussion of integration.*

Finally, we introduce the related notions of injectivity *and* strict monotonicity *for functions. We show that:*

Injectivity is a necessary and sufficient condition for a function to have an inverse.

If the function is injective and is continuous as well, then the inverse is also continuous.

Strict monotonicity is a sufficient condition for injectivity.

FUNCTIONS

Domain and Range

A real valued *function* of a single real variable can be defined as a rule which assigns to each real number x in a subset D of the reals, a uniquely determined real number f(x). The set D is called the *domain* of the function and the set of values y = f(x) obtained as x varies over D is called the *range* of the function f(x). A function can be more precisely defined as a set of ordered pairs (x, y) such that no two distinct pairs have the same first element. The set D of all x values in the collection is the domain of the function and the set of all values $y = f(x)$ forms the range of the function. The *graph* of the function is the set of all points $(x, f(x))$ in the plane.

Example 3.1 Functions

3.1 (a) The function $f(x) = x^2$ with domain D equal to the closed interval $[0, 2]$ assigns to each x in D, the real number x^2. The range of this function is then the set $[0,4]$. The graph of the function is the collection of points, (x, x^2), $0 \le x \le 2$, in the plane. These points form an arc of a parabola. Note that the set $[0,2]$ is not the largest possible domain for this function.

3.1 (b) The function $f(x) = \sqrt{x - 3}$ with domain D equal to the unbounded interval, $x \ge 3$, has for its range the unbounded interval $y \ge 0$. Note that for $x < 3$, this function does not produce real values. Thus the set $x \ge 3$ is the largest possible domain for this function.

3.1 (c) The function defined by

$$f(x) = \begin{cases} -1 & \text{for } -1 \le x < 0 \\ 1 & \text{for } 0 \le x \le 1 \end{cases}$$

for x in the domain $D = \{x : -1 \le x \le 1\}$ has for its range the set consisting of the two points $y = 1$ and $y = -1$. The graph of this function is two disconnected horizontal line segments. Since the function is not given by a single formula that applies over its entire range, we say this function is *piecewise defined.*

CONTINUITY

Definition The function $f(x)$ with domain D is *continuous at the point* c in D if, for every $\in > 0$ there exists a $\delta > 0$ such that $f(x)$ belongs to $N_\delta[f(c)]$ whenever x belongs to $N_\in[c] \cup D$.

If $f(x)$ is continuous at every point c of D, we say $f(x)$ is *continuous*, or more precisely, $f(x)$ is *continuous on* D. A function that is not continuous is said to be *discontinuous*.

Example 3.2 Continuous Functions

3.2 (a) For fixed constants $a \ne 0$ and b, consider the linear function $f(x) = ax + b$ with domain $D = \mathbb{R}$. We can use the definition of continuity to show that this function is continuous at every point c in $\mathbb{R}$.

Let c in $\mathbb{R}$ be fixed but arbitrary. Then

$$|f(x) - f(c)| = |ax + b - (ac + b)| = |a|\,|x - c|$$

and for $\in > 0$ given, it is easy to see that

$$|f(x) - f(c)| < \epsilon \text{ for x in D such that } |x - c| < \frac{\epsilon}{|a|}$$

Therefore, for each $\epsilon > 0$, there exists a $\delta = \epsilon/|a| > 0$ such that f(x) belongs to $N_\epsilon[f(c)]$ whenever x belongs to $N_\delta[c]$. Then by definition, f(x) is continuous at c. Since the proof did not depend on c having any particular value it follows that f(x) is continuous at every point c in $\mathbb{R}$.

3.2 (b) In a similar fashion we can show that the function $f(x) = x^2$ with $D = [0, 2]$ is continuous at each point c in D. Suppose c in D is fixed but arbitrary. Then

$$|f(x) - f(c)| = |x^2 - c^2| = |x - c||x + c|$$

and for $\epsilon > 0$, given, $|f(x) - f(c)| < \epsilon$ if x in D satisfies

$$|x - c| < \frac{\epsilon}{|x + c|}.$$

But for x in D=[0, 2], and c fixed in D

$$\frac{\epsilon}{2 + c} \leq \frac{\epsilon}{|x + c|} \leq \frac{\epsilon}{c}$$

and it follows that

$$|f(x) - f(c)| < \epsilon \text{ for all x in D such that } |x - c| < \delta = \frac{\epsilon}{2 + c}$$

Thus for any $\epsilon > 0$, and $\delta = \delta(e)$ as above, f(x) belongs to $N_\epsilon[f(c)]$ whenever x belongs to $N_\delta[c]$. The definition of continuity implies that f(x) is continuous at the point c in D. Since we have assumed nothing special about c for this proof it follows that f(x) is continuous at each c in D.

3.2 (c) Each of the following functions can be shown to be continuous at every real value x:

polynomials $P(x)=a_nx^n + a_{n-1}x^{n-1} + \cdots + a_1x + a_0$

Sin x, Cos x, e^x

Each of the following functions is continuous at all points where it is defined:

$r(x) = \frac{P(x)}{Q(x)}$ for polynomials P, Q is continuous except where $Q(x) = 0$

Tan x, Sec x are continuous except at $x_n = (2n + 1)\pi/2 \; n \in \mathbb{Z}$
Cot x, Csc x are continuous except at $x_n = n\pi \; n \in \mathbb{Z}$

Lnx is continuous except at $x = 0$

Proving continuity or discontinuity for a function at a point in its domain by means of the definition is often cumbersome. For that reason we need alternative characterizations for continuity. Function limits and sequences lead to useful alternative descriptions for continuity.

FUNCTION LIMITS

Definition of function limit

In the previous chapter we considered limits of sequences. Now we shall introduce the notion of *function limits*. Let f(x) be a real valued function with domain D in $\mathbb{R}$, and let c be an accumulation point in D. We say that the limit of f(x) as x approaches c exists and equals L if, for every $\in > 0$, there exists a $\delta > 0$ such that $|f(x) - L| < \in$ whenever $0 < |x - c| < \delta$. We indicate this by writing

$$\lim_{x \to c} f(x) = L.$$

Note that c is an accumulation point of D but need not belong to D. Note also that we allow x to *approach* the value c but we do not allow x to become *equal to* c.

Limit as x tends to infinity

We can also define the limit as x tends to infinity,

$$\lim_{x \to \infty} f(x) = L$$

to mean that for every $\in > 0$ there exists a $B > 0$ such that $|f(x) - L| < \in$ for all x such that $x > B$. The limit as x approaches $-\infty$ is defined in an analogous fashion.

Example 3.3 Function Limits via the Definition

Consider the following limit

$$\lim_{x \to 4} \sqrt{x}.$$

Then $c = 4$ lies in the allowable domain of the function $f(x) = \sqrt{x}$ and we can use the definition to show that the limit of f(x) as x approaches 4 exists and equals 2. Let $\in > 0$ be given. Then we must find $\delta = \delta(\in)$ such that $|\sqrt{x} - 2| < \in$ whenever, $0 < |x - 4| < \delta$. If we write

$$|\sqrt{x} - 2| = \frac{|x - 4|}{\sqrt{x} + 2}$$

and note that $\sqrt{x} + 2 > 3$ if $|x - 4| < 1$, then it follows that

$$|\sqrt{x} - 2| < |x - 4|/3 \text{ if } |x - 4| < 1. \tag{1}$$

Then if we choose $\delta(\in)$ to be the smaller of the two numbers 1 and $3\in$, we see at once from (1) that

$$|\sqrt{x} - 2| < \in \text{ whenever } |x - 4| < \delta.$$

Function limits are closely related to limits of sequences.

Theorem 3.1 Function Limits and Sequences

Theorem 3.1 Suppose f(x) is a real valued function with domain D in $\mathbb{R}$ and let c be an accumulation point of D. Then the following are equivalent,

(i) $\lim_{x \to c} f(x) = L$

(ii) if the sequence $\{a_n\}$ in D converges to c then the sequence $\{f(a_n)\}$ converges to L.

Theorem 3.1 is often useful for identifying function limits that do *not* exist.

Example 3.4 Function Limits

3.4(a) Consider the function

$$f(x) = \begin{cases} -1 & \text{for } -1 \leq x < 0 \\ 1 & \text{for } 0 \leq x \leq 1 \end{cases}$$

We can use Theorem 3.1 to show that the limit of f(x) as x approaches 0 does not exist. The sequence

$$a_n = \frac{(-1)^n}{n} \qquad n \in \mathbb{N}$$

converges to the limit point $c = 0$, but $a_n < 0$ for odd n and $a_n > 0$ for n even. This leads to the following corresponding sequence of function values

$$f(a_n) = \begin{cases} -1 & \text{for n odd} \\ 1 & \text{for n even} \end{cases}$$

Since this sequence has two limit points it is not convergent. We have found a sequence $\{a_n\}$ converging to c for which $\{f(a_n)\}$ does not converge to any L; i.e., for this f(x) and limit point c and any L, statement (ii) of Theorem 3.1 fails. Since (i) and (ii) are equivalent (i) must also fail for every L.

3.4 (b) Consider the function

$$f(x) = \frac{1}{x} \text{ with } D=(0, \infty).$$

Then $c = 0$ is an accumulation point of D and we can use Theorem 3.1 to show that the limit of f(x) as x approaches 0 does not exist.

The sequence $a_n = 1/n$, $n \in \mathbb{N}$, is a sequence in D that tends to the limit $c = 0$. However the sequence $\{f(a_n) = n\}$ of corresponding function values is not bounded. Hence the sequence $\{f(a_n)\}$ does not converge and Theorem 3.1 implies then that the function limit does not exist.

Analogous to Theorems 2.4 and 2.5 for sequences, we have the following theorem for function limits:

Theorem 3.2 Properties of Limits

Theorem 3.2 Suppose f(x) and g(x) are real valued functions, each with domain D in $\mathbb{R}$. Suppose also that c is an accumulation point of D and that

$$\lim_{x \to c} f(x) = L \qquad \lim_{x \to c} g(x) = K.$$

Then

(i) $\lim_{x \to c} (\alpha f(x) + \beta g(x)) = \alpha L + \beta K$ for any α, β in R

(ii) $\lim_{x \to c} f(x)\,g(x) = LK$

(iii) $\lim_{x \to c} f(x)/g(x) = L/K$ if $K \neq 0$.

(iv) If $K = L$ and if $f(x) \leq h(x) \leq g(x)$ for x in D, $x \neq c$, then $\lim_{x \to c} h(x) = L$.

EQUIVALENT DEFINITIONS OF CONTINUITY

We can define continuity equivalently in terms of function limits.

Theorem 3.3 Limit definition of continuity

Theorem 3.3 The function f(x) with domain D is continuous at the point c in D if and only if,

$$\lim_{x \to c} f(x) = f(c).$$

Note that Theorem 3.3 states that c *must belong* to D so that f(c) is defined; it requires further that f(x) must tend to some limit L, as x tends to c; finally it requires that $f(c) = L$. If any of these conditions fails then f is not continuous at the point $x = c$.

Theorems 3.3 and 3.1, taken together give one more equivalent definition for continuity.

Theorem 3.4 Sequence definition of

Theorem 3.4 The function f(x) with domain D is continuous at the point c in D if and only if, for every sequence $\{a_n\}$ in D that converges to c, the sequence $\{f(a_n)\}$ converges to f(c).

Theorem 3.4 is used most often to show that a function is *not* continuous. All that is required is to find a sequence $\{a_n\}$ in D that converges to c for which $\{f(a_n)\}$ does not converge.

Continuous functions can be combined in various ways to form new functions that are also continuous:

Theorem 3.5

Theorem 3.5 Suppose f(x) and g(x) are real valued functions with domain D. Suppose also that f and g are each continuous at the point c in D. Then each of the functions $\alpha f(x) + \beta g(x)$, $f(x)\,g(x)$, is continuous at c, and if $g(c) \neq 0$ then so is the function $f(x)/g(x)$. Finally, if F(x) is a function that is defined and continuous in a neighborhood of the point $b = f(c)$, then the composed function $F(f(x))$ is continuous at c.

CONSEQUENCES OF CONTINUITY

There are a number of function properties that follow from continuity. We list the most important of these now.

Theorem 3.6

Theorem 3.6 A function that is continuous on a compact domain has a compact range.

COROLLARY 3.7 BOUNDED RANGE THEOREM

A function that is continuous on a compact domain is bounded.

COROLLARY 3.8 EXTREME VALUE THEOREM

If $f(x)$ is continuous on a compact domain D, then there exist points c and d in D such that $f(c) \le f(x) \le f(d)$ for all x in D

Theorem 3.9

Theorem 3.9 A function that is continuous on a connected domain D has a connected range.

COROLLARY 3.10 INTERMEDIATE VALUE THEOREM

Suppose $f(x)$ is continuous on an interval I. Then for any $p<q$ in I, and any real number S lying between $f(p)$ and $f(q)$, there exists an s in I such that $p \le s \le q$ and $f(s)=S$.

UNIFORM CONTINUITY

A function is said to be *uniformly continuous* on D if, for every $\in > 0$ there exists a $\delta > 0$ such that for all x, y in D

$$|x - y| \le \delta \text{ implies } |f(x) - f(y)| \le \in.$$

Every function that is uniformly continuous on D is continuous at every point of

D but the converse is false unless the domain D is compact (see Problems 3.23 and 3.24).

Theorem 3.11

Theorem 3.11 A function that is continuous at each point of a compact domain D is uniformly continuous on D.

Theorem 3.12

Theorem 3.12 Suppose f(x) is defined and uniformly continuous on domain D. Then whenever $\{a_n\}$ is a Cauchy sequence in D, it follows that $\{f(a_n)\}$ is a Cauchy sequence. If D is bounded, the converse is also true.

INVERSE FUNCTIONS

Injective Functions

A real valued function f(x) on domain D is said to be *one to one* or *injective* if

$$\text{for all } x_1, x_2 \text{ in D, } x_1 \neq x_2 \text{ implies } f(x_1) \neq f(x_2).$$

If f(x) is a function on D then the set of ordered pairs $\{(x,y) : x \in D, y = f(x)\}$ is such that no two pairs have the same first element; if f(x) is injective then no two pairs have the same second element. In this case the set of reversed pairs $\{(y,x) : x \in D, y = f(x)\}$ defines a function x = g(y) with the property,

y=f(x) if and only if x=g(y)

We say that x = g(y) is the *inverse* of the function y = f(x) and we write $g = f^{-1}$

Monotone Functions

A real valued function f(x) whose domain is the interval D is said to be:

increasing on D if $x_1 < x_2$ implies $f(x_1) < f(x_2)$

decreasing on D if $x_1 < x_2$ implies $f(x_1) > f(x_2)$.

We say f(x) is monotone on D if f(x) is either increasing or decreasing on D. If the inequality $f(x_1) \leq f(x_2)$ is replaced by the *strict* inequality $f(x_1) < f(x_2)$, then we say f(x) is *strictly increasing* on D. Strictly decreasing is defined analogously and the function is said to be *strictly monotone* if it is either strictly increasing or strictly decreasing. Evidently, a function f(x) is injective if it is strictly monotone. Conversely we have

Theorem 3.13

Theorem 3.13 If f(x) is injective and continuous on the interval I, then f(x) is strictly monotone on I.

Theorem 3.14 Continuity of the inverse

Theorem 3.14 If f(x) is strictly monotone and continuous on the interval I, then $g = f^{-1}$ is strictly monotone and continuous on the interval J=f[I].

Example 3.3 Injectivity and

3.3 (a) The continuous function $f(x) = x^2$ is strictly increasing on (0,b) for $b > 0$. The continuous function

$$g(x) = \begin{cases} x & \text{for } 0 \le x \le 1 \\ 1 & \text{for } x > 1 \end{cases}$$

is increasing but not strictly increasing on (0, b) for $b > 1$. Then f(x) has a continuous inverse given by $f^{-1}(x) = \sqrt{x}$ but g(x) is not injective and has no inverse.

3.3 (b) The function $f(x) = x^2$ is continuous on (−3, 3) but it is not monotone on this interval. Then Theorem 3.13 implies that f(x) is not injective. Of course this is clear since for each x, $0 < x < 3$, $f(x) = f(-x)$.

3.3 (c) An example of an injective function which is not monotone is

$$g(x) = \begin{cases} x & \text{for } 0 \le x \le 1 \\ 1-x & \text{for } x > 1 \end{cases}$$

Of course this function is discontinuous as predicted by Theorem 3.13.

SOLVED PROBLEMS

CONTINUITY

PROBLEM 3.1

Show that the function $f(x) = 1/x$ with domain $D=(0, \infty)$ is continuous at each point of D.

SOLUTION 3.1

Let c in D be fixed but arbitrary and choose an $\in > 0$. Then let δ be the smaller of the two numbers, $c/2$ and $\in c^2/2$. Then $\delta \le c/2$ implies

$$c/2 < x < 3c/2 \text{ for all x in } N_\delta[c]$$

and

$$|1/x - 1/c| = \frac{|x-c|}{xc} < \frac{|x-c|}{c \cdot c/2} = (2/c^2)|x-c| \text{ for x in } N_\delta[c]\cdot$$

But then since δ is also less than or equal to $\in c^2/2$, this last estimate implies that

$$|1/x - 1/c| < \in \text{ for all x such that } |x - c| < \delta;$$

i.e.

$$f(x) \in N_{\in}[f(c)] \text{ if } x \in N_\delta[c].$$

PROBLEM 3.2

Show that the functions Sin x and Cos x are continuous at each real x.

SOLUTION 3.2

We shall make use of the following facts about the functions Sin x and Cos x:

$$|\text{Sin } x| \le |x| \text{ and } |\text{Sin } x|, |\text{Cos } x| \le 1 \text{ for all real x.} \quad (1)$$

We shall establish the estimates in (1) later. We have trigonometric identities that state that for all real x and y,

$$\text{Sin } x - \text{Sin } y = 2 \text{ Sin}((x - y)/2)(\text{Cos}((x + y)/2) \quad (2)$$

$$\text{Cos } x - \text{Cos } y = 2 \text{ Sin}((x + y)/2)(\text{Sin}((y - x)/2) \quad (3)$$

Using (1) in (2) and (3) leads to the estimates

$$|\text{Sin } x - \text{Sin } y| \le 2 |(x - y)/2| = |x - y| \quad (4)$$

$$|\text{Cos } x - \text{Cos } y| \le 2 |(y - x)/2| = |y - x| \quad (5)$$

It follows at once from (4) and (5) that for all $\in > 0$ we can choose $\delta = \in$ and have

$$f(x) \in N_{\in}[f(y)] \quad \text{whenever } x \in N_\delta[y]$$

in either of the cases f(x) = Sin x or f(x) = Cos x.

PROBLEM 3.3

Use the definition of continuity to show that the function defined by

$$f(x) = \begin{cases} -1 & \text{for } -1 \le x < 0 \\ 1 & \text{for } 0 \le x \le 1 \end{cases}$$

is discontinuous at x=0.

SOLUTION 3.3

For any $\in > 0$, the point f(x) belongs to $N_{\in}[f(0)]$ if and only if

$$|f(x) - 1| < \in.$$

But for any $\delta > 0$, the neighborhood $N_\delta[0]$ contains both positive and negative values for x. For the positive x we have f(x) = 1, while for the negative values of x we have f(x) = –1. Then for any $\in < 2$, there can be no $\delta > 0$ for which x in $N_\delta[0]$ implies f(x) is in $N_{\in}[1]$. Thus f is not continuous at x = 0.

FUNCTION LIMITS

PROBLEM 3.4

Show that

$$\lim_{x \to c} \frac{x+1}{x^2+1} = \frac{2}{3}$$

SOLUTION 3.4

We must show that for a given $\in > 0$, we have

$$\left| \frac{x+1}{x^2+1} - \frac{2}{3} \right| < \in$$

when $|x-1|$ is sufficiently small. In order to see what "sufficiently small" must mean, we write

$$\left| \frac{x+1}{x^2+1} - \frac{2}{3} \right| = \left| \frac{3(x+1)-2(x+2)}{3(x^2+2)} \right| = \left| \frac{(2x-1)(x-1)}{3(x^2+2)} \right|$$

For x such that $|x-1| < 1$, it is clear that

$$|2x-1| < |2x+1| < 5 \quad \text{and} \quad |3(x^2+2)| = 3x^2+6 > 6$$

hence

$$\left| \frac{2x-1}{3(x^2+2)} \right| < \frac{5}{6} \quad \text{for } |x-1| < 1.$$

i.e., we obtained an upper bound for the quotient by finding an upper bound for the numerator and a lower bound for the denominator. Then the quotient is not greater than the quotient of these bounds. It follows that

$$\left| \frac{x+1}{x^2+1} - \frac{2}{3} \right| < \frac{5}{6}|x-1| \quad \text{for } |x-1| < 1.$$

Then

$$\left| \frac{x+1}{x^2+1} - \frac{2}{3} \right| < \in \quad \text{for all x such that } |x-1| < \delta$$

if we choose δ to be the smaller of the two numbers, 1 and $6 \in/5$.

PROBLEM 3.5

Show that (a) $\lim_{x \to \infty} 1/(1+x^2) = 0$ and (b) $\lim_{x \to \infty} x^2/(1+x^2) = 1$.

SOLUTION 3.5

We show first that $1/x$ tends to zero as x tends to infinity. For any fixed $\in > 0$, we simply choose $B > 1/\in$. Then $|1/x| < \in$ whenever $x > B$. Part (ii) of Theorem 3.2 implies that $1/x^2$ also tends to zero as x tends to infinity.

3.5(a) Since

$$0 < \frac{1}{1+x^2} < \frac{1}{x^2}$$

we use the final statement in Theorem 3.2 to conclude that $1/(1+x^2)$ tends to zero as x tends to infinity.

3.5 (b) The second limit follows by writing

$$\frac{x^2}{1+x^2} = \frac{1}{1+1/x^2}$$

and using part (iii) of Theorem 3.2.

PROBLEM 3.6

Show that Sin x tends to zero and Cos x tends to 1 as x tends to zero.

SOLUTION 3.6

In Figure 3.1 we see a circle of radius 1 and center O with angle POR denoted by x. If PQ is perpendicular to the radius OR then it follows that

$$PQ = \text{Sin } x,\ OQ = \text{Cos } x, \text{ and } QR = 1 - \text{Cos } x.$$

The chord PR is less than the circular arc PR and since the circle has radius 1, the arc PR equals the angle x in radian measure. Then $|PR| < x$ so by the Pythagorean theorem

$$|PQ|^2 + |QR|^2 = |PR|^2 < x^2;$$

i.e.,

$$\text{Sin}^2 x + (1 - \text{Cos } x)^2 < x^2.$$

Then for $x > 0$

$$0 < \text{Sin } x < x, \quad 0 < 1 - \text{Cos } x < x$$

and it follows by the last statement in Theorem 3.2 that

$$\lim_{x\to 0} \text{Sin } x = 0 \text{ and } \lim_{x\to 0} \text{Cos } x = 1$$

PROBLEM 3.7

Show that $\lim_{x\to 0} \text{Sin } x/x = 1$.

SOLUTION 3.7

Since $\text{Sin}(-x)/(-x) = \text{Sin } x/x$ it will be sufficient to consider only positive values for x. Referring again to Figure 3.1 we see an arc PR of a circle of radius 1 with center at O. PQ is perpendicular to the radius OR and PS is perpendicular to the radius OR. Then it is evident from the picture that

$$\text{Area } \Delta OPQ < \text{Area of sector } OPR < \text{Area } \Delta OPS,$$

that is,

$$\frac{1}{2}\text{Cos x Sin x} < x/2 < \frac{1}{2}\text{Tan x}. \quad (1)$$

The areas have been expressed in terms of the angle POR which we denote by x. For x > 0 we can divide (1) by 1/2 Sin x to obtain

$$\text{Cos x} < \frac{x}{\text{Sin x}} < \frac{1}{\text{Cos x}}$$

or

$$\text{Cos x} < \frac{\text{Sin x}}{x} < \frac{1}{\text{Cos x}}. \quad (2)$$

Since we know from the previous problem that Cos x tends to 1 as x tends to zero, it follows from the final part of Theorem 3.2 that Sin x/x tends to 1 as x tends to zero.

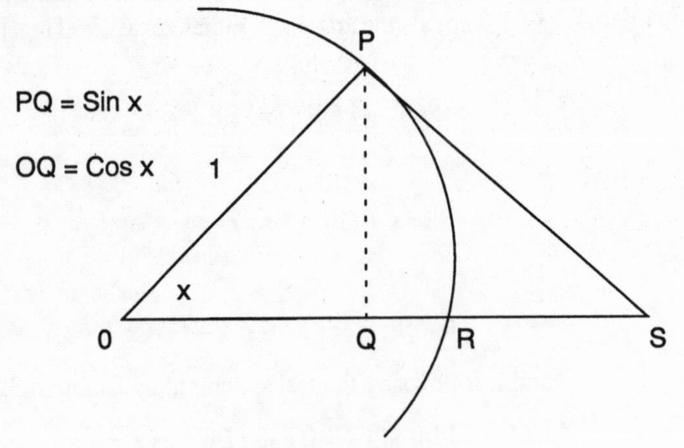

Figure 3.1
Sketch for Problem 3.6

PROBLEM 3.8

Evaluate the limits,

(a) $\lim\limits_{x\to 0} \dfrac{1 - \text{Cos x}}{x}$ (b) $\lim\limits_{x\to 0} \dfrac{1 - \text{Cos x}}{x^2}$

SOLUTION 3.8

We can write

$$\frac{1 - \text{Cos } x}{x} = \frac{\text{Sin}^2 x}{x(1 + \text{Cos } x)} = \left[\frac{\text{Sin } x}{x}\right]^2 \frac{x}{1 + \text{Cos } x}$$

and

$$\frac{1 - \text{Cos } x}{x^2} = - \frac{\text{Sin}^2 x}{x^2(1 + \text{Cos } x)} = \left[\frac{\text{Sin } x}{x}\right]^2 \frac{1}{1 + \text{Cos } x}$$

Then we can use parts (ii) and (iii) of Theorem 3.2, together with the results of the previous two problems to conclude that

$$\text{(a) } \lim_{x \to 0} \frac{1 - \text{Cos } x}{x} = 0 \qquad \text{(b) } \lim_{x \to 0} \frac{1 - \text{Cos } x}{x^2} = 1/2$$

Equivalent Definitions of Continuity

PROBLEM 3.9 REMOVABLE DISCONTINUITIES

Let f(x) be defined on the inteval [–1, 1] by

$$f(x) = \begin{cases} x & \text{if } x \neq 0 \\ 1 & \text{if } x = 0 \end{cases}$$

Show that f(x) is not continuous at the point x = 0.

SOLUTION 3.9

We can show that

$$\lim_{x \to 0} f(x) = 0. \tag{1}$$

Then the function limit at x = 0 exists but since f(0) = 1 we see that the function limit does not equal the function value at x = 0. Then it follows from Theorem 3.3 that f(x) is not continuous at x = 0. This is an example of a *removable discontinuity*. That is, the discontinuity can be removed by redefining the function at the point of discontinuity. In this case if we define f(0) to be equal to zero, then f(x) becomes continuous.

To prove (1) we must show that for every $\in > 0$, there exists a $\delta > 0$ such that $|f(x) - 0| < \in$ whenever $0 < |x - 0| < \delta$. But since f(x) = x for $x \neq 0$, we can choose $\delta = \in$ and the result follows.

PROBLEM 3.10 JUMP DISCONTINUITIES

Let f(x) be defined on D = (–1, 1) by

$$f(x) = \begin{cases} 0 & \text{if } -1 < x < 0 \\ 1 + x^2 & \text{if } 0 \leq x < 1 \end{cases}$$

Then show that f(x) is not continuous at x = 0.

SOLUTION 3.10

We can show that f(x) tends to no limit as x tends to zero. Then Theorem 3.3 implies that f(x) is not continuous at x = 0.

We define a sequence of points a_n in D by

$$a_n = \frac{(-1)^n}{n+1} \quad n = 1, 2, \ldots$$

Then a_n converges to 0 but $a_n < 0$ when n is odd and $a_n > 0$ when n is even. The sequence of function values $\{f(a_n)\}$ does not converge since

$$f(a_n) = \begin{cases} 0 & \text{if n is odd} \\ 1 + a_n{}^2 & \text{if n is even} \end{cases}$$

i.e., the subsequence $\{f(a_{2n})\}$ converges to 1 while the subsequence $\{f(a_{2n-1})\}$ converges to 0 hence it follows from Theorem 2.6 that the sequence $\{f(a_n)\}$ does not converge. Since the sequence does not converge, Theorem 3.1 implies that the function f(x) tends to no limit as x tends to zero.

The function f(x) is said to have a *finite jump discontinuity* at x=0.

PROBLEM 3.11 ONE SIDED LIMITS

Show that the function in the previous problem has *one sided limits* from the left and right at x = 0.

SOLUTION 3.11

Let f(x) be a real valued function with domain D in $\mathbb{R}$, and let c be an accumulation point for the set $D^+ = \{x \in D : x > c\}$. Then we say that the *one sided limit* of f(x) as x approaches c from the right exists and equals L if, for every $\in > 0$ there exists a $\delta > 0$ such that $|f(x) - L| < \in$ for all x such that, $0 < x - c < \delta$. We write $f(c+) = L$ in this case. The one sided limit for f(x) as x approaches c from the left is defined similarly. If this limit exists and equals L, we write $f(c-) = L$.

For the function f(x) from Problem 3.10, we have $f(0-) = 0$ and $f(0+) = 1$. To see this note that for all x in $D- = \{x \in D : x < 0\}$ we have $f(x) = 0$. Then for any $\in > 0$

$$|f(x) - 0| = 0 < \in \quad \text{for all x such that } 0 < 0 - x < \delta.$$

In fact, this holds for any $\delta > 0$. This proves that $f(0-) = 0$. For x in the set $D+ = \{x \in D : x > 0\}$ we have $f(x) = 1 + x$ and hence if we choose $\delta = \sqrt{\in}$, then

$$|f(x) - 1| = |x| < \in \quad \text{for all x such that } 0 < x - 0 < \delta.$$

This proves that $f(0+) = 1$. Since $f(0-)$ and $f(0+)$ both exist but are not equal, we say that f(x) has a *jump discontinuity* at x = 0. The difference $f(0+) - f(0-)$ is equal to the jump in f(x) at x = 0.

PROBLEM 3.12 AN EVERYWHERE DISCONTINUOUS FUNCTION

Show that the function defined by

$$f(x) = \begin{cases} 0 & \text{if x is an irrational number} \\ 1 & \text{1 if x is a rational number} \end{cases}$$

is not continuous at any point. This function is known as the Dirichlet function, named for the mathematician who invented it.

SOLUTION 3.12

For any point c in the domain of this function, we can exhibit a sequence $\{a_n\}$ in D converging to c for which $\{f(a_n)\}$ fails to converge. Then Theorem 3.4 implies f(x) is not continuous at c. Since this can be done for every c in D, it follows that this function is nowhere continuous.

Let c denote any fixed real number. Then it follows from Corollary 1.6 that there exists a sequence of *irrational* numbers $\{a_n\}$ converging to c. In the same way, by Corollary 1.5 there is a sequence $\{b_n\}$ of *rational* numbers that converges to c. We define a new sequence $\{c_n\}$ by letting $c_{2n-1} = a_n$ and $c_2 = b_n$ for $n \in \mathbb{N}$. Then $\{c_n\}$ converges to c, but the sequence of function values $f(c_n)$ does not converge since the subsequence $f(c_{2n-1})$ converges to 0, while the subsequence $\{f(c_{2n})\}$ converges to 1.

Essentially the same argument shows that this function approaches no limit at any point in its domain. Then as an alternative method of proof we could use Theorem 3.3 to conclude that the function is discontinuous at every point.

PROBLEM 3.13

Let the function f(x) be defined on $D =[0, \infty)$ in the following way,

$$f(x) = \begin{cases} 1/x & \text{if } x > 0 \\ a & \text{if } x = 0 \quad \text{where a is a real number.} \end{cases}$$

Show that f(x) is not continuous at $x = 0$ for any choice of the constant, a.

SOLUTION 3.13

In Example 3.4(b) we showed that $f(x) = 1/x$ tends to no limit as x approaches zero. Since there is no limit, the limit cannot equal f(0) for any choice of the value a. Then by Theorem 3.3 f(x) is not continuous at $x = 0$. The function $f(x)=1/x$ has an *infinite jump discontinuity* at $x = 0$.

PROBLEM 3.14

Let the function f(x) be defined on $D=[0, \infty)$ in the following way,

$$f(x) = \begin{cases} \text{Sin}(1/x) & \text{if } x > 0 \\ a & \text{if } x = 0 \quad \text{where } a \in \mathbb{R}. \end{cases}$$

Show that f(x) is not continuous at $x = 0$ for any choice of the constant, a.

SOLUTION 3.14

We can exhibit a sequence $\{a_n\}$ in D tending to zero for which $f(a_n)$ fails to converge. Then Theorem 3.4 implies that f(x) is not continuous at x=0.

The function f(x) oscillates very rapidly between the values +1 and −1 in a neighborhood of x = 0. Thus we can define the following sequence in D:

$$a_n = 1/z_n \in \mathbb{N} \quad \text{where } z_n = \frac{\pi}{2} + n\pi .$$

Then $\{a_n\}$ converges to 0 and

$$f(a_n) = \text{Sin } z_n = \begin{cases} -1 & \text{if n is odd} \\ 1 & \text{if n is even} \end{cases}$$

Since the sequence $\{f(a_n)\}$ has two limit points, it is divergent by Theorem 2.7.

This discontinuity is neither removable nor a jump discontinuity. It is sometimes referred to as an *oscillatory discontinuity.*

Consequences of Continuity

PROBLEM 3.15

Prove Theorem 3.6, a continuous function with a compact domain has a compact range.

SOLUTION 3.15

Suppose f is continuous on the compact domain D, and that the range of f(x) is denoted by B. We shall suppose that B is not compact and show that this leads to a contradiction.

If B is not compact, then by Theorem 1.13, B contains a sequence $\{b_n\}$ with the property that there is no subsequence of $\{b_n\}$ that converges to a point of B. For each n, we let a_n denote a point of D such that $f(a_n) = b_n$. Since D is compact, $\{a_n\}$ contains a subsequence $\{a_{n'}\}$ converging to α in D. But f(x) is continuous on D so Theorem 3.4 implies that $f(a_{n'})$ converges to $f(\alpha)$. Then $\{b_{n'}\} = \{f(a_{n'})\}$ is a subsequence of $\{b_n\}$ and $b_{n'}$ converges to $\beta = f(\alpha)$ in B. This contradicts the assumption that B is not compact and the theorem is proved.

PROBLEM 3.16 THE BOUNDED RANGE THEOREM

Prove Corollary 3.7 from Theorem 3.6

SOLUTION 3.16

Suppose f(x) is continuous on the compact domain D. Then Theorem 3.6 implies that f(x) has compact range. In particular, if D is a closed bounded interval [a, b] then f(x) has compact range.

Since the range of f(x) is compact, it is bounded; i.e. there exists a constant, $M > 0$ such that $|f(x)| \le M$ for all x in D. This is the same thing as saying f(x) is bounded on D. Thus Corollary 3.7 follows at once from Theorem 3.6. This is sometimes called the *bounded range* theorem. It says that a continuous function on a bounded domain must have a bounded range.

PROBLEM 3.17 THE EXTREME VALUE THEOREM

Prove Corollary 3.8 from Theorem 3.6

SOLUTION 3.17

Suppose f has a domain D which is compact. Then according to Theorem 3.6 the range of f must also be compact.

To prove Corollary 3.8, we see by Theorem 1.12, that the compact range contains its least upper bound F^* and its greatest lower bound, F_*. Since F^* and F_* belong to the range, there exist points c and d in D such that $f(c) = F_*$ and $f(d) = F^*$. Then $f(c) \leq f(x) \leq f(d)$ for all x in D. This is Corollary 3.8, sometimes referred to as the *extreme value* theorem. It says that if f(x) is continuous on a closed, bounded domain, then there are points in the domain where f(x) attains its maximum and minimum values.

PROBLEM 3.18

Prove Theorem 3.9, a function that is continuous on a connected domain has a connected range.

SOLUTION 3.18

Suppose f(x) is continuous at each point c of the connected domain D. Let the range of f(x) be denoted by B and suppose that B is not connected. We shall show that this leads to a contradiction.

If B is not connected, then B can be represented as the union of sets B_1 and B_2, that are nonempty, disjoint and neither contains a limit point of the other. Now let $D_1 = \{x \in D : f(x) \in B_1\}$ and $D_2 = \{x \in D : f(x) \in B_2\}$. Then the sets D_1, D_2 are nonempty, disjoint and, since D is connected, at least one of them must contain an accumulation point of the other. In particular, suppose c in D_1 is an accumulation point for D_2. Then there exists a sequence $\{c_n\}$ in D_2 converging to c in D_1. But f(x) is continuous at each point of D and therefore $f(c_n)$ converges to f(c). But then f(c) belongs to B_1 and is an accumulation point for B_2. This contradicts the assumptions that B_1 and B_2 are nonempty, disjoint and neither contains a limit point of the other. This proves that B is connected.

PROBLEM 3.19 THE INTERMEDIATE VALUE THEOREM

Prove Corollary 3.10.

SOLUTION 3.19

Theorem 3.9, taken together with Theorem 1.14, implies that if f(x) is continuous on domain D, and D is the interval I, then the range of f(x) is also an interval. We agree that a point is a degenerate case of an interval. The range of f(x) is the set of points $J = f[I] = \{y : y = f(x) \text{ for } x \in I\}$, so as x moves across the interval I, $y = f(x)$ sweeps out the interval J. One consequence of this observation is the *intermediate value theorem* which states that

if f(x) is continuous on the interval I, then for $p < q$ in I and S lying between the values f(p) and f(q), there exists an s such that $p \le s \le q$ and $f(s) = S$.

This is Corollary 3.10.

PROBLEM 3.20

Use the intermediate value theorem to prove that every polynomial of odd degree must have at least one real root.

SOLUTION 3.20

If P(x) is a polynomial of *odd* degree, then the sign of P(x) as x tends to plus infinity is opposite to the sign of P(x) when x tends to minus infinity. Then it is always possible to find real numbers $p < q$ such that 0 lies between f(p) and f(q). Then the intermediate value theorem implies the existence of a real number s between p and q such that $f(s) = 0$.

In particular suppose

$$P(x) = a_n x^n + a_{n-1}x^{n-1} + \cdots + a_1 x + a_0$$

for n and odd integer and $a_n > 0$. Then $P(p) < 0 < P(q)$ for p sufficiently large, negative and q sufficiently large positive. It follows that for some s between p and q we have $P(s) = 0$.

PROBLEM 3.21 PERSISTENCE OF SIGN

Let f(x) be defined and continuous on an open interval I containing the point c. If $f(c) \neq 0$ then there is an interval about the point c on which f(x) has the same sign as f(c).

SOLUTION 3.21

Since I is an open interval containing the point c, there exists $h>0$ such that $(c-h, c+h)$ is contained in I. Suppose for the sake of discussion that $f(c) > 0$. Then for $\epsilon = f(c)/2 > 0$ there is a δ, $0 < \delta \le h$, such that

$$|f(x) - f(c)| < \epsilon = f(c)/2 \quad \text{for all } x \text{ such that } |x - c| < \delta.$$

Thus,

$$0 < f(c)/2 < f(x) < 3f(c)/2 \quad \text{for all } x \text{ such that } c - \delta < x < c + \delta.$$

Since $\delta \le h$, the interval $c - \delta < x < c + \delta$ is contained in I. The proof when f(c) is negative is similar. A related result holds when the interval I is allowed to be closed and c an endpoint. In this case the interval on which f has the same sign as f(c) will lie on one side of c.

PROBLEM 3.22 A FIXED POINT THEOREM

Suppose f(x) is continuous on the interval I= [0, 1] and that $0 \le f(x) \le 1$ for x in I. Then show that for some c in I, we have f(c) = c; i.e. if f maps I into I continuously, then f has a *fixed point*.

SOLUTION 3.22

If f(x) is continuous on I, then the function g(x) = f(x) – x is also continuous on I. Moreover, $g(0) = f(0) \ge 0$ and $g(1) = f(1) - 1 \le 0$. If either of the values g(0) or g(1) is zero then the fixed point is an end point of [0, 1]. If neither g(0) nor g(1) is zero, then we have $g(1) < 0 < g(0)$ and the intermediate value theorem implies the existence of a c in I such that g(c) = 0; i.e. such that f(c) = c.

Uniform Continuity

PROBLEM 3.23

Show that f(x) = 1/x is not uniformly continuous on D = (0,1).

SOLUTION 3.23

In order to show that f(x) = 1/x is not uniformly continuous on (0,1) we will find an $\epsilon > 0$ such that for all $\delta > 0$ there exist a, b in D with $|a - b| < \delta$ but $|f(a) - f(b)| > \epsilon$.

If we fix $\epsilon = 1/2$, then it will be sufficient to show we can always find a and b in D such that

$$|a - b| < \delta \text{ but } |f(a) - f(b)| = \left|\frac{1}{a} - \frac{1}{b}\right| > \frac{1}{2}.$$

Since the function f(x) = 1/x grows without bound as x tends toward zero, it seems reasonable that we should be able to find a and b near zero such that even if a and b are close together, f(a) and f(b) remain separated by at least ϵ. In particular, for δ given, $0 < \delta < 1$, we can choose $a = \delta$ and $b = \delta - \delta^2/2$. Then since $\delta^2 < \delta$ for $0 < \delta < 1$, we have $0 < b < a < 1$ and,

$$|f(a) - f(b)| = \left|\frac{b - a}{ab}\right| > \frac{\delta^2/2}{\delta \cdot \delta} = \frac{1}{2}$$

It follows that f(x)=1/x is not uniformly continuous on (0, 1).

PROBLEM 3.24

Prove Theorem 3.11, a function continuous on a compact domain is uniformly continuous.

SOLUTION 3.24

Suppose f(x) is continuous on a closed, bounded interval I. If f is not uniformly continuous on I then there exists an $\epsilon > 0$ and sequences $\{a_n\}$ and $\{b_n\}$ in I such that for all n in N,

$$|a_n - b_n| < 1/n \qquad (1)$$

and

$$| f(a_n) - f(b_n) | > \in \quad (2)$$

Since I is compact, it follows from Theorem 1.13 that $\{a_n\}$ contains a convergent subsequence $\{a_{n'}\}$ converging to limit a in I. Since

$$| b_{n'} - a | < | b_{n'} - a_{n'} | + | a_{n'} - a |$$

it follows from (1) that the subsequence $\{b_{n'}\}$ of the sequence $\{b_n\}$ must also converge to the limit a. Since f(x) is continuous at each point of I, both of the subsequences $\{f(a_{n'})\}$ and $\{f(b_{n'})\}$ must converge to the value f(a). But this is in contradiction to (2). Then if f is continuous on a compact domain, it must be uniformly continuous on that domain.

PROBLEM 3.25

Use Theorem 3.12 to prove that f(x) = Sin1/x is not uniformly continuous on the interval (0,1).

SOLUTION 3.25

The sequence $a_n = 2/(n\pi)$ is a Cauchy sequence in the bounded interval (0, 1) but

$$f(a_n) = \begin{cases} \pm 1 & \text{if n is odd} \\ 0 & \text{if n is even} \end{cases}$$

so $f(a_n)$ is not a Cauchy sequence. Then Theorem 3.12 implies that f(x) is not uniformly continuous on (0, 1).

Inverse Functions

PROBLEM 3.26

Prove that if f(x) is continuous and injective on the interval I = [a, b] then f(x) is strictly monotone on I.

SOLUTION 3.26

Since f(x) is injective on I, f(a) ≠ f(b). Suppose f(b) > f(a). Then we shall show that f(x) is strictly increasing.

Choose an x, a < x < b. Then we must have f(a) < f(x) < f(b). To see this, suppose we have f(x) < f(a) < f(b). Then we can apply Corollary 3.10, the intermediate value theorem, to the interval [x, b] to conclude that there exists a point c, x < c < b, with f(c) = f(a). But this contradicts the assumption that f(x) is injective. Similarly, if we suppose f(a) < f(b) < f(x), then we obtain in the same way, a point c, a < c < x such that f(c) = f(b); again this contradicts the assumption of injectivity for f(x). It follows that f(a) < f(x) < f(b) for a < x < b.

Now choose y, a < x < y < b. Just as we did above, we can show we must have f(a) < f(y) < f(b). If we suppose that f(a) < f(y) < f(x), then the same sort of argument leads to the existence of a point c, a < c < x such that f(c) = f(y). Since this contradicts the injectivity assumption on f(x), we conclude $f(x) < f(y)$ for $x < y$; i.e., f is strictly increasing.

PROBLEM 3.27 THE CONTINUOUS INVERSE THEOREM

Prove that if f(x) is strictly monotone and continuous on $I = [a, b]$, then $g = f^{-1}$ is continuous on the interval $J = [c, d]$.

SOLUTION 3.27

We shall suppose f(x) is strictly increasing on I. The proof for the case where f(x) is strictly decreasing is similar.

In order to prove g(y) is continuous, we shall suppose that g(y) is discontinuous at some point y_0 in J and show that this leads to a contradiction. If g(y) is discontinuous at y_0 in J, it means that there exists an $\in > 0$ such that for every $\delta > 0$ there is a y in J such that $|y - y_0| < \delta$ while at the same time, $|g(y) - g(y_0)| > \in$. In particular we can find a sequence $\{y_n\}$ in J such that

$$|y_n - y_0| < 1/n \text{ for n in N} \tag{1}$$

and

$$|x_n - x_0| > \in. \tag{2}$$

Here $x_n = g(y_n)$; i.e. $y_n = f(x_n)$.

It follows from (1) that y_n converges to y_0. In addition, the sequence $\{x_n\}$ is bounded since it is contained in I. Then $\{x_n\}$ contains a convergent subsequence $\{x_{n'}\}$ with limit x^0 in I. However, it follows from (2) that x^0 cannot be equal to x_0.

Since f(x) is continuous on I, it is continuous at x^0 and therefore $x_{n'}$ converging to x^0 implies $f(x_{n'})$ converges to $f(x^0)$. But y_n converges to $y_0 = f(x_0)$ and so the subsequence $y_{n'}$ must also converge to $y_0 = f(x_0)$; that is,

$$f(x_{n'}) = y \to f(x^0) \quad \text{and} \quad y_{n'} \to f(x_0).$$

By the uniqueness of the limit, Theorem 2.3, it follows that $f(x^0) = f(x_0)$. Then we have

$$x^0 \neq x_0 \text{ and } f(x^0) = f(x_0),$$

which contradicts the assumption that f(x) is strictly monotone and is therefore injective. We conclude that there can be no discontinuity y_0 for g(y) and the result is proved.

Let f(x) be a real valued function with domain D in $\mathbb{R}$ and let c denote an accumulation point of D. Then the following statements are equivalent (i.e., each implies all of the others):

(a) *the limit of $f(x)$ as x approaches c exists and equals L*

(b) $\forall \in > 0\ \exists \delta > 0$ *such that* $|f(x) - L| < \in$ *if* $0 < |x - c| < \delta$

(c) *if the sequence* $\{a_n\}$ *in D converges to c, then* $\{f(a_n)\}$ *converges to L*

Theorem 3.2 states that the limits of various combinations of functions are equal to the corresponding combinations of the limits of the functions. We can now characterize continuity in terms of limits.

Let $f(x)$ be defined on domain D in $\mathbb{R}$. Then the following properties are equivalent:

(a) *$f(x)$ is continuous at the point c in D*

(b) $\forall \in > 0\ \exists \delta > 0$ *such that* $f(x) \in N_{\in}[f(c)]$ *if* $x \in N_{\delta}[c]$

(c) $\lim_{x \to c} f(x) = f(c)$

(d) *for every sequence a_n in D converging to c, $f(a_n)$ converges to $f(c)$*

The following functions are continuous at all points of the real line: polynomials of all degrees, Sin x, Cos x, e^x. Theorem 3.5 contains the information that certain combinations of continuous functions are continuous.

There are a number of important consequences of continuity. These include the following results:

(a) *Bounded Range Theorem: If $f(x)$ is continuous on the closed, bounded interval $I = [a,b]$, then there exists $M > 0$ such that $|f(x)| \le M$ for all x in I*

(b) *Extreme Value Theorem: If $f(x)$ is continuous on the closed bounded interval $I=[a,b]$, then there exist real numbers F^* and F_* such that $F_* \le f(x) \le F^*$ for all x in I; moreover, there exist points c and d in I such that $f(c) = F_*$ and $f(d) = F^*$. Here F_* is the Glb of the range and F^* is the Lub of the range.*

(c) *Intermediate Value Theorem: If $f(x)$ is continuous on an interval I, then for any two points $p < q$ in I, $f(x)$ assumes every value between $f(p)$ and $f(q)$ as x ranges from p to q.*

A function is said to be uniformly continuous on a domain D if, for each $\in > 0$ there exists a $\delta > 0$ such that for all x, y in D, $|x - y| < \delta$ implies that $|f(x) - f(y)| < \in$. Clearly if $f(x)$ is uniformly continuous on D, then $f(x)$ is continuous at each point of D. The converse is false unless D is compact.

A function $f(x)$ on domain D is said to be injective if for all x ,x in D,

$$x_1 \neq x_2 \quad \text{implies that } f(x_1) \neq f(x_2).$$

The function $f(x)$ is said to be:

strictly increasing on D if $x_1 < x_2$ implies $f(x_1) < f(x_2)$

strictly decreasing on D if $x_1 < x_2$ implies $f(x_1) > f(x_2)$

We say f(x) is strictly monotone on D if f(x) is either strictly increasing or strictly decreasing on D. If f(x) is strictly monotone then f(x) is injective. If f(x) is injective and continuous on an interval I, then f(x) is strictly monotone on I. If f(x) is injective then f has an inverse, and if f(x) is also continuous, then the inverse is continuous.

SUPPLEMENTARY PROBLEMS

Use the definition of continuity to prove the continuity or discontinuity of the following functions at the indicated points.

3.1 $f(x) = x^3$ at $x = c$ for an arbitrary point c.

3.2 $f(x) = 1/x$ at $x = 2$

3.3 $f(x) = \begin{cases} 2 & \text{if } x < 2 \\ 1 & \text{if } x > 2 \end{cases}$ at the point $x = 2$

Prove the following function limits exist and evaluate the limit or prove the limit fails to exist.

3.4 $\lim_{x\to 3} 2x^2 - 1$

3.5 $\lim_{x\to 0} \sin(1/x)$

3.6 $\lim_{x\to 0} x \cos(1/x)$

3.7 $\lim_{x\to 0} 1/\sqrt{x}$

Use the sequential characterization of continuity to show that the following functions are discontinuous at the indicated point.

3.8 $f(x) = \sin(1/x)$ for $x \neq 0$ and $f(0) = 1$, at the point $x = 0$.

3.9 $f(x) = \dfrac{1}{x-2}$ at the point $x = 2$.

3.10 $f(x) = \begin{cases} 2 & \text{if } x \leq 2 \\ 1 & \text{if } x \geq 2 \end{cases}$ at the point $x = 2$.

3.11 $f(x) = \begin{cases} x^2 & \text{if } x \neq 2 \\ 1 & \text{if } x = 2 \end{cases}$ at the point $x = 2$.

3.12 Suppose $(1 + x^4)g(x)$ is continuous on [0, 1]. Does it follow that g(x) is continuous on [0, 1]?

3.13 Suppose f(x) + 2g(x) and g(x) – 4f(x) are both continuous on [0, 1]. Does it follow that f(x) and g(x) are both continuous on [0, 1]?

3.14 If f(x) is continuous and x = a is an accumulation point of dom f then is f(a) necessarily an accumulation point of the range of f?

3.15 Is $f(x) = x^{1/3}$ continuous on [0, 10]? Is f(x) uniformly continuous on [0, 10]?

3.16 Is $f(x) = x^{-1/3}$ continuous on (0, 10)? Is f(x)uniformly continuous on (0,10)?

3.17 Give an example of a function f(x) that is continuous but not uniformly continuous on (1, 2).

3.18 Give an example of a function f(x) that is uniformly continuous on (1,2). Is f(x) then continuous on [1, 2]?

3.19 Give an example of a function f(x) such that dom f is closed and bounded but the range of f is not bounded.

3.20 Give an example of a continuous function whose domain is bounded and whose range is also bounded but the function does not assume its maximum value at any point of the domain.

4

Differentiation

In the study of the calculus of functions of one real variable, the notions of continuity, differentiability, *and* integrability *play a central role. The previous chapter was devoted to continuity and its consequences and the next chapter will focus on integrability. In this chapter we define the* derivative *of a function of one variable and discuss several important consequences of differentiability. In particular, we show that differentiability implies continuity.*

We use the definition of the derivative to derive a few differentiation formulas but we assume that the formulas for differentiating the most common elementary functions are known. Similarly, we assume that the rules of differentiation are already known although the chain rule and some of its corollaries are proved in the solved problems.

We shall not discuss the various geometrical and physical applications of the derivative but concentrate instead on the more mathematical aspects of differentiation. We present several forms of the mean value theorem for derivatives, including the Cauchy mean value theorem which leads to L'Hospital's rule. This latter result is useful in evaluating so called "indeterminate" limits. Finally, we discuss the representation of a function by Taylor polynomials.

THE DERIVATIVE

DIFFERENCE QUOTIENTS

Let $f(x)$ be a real valued function with domain D containing an $\in$–neighborhood of the point x_0; i.e, for some $\in > 0$, $N_\in[x_0]$ is contained in D. We say that x_0 is an *interior point* of D. Then for any h such that $0 < |h| < \in$, we can define the *difference quotient* for $f(x)$ near x_0

$$D_h f(x_0) = \frac{f(x_0 + h) - f(x_0)}{h} \tag{4.1}$$

It is well known from elementary calculus that $D_hf(x)$ represents the slope of the secant line through the points $(x_0, f(x_0))$ and $(x_0 + h, f(x_0 + h))$. Then we have

Definition of the Derivative

Definition The function f(x) is said to be *differentiable* at the interior point c in D if

$$\lim_{h \to 0} D_h f(c) \text{ exists.} \tag{4.2}$$

We denote the value of this limit by f′(c) and refer to this as the *derivative* of f(x) at x = c.

NOTATION FOR THE DERIVATIVE

We may also define the derivative of f(x) at the endpoints of an interval by computing one sided limits for $D_hf(x)$. The set of points in D_h where the limit (4.2) exists is the domain of a new function, f′(x), called the derivative of f(x). We shall also use the notation dy/dx for the derivative of y = f(x). The derivative dy/dx = f′(x) may be variously interpreted as:

the slope of the TANGENT LINE at (x,f(x)) to the graph of y = f(x),

the instantaneous RATE OF CHANGE of y = f(x) with respect to x

HIGHER DERIVATIVES

If the function f′(x) is differentiable, then its derivative is denoted by f″(x) or by d^2y/dx^2. This is called the *second derivative* of y = f(x). We denote derivatives of order higher than two by $f^{(n)}(x)$ or d^ny/dx^n for $n \in \mathbb{N}$.

Example 4.1 Derivatives

4.1 (a) In an elementary calculus course we derive formulas for the derivatives of many elementary functions. The following functions are differentiable at each point where they are defined:

f(x)	f′(x)
x^p	px^{p-1}
Sin x	Cos x
Cos x	–Sin x
$A^x \quad A > 0$	$A^x \operatorname{Log}_e A$
$\operatorname{Log}_e x$	1/x

4.1 (b) The following functions are continuous for all x, but the derivative fails to exist at the indicated point:

f(x) = | x | is not differentiable at x = 0 since $\lim_{h \to 0} D_hf(0)$ fails to exist (see Problem 4.4)

$f(x) = x^{2/3}$ is not differentiable at x = 0 since $D_hf(0)$ tends to infinity as h tends to zero

Theorem 4.1

Theorem 4.1 If f(x) is differentiable at the point c in D then f(x) is continuous at c.

In particular, Theorem 4.1 implies that continuity at point is a necessary condition for differentiability at the point. Example 4.1(b) shows that continuity is not sufficient for differentiability.

RULES FOR DIFFERENTIATION

In elementary calculus we learn differentiation formulas for commonly occuring functions. In addition, we learn differentiation *rules* which allow us to compute derivatives of various combinations of functions when the derivatives of the separate functions are known.

Theorem 4.2 Derivatives of sums, products and quotients

Theorem 4.2 Suppose f(x) and g(x) are differentiable at a point c in an interval D. Then

(a) $[\alpha f + \beta g]'(c) = \alpha f'(c) + \beta g'(c)$ for all α, β in $\mathbb{R}$

(b) $[f \cdot g]'(c) = f'(c)g(c) + f(c).g'(c)$

(c) $[f/g]'(c) = (f'(c)g(c) - f(c)g'(c))/g^2(c)$ if $g(c) \neq 0$

Theorem 4.3 Chain Rule

Theorem 4.3 Suppose f(x) is differentiable at a point c and that g(x) is differentiable at the point f(c). Then the composed function F(x) = g[f(x)] is differentiable at c. The derivative at c is equal to

$$F'(c) = g'[f(c)] \cdot f'(c).$$

COROLLARY 4.4 DERIVATIVE OF THE INVERSE

Suppose f(x) is strictly monotone and continuous on an interval I. Then the inverse function $g(x) = f^{-1}(x)$ is strictly monotone and continuous on the interval J = f[I] and if f(x) is differentiable at c in I, and if $f'(c) \neq 0$, then g(x) is differentiable at b = f(c) and $g'(b) = 1/f'(c)$.

COROLLARY 4.5

Suppose x = f(t) and y = g(t) are differentiable functions of t for $a \leq t \leq b$, and that $f'(t) \neq 0$ for $a \leq t \leq b$. Then

$$y'(x) = \frac{dy}{dx} = \frac{dy/dt}{dx/dt} = \frac{g'(t)}{f'(t)}$$

CONSEQUENCES OF DIFFERENTIABILITY

Just as there were a number of useful facts resulting from the property of continuity, there are a similar number of important consequences of differentiability.

Local Extreme Points

A point c in the domain D of a function f(x) is said to be a local maximum for f(x) if for some $\delta > 0$, $f(c) \geq f(x)$ for all x in $N_\delta[c] \cap D$. If $f(c) \leq f(x)$ for all x in $N_\delta[c] \cap D$, then c is said to be a local minimum for f(x). The point c is said to be a *local extreme point* for f(x) if it is either a local maximum or a local minimum.

Theorem 4.6 Extreme Points

Theorem 4.6 Let f(x) be defined and continuous on the interval I and suppose c in I is a local extreme point for f(x). Then exactly one of the following must hold:

i) c is an endpoint for I
ii) c is an interior point of I and $f'(c) = 0$
iii) c is an interior point of I but $f'(c)$ does not exist

Mean Value Theorems for Derivatives

There are several useful results, each of which can be described as a mean value theorem for derivatives. The first of these is Rolle's theorem.

Theorem 4.7 Rolle's theorem

Theorem 4.7 Suppose f(x) is continuous on the closed interval [a, b], and is differentiable at each point of the open interval (a, b). Suppose also that $f(a) = f(b) = 0$. Then there exists a point c in (a, b) where $f'(c) = 0$.

COROLLARY 4.8 THE MEAN VALUE THEOREM FOR DERIVATIVES

Suppose f(x) is continuous on the closed interval [a, b], and is differentiable on the open interval (a,b). Then there exists a point c in (a, b) where

$$f(b) - f(a) = f'(c)(b - a).$$

COROLLARY 4.9

Suppose f(x) is continuous on the closed interval [a, b], and is differentiable on the open interval (a, b). Suppose also that $f'(x) = 0$ for all x in (a, b). Then f is constant on [a, b].

Theorem 4.10 The Cauchy Mean Value Theorem

Theorem 4.10 Suppose f(x) and g(x) are continuous on the closed interval [a,b] and are differentiable on the open interval (a,b). Suppose also that $g(a) \neq g(b)$ and $f'(x)$ and $g'(x)$ never vanish simultaneously. Then there exists a point c in (a, b) such that

$$\frac{f(b) - f(a)}{g(b) - g(a)} = \frac{f'(c)}{g'(c)}$$

INDETERMINATE FORMS

A limit of a quotient in which the numerator and denominator tend to zero simultaneously is said to be an indeterminate limit of the form 0/0. Often such limits can be evaluated by means of a corollary of the Cauchy mean value theorem known as L'Hospital's rule.

COROLLARY 4.11 L'HOSPITAL'S RULE

Suppose f(x) and g(x) are continuous on the closed interval [a,b] and are differentiable on the open interval (a,b). Suppose also that $g(x) \neq 0$ and $g'(x) \neq 0$ for all x in (a,b) and that $f(a) = g(a) = 0$. Then either the one sided limits

$$\lim_{x \to a} \frac{f(x)}{g(x)} \quad \text{and} \quad \lim_{x \to a} \frac{f'(x)}{g'(x)} \tag{4.3}$$

both exist and they are equal, or both limits fail to exist.

The left hand limits in (4.3) can be replaced by right hand limits as x approaches b and the conclusion still holds. In addition, the conclusions still hold when the limits are replaced by limits as x tends to $+\infty$ or $-\infty$. Finally, if f(x) and g(x) both tend to infinity as x tends to a, the conclusions continue to hold in this case as well. All of these limits are various types of *indeterminate forms*.

Taylor's Theorem

Corollary 4.8 is often referred to as the mean value theorem for derivatives. This result can be extended to derivatives of order higher than one. The result is known as Taylor's theorem.

Theorem 4.12 Taylor's theorem

Theorem 4.12 Suppose that f(x) together with its derivatives up to order n, $f'(x), f''(x), \ldots, f^{(n)}(x)$ are all continuous on the interval [a, b]. Suppose also that $f^{(n+1)}(x)$ exists at each point of (a,b). Then for each fixed x_0 in [a, b],and all x in the interval, there exists a value c between x_0 and x such that

$$f(x) = f(x_0) + f'(x_0)(x - x_0) + \frac{f''(x_0)}{2!}(x - x_0)^2 + \ldots + \frac{f^{(n)}(x_0)}{n!}(x - x_0)^n + \frac{f^{(n+1)}(c)}{(n+1)!}(x - x_0)^{n+1} \tag{4.4}$$

The sum of the first n + 1 terms in the expression (4.4) is called the nth degree Taylor polynomial for f(x), expanded about the point $x = x_0$. The final term in the expression is called the *Lagrange form* of the *remainder* term. An integral form for the remainder term will be given in the next chapter. In the special case $x_0 = 0$, we refer to (4.4) as the *Maclaurin expansion* for f(x).

Example 4.2 Maclaurin Expansions and Taylor

4.2 (a) The following are some frequently used Maclaurin expansions,

$$e^x = 1 + x + x^2/2! + x^3/3! + \cdots.$$

$$\text{Sin } x = x - x^3/3! + x^5/5! - x^7/7! + \cdots$$

$$\text{Cos } x = 1 - x^2/2! + x^4/4! - x^6/6! + \cdots$$

$$\ln(1 + x) = x - x^2/2 + x^3/3 - x^4/4 + \cdots$$

$$\text{Arctan } x = x - x^3/3 + x^5/5 - x^7/7 + \cdots$$

4.2 (b) The Taylor series expansion for lnx about the point $x_0 = 1$ takes the form,

$$\ln x = \ln(1 + x - 1) = (x - 1) - 1/2\,(x - 1)^2 + 1/3\,(x - 1)^3 - 1/4\,(x - 1)^4 + \cdots$$

The Taylor and Maclaurin expansions are examples of infinite series of functions. This important topic is discussed in detail in Part IV of this text.

SOLVED PROBLEMS

The Derivative

PROBLEM 4.1

Use the definition of derivative to compute the derivative of $f(x) = \sqrt{x}$ for $x > 0$.

SOLUTION 4.1

For $x > 0$ we have from (4.1)

$$D_h f(x) = \frac{\sqrt{x+h} - \sqrt{x}}{h} = \frac{x + h - x}{h(\sqrt{x+h} + \sqrt{x})} = \frac{1}{\sqrt{x+h} + \sqrt{x}}$$

Letting h tend to zero in this last expression, we can see that

$$\lim_{h \to 0} D_h f(x) \text{ exists and equals } \frac{1}{2\sqrt{x}}$$

Note, however, that for x=0 we have

$$D_h f(0) = \frac{\sqrt{|0 + h|} - 0}{h} = |h|^{-1/2}$$

hence the limit as h tends to zero of $D_h f(0)$ does not exist. Then f(x) is not differentiable at x=0.

PROBLEM 4.2

Use the definition of derivative to compute the derivative of f(x) = Sin x.

SOLUTION 4.2

In this case, we have from (4.1),

$$D_hf(x) = \frac{\text{Sin}(x + h) - \text{Sin } x}{h} = \frac{\text{Sin } x \text{ Cos } h + \text{Sin } h \text{ Cos } x - \text{Sin } x}{h}$$

$$= \frac{\text{Sin } x(\text{Cos } h - 1)}{h} + \frac{\text{Cos } x \text{ Sin } h}{h} \qquad (1)$$

In Problem 3.8 we showed (1 – Cos h)/h tends to zero as h tends to zero and in Problem 3.6 we showed that Sin h/h tends to 1 as h tends to zero. Then it follows from (1) that

$$\lim_{h \to 0} D_hf(x) \text{ exists and equals Cos } x$$

PROBLEM 4.3 THE EXPONENTIAL FUNCTION

Use the definition of derivative to compute the derivative of $f(x) = e^x$.

SOLUTION 4.3

Let us first indicate how the exponential function is defined. For A > 0 and positive integers m, n we define

$$A^{m/n} = \sqrt[n]{A^m} = (\sqrt[n]{A})^m.$$

This defines A^x for A > 0 and x a rational number. Now for A > 1, and any real number x, let S denote the following set,

$$S = \{A^r : r < x \text{ and } r \in \mathbb{Q}\}.$$

Then S has an upper bound (i.e., take an integer n > x; then A > 1 implies $A^r \leq A^n$). It follows that S has a least upper bound and we then define

Definition For A, x real numbers with A > 0,

i) if A > 1 then A^x = least upper bound of S

ii) if A = 1 then $A^x = 1$

iii) if A < 1 then 1/A > 1 and $A^x = 1/(1/A)^x$

Thus A^x has been defined for A > 0 and any real x.

Now let e denote the number between 2 and 3, found in Problem 2.14 as the following limit,

$$e = \lim_{n \to \infty} (1 + 1/n)^n. \qquad (1)$$

Now for $f(x) = e^x$ we have from (4.1),

$$D_h f(x) = \frac{e^{x+h} - e^x}{h} = e^x \frac{e^h - 1}{h}$$

If we can show

$$\lim_{h \to 0} \frac{e^h - 1}{h} = 1 \tag{2}$$

then it will follow that $f'(x) = e^x$.

In order to show (2) it will suffice to consider h such that $|h| \le 1/2$. For $0 < h < 1/2$, there exists an integer m such that

$$m \le 1/h < m + 1 \quad \text{i.e., } 1/(m+1) < h \le 1/m \tag{3}$$

Then

$$e^{1/(m+1)} < e^h \le e^{1/m} \tag{4}$$

and, for e given by (1),

$$1 + 1/n \le e^{1/n} \le 1 + 1/(n-1) \tag{5}$$

Using (4) and (5) together, we get

$$1 + 1/(m+1) < e^h < 1 + 1/(m-1) \tag{6}$$

We conclude from (3) that

$$h/(1+h) < 1/(m+1) \quad \text{and} \quad 1/(m-1) < h/(1-2h)$$

and hence

$$\frac{1}{1+h} \le \frac{e^h - 1}{h} \le \frac{1}{1-2h} \qquad 0 < h < 1/2 \tag{7}$$

Similarly

$$\frac{1}{1+h} \ge \frac{e^h - 1}{h} \ge \frac{1}{1-2h} \qquad -1/2 < h < 0 \tag{8}$$

Then (7) and (8), used with Theorem 3.2 part iv) , leads to (2).

PROBLEM 4.4

Show that $f(x) = |x|$ is not differentiable at $x = 0$.

SOLUTION 4.4

We have for this function

$$D_h f(0) = \frac{|0 + h| - 0}{h} = \frac{|h|}{h} = \begin{cases} 1 & \text{if } h > 0 \\ -1 & \text{if } h < 0 \end{cases}$$

Then the limit of $D_h f(0)$ as h tends to zero clearly does not exist and by the definition, the function is not differentiable at $x = 0$.

PROBLEM 4.5

Prove that if f(x) is differentiable at the point c in D, then f(x) is continuous at the point c.

SOLUTION 4.5

According to Theorem 3.3, f(x) is continuous at the point c in D if

$$\lim_{h \to 0} (f(c+h) - f(c)) = 0. \qquad (1)$$

But

$$f(c+h) - f(c) = D_h f(c)(x-c) \qquad (2)$$

and by Theorem 3.2, part ii), we have

$$\lim_{h \to 0} (D_h f(c)(x-c)) = \lim_{h \to 0} D_h f(c) \lim_{h \to 0} (x-c) = f'(c) \cdot 0 = 0 \qquad (3)$$

Then (3) and (2) together give (1).

PROBLEM 4.6

Show that the function

$$f(x) = \begin{cases} x \operatorname{Sin}(1/x) & \text{if } x \neq 0 \\ 0 & \text{if } x=0 \end{cases}$$

is continuous at $x = 0$ but is not differentiable there. This is another example showing that the converse of Theorem 4.1 is false.

SOLUTION 4.6

Since $|\operatorname{Sin}(1/x)| < 1$, it follows that $|f(x)| < |x|$. Then it is evident that

$$\lim_{x \to 0} f(x) = 0 = f(0);$$

i.e., f(x) is continuous at $x = 0$. On the other hand.

$$D_h f(0) = \frac{h \operatorname{Sin}(1/h) - 0}{h} = \operatorname{Sin}(1/h).$$

A slight modification of the argument used in Problem 3.14 shows that Sin(1/h) tends to no limit as h tends to zero. Then f(x) is not differentiable at $x = 0$.

However, f(x) is differentiable at x different from zero. We can use the differentiation formulas from Example 4.1 together with the differentiation rules from Theorem 4.2 to compute

$$f'(x) = \operatorname{Sin}(1/x) + x \operatorname{Cos}(1/x) \cdot (-1/x^2) \quad \text{for } x \neq 0 .$$

Rules for Differentiation

PROBLEM 4.7

Prove Theorem 4.3, the chain rule.

SOLUTION 4.7

Suppose that g(x) is differentiable at the point x = c and that f(y) is differentiable at the point y = d = g(c). Then if we denote the composed function by F, F(x) = f[g(x)], the difference quotient for F has the form

$$D_hF(c) = \frac{f[g(c+h)] - f[g(c)]}{h} = \frac{f[g(c+h)] - f[g(c)]}{g(c+h) - g(c)} \frac{g(c+h) - g(c)}{h} \quad (1)$$

Since g(x) is differentiable at x = c, it is continuous there by Theorem 4.1. Thus as h tends to zero, g(c + h) tends to g(c) = d and so long as g(c + h) – g(c) does not vanish in a neighborhood of d, we have

$$\lim_{h \to 0} D_hF(c) \text{ exists and equals } f'(d) \cdot g'(c)$$

We can give a more precise argument which avoids the necessity of assuming g(c + h) – g(c) does not vanish. Define a function G(y) by

$$G(y) = \begin{cases} \dfrac{f(y) - f(d)}{y - d} & \text{if } y \neq d = g(c) \\ f'(d) & \text{if } y = d \end{cases} \quad (2)$$

Then

$$\lim_{y \to d} G(y) = f'(d) = G(d)$$

so G(y) is continuous at y = d. Then it follows from Theorem 3.5 that

$$\lim_{x \to c} G(g(x)) = \lim_{y \to d} G(y) = G(d) = f'[g(c)] \quad (3)$$

Now (2) implies

$$f[g(c+h)] - f[g(c)] = G(y)(g(c+h) - g(c)) \quad (4)$$

and this, together with (1), leads to

$$D_hF(c) = G(g(c+h))D_hg(c) \quad (5)$$

Then

$$\lim_{h \to 0} D_hF(c) = \lim_{h \to 0} G(g(c+h)) \lim_{h \to 0} D\, g(c) = f'[g(c)]g'(c).$$

PROBLEM 4.8

Prove Corollary 4.4.

SOLUTION 4.8

Suppose f(x) is strictly monotone and continuous on an interval I. Then by Theorem 3.14, $g(x) = f^{-1}(x)$ is also strictly monotone and continuous on the interval J = f[I]. Now if f(x) is differentiable at c in I and if $f'(c) \neq 0$, then g(x) is differentiable at b = f(c) and $g'(b) = 1/f'(c)$. To see this, let F(x) denote the composed function,

$$F(x) = g[f(x)\} = x \quad \text{for each x in I.} \tag{1}$$

The chain rule implies $F'(c) = g'[f(c)] \cdot f'(c) = g'(b)f'(c)$. But it follows from (1) and the first differentiation formula in Example 4.1(a), that $F'(x) = 1$ for all x in I. Then $g'(b)f'(c)=1$.

Consequences of Differentiability

PROBLEM 4.9

Suppose f(x) is continuous on the closed interval [a, b] and differentiable at each point of the open interval (a,b). Suppose also that f(x) has a local extreme point at x = c for a < c < b. Then prove that $f'(c) = 0$.

SOLUTION 4.9

We shall suppose that x = c is a local maximum for f(x). The proof for the case of a local minimum is similar.

If x = c is a maximum, then $f(c + h) - f(c) < 0$ for all h sufficiently small and it follows that

$$D_h f(c) \leq 0 \text{ for } h > 0 \tag{1}$$

and

$$D_h f(c) \geq 0 \text{ for } h < 0. \tag{2}$$

Since

$$\lim_{h \to 0} D_h f(c) = f'(c)$$

it follows from (1) that $f'(c) \leq 0$, while (2) implies that $f'(c) \geq 0$. But then it must be the case that $f'(c) = 0$.

PROBLEM 4.10

Prove Rolle's theorem, Theorem 4.7.

SOLUTION 4.10

If f(x) is identically zero, then at each point c in (a, b) we have $f'(c) = 0$. If f(x) is not identically zero, then the extreme value theorem, Corollary 3.8, implies that f(x) assumes its maximum and minimum values at points of [a, b]. Since f(x) is not identically zero, at least one of these extreme values is not zero. Since f(x) vanishes at x = a and x = b, the nonzero extreme value must occur at a point c that lies *between* x = a and x = b. But then, by the result of the previous problem, we see that $f'(c) = 0$.

PROBLEM 4.11

Prove Corollary 4.8, the *mean value theorem for derivatives.*

SOLUTION 4.11

Suppose the function f(x) is continuous on the closed interval [a,b] and is differentiable at each point of the open interval (a, b). Then let

$$g(x) = f(b) - f(x) - \frac{f(b) - f(a)}{b - a}(b - x).$$

According to Theorems 3.5 and 4.2, the function g(x) is continuous on the closed interval [a,b] and is differentiable at each point of the open interval (a,b). In addition it is easy to see that $g(a) = g(b) = 0$. Then Rolle's theorem implies that $g'(c) = 0$ at some point c in (a,b). But

$$g'(c) = -f'(c) + \frac{f(b) - f(a)}{b - a} = 0,$$

from which it follows that

$$f(b) - f(a) = f'(c) \cdot (b - a).$$

PROBLEM 4.12

Prove Corollary 4.9.

SOLUTION 4.12

Suppose the function f(x) is continuous on the closed interval [a,b] and is differentiable at each point of the open interval (a,b). Suppose also that $f'(x) = 0$ for all x in (a,b). We shall assume that f(x) is not constant and show that this leads to a contradiction. Then this will imply that f(x) must be constant.

If f(x) is not a constant, then there exist points α and β in [a,b] where $f(\alpha) \neq f(\beta)$. We apply Corollary 4.8 to f(x) on the interval $[\alpha, \beta]$ to conclude that there exists a point c in (α, β) where

$$f'(c) = \frac{f(\beta) - f(\alpha)}{\beta - \alpha} \neq 0.$$

But this is in contradiction to the assumption that $f'(x) = 0$ for all x in (a, b). Then it follows that f(x) must be constant on [a, b].

PROBLEM 4.13

Prove that if f(x) and g(x) are continuous on [a, b], differentiable on (a, b) and $f'(x) = g'(x)$ for all x in (a, b), then $f(x) = g(x) + C$ for some constant C.

SOLUTION 4.13

The hypotheses imply that we can apply Corollary 4.9 to the function $F(x) = f(x) - g(x)$ to conclude that $F(x) = C$.

PROBLEM 4.14

Suppose the function $f(x)$ is continuous on the closed interval $[a,b]$ and is differentiable at each point of the open interval (a, b). Prove that if $f'(x)$ does not change sign on (a, b), then $f(x)$ is monotone on $[a, b]$.

SOLUTION 4.14

Suppose $f'(x) \geq 0$ at each point x in (a, b). Then for arbitrary points α and β such that $a \leq \alpha < \beta \leq b$, the mean value theorem implies the existence of a point c in (α, β) such that

$$f(\beta) - f(\alpha) = f'(c)(\beta - \alpha) \geq 0.$$

This shows $\beta - \alpha \geq 0$ implies $f(\beta) - f(\alpha) \geq 0$; i.e., $f(x)$ is *increasing* on $[a ,b]$. Similarly, if $f'(x) \leq 0$ at each point x of (a, b), we can show that $f(x)$ is *decreasing* on $[a, b]$. Thus if $f'(x)$ is of one sign on (a, b), $f(x)$ must be monotone on $[a, b]$.

PROBLEM 4.15

Prove that $|\text{Sin } x| \leq |x|$ with equality only for $x = 0$.

SOLUTION 4.15

We shall prove that $\text{Sin } x \leq x$ for $x \geq 0$. The other half of this inequality is proved in a similar fashion.

Clearly $\text{Sin } 0 = 0$. For $x > 0$, apply the mean value theorem to the function $f(x) = \text{Sin } x$ to conclude that there exists some point c, $0 < c < x$, such that

$$f(x) = f(0) + f'(c)(x - 0) = x \text{ Cos } c;$$

i.e.

$$\text{Sin } x = x \text{ Cos } c.$$

If $x \leq 1$, then $0 < c < 1$ and $0 < \text{Cos } c < 1$. In this case we have $\text{Sin } x < x$. If $x > 1$, then $\text{Sin } x \leq 1 < x$. In either case we have $\text{Sin } x < x$ for $x > 0$.

PROBLEM 4.16

Prove that

$$e^x > 1 + x \quad \text{for all } x \tag{1}$$

and that equality occurs only for $x = 0$.

SOLUTION 4.16

Clearly we have equality in (1) in the case $x = 0$. For $x > 0$, we apply the mean value theorem to $f(x) = e^x$ on the interval $[0, x]$. We conclude that for some value c, $0 < c < x$

$$e^x - e^0 = e^c (x - 0);$$

i.e.,

$$e^x - 1 = e^c x > x \quad \text{for } x > 0$$

since $e^c > 1$ for $c > 0$. For $x < 0$, we apply the same argument on the interval $[x, 0]$ to obtain (1).

PROBLEM 4.17 INTERMEDIATE VALUE PROPERTY FOR DERIVATIVES

Suppose that $f(x)$ is differentiable on the interval $[a, b]$ and that $\alpha \in \mathbb{R}$ is between $f'(a)$ and $f'(b)$. Show there exists at least one point c in $[a,b]$ such that $f'(c) = \alpha$. In other words, in order for a function to be the derivative of some differentiable function, it is necessary for it to have the *intermediate value property*.

SOLUTION 4.17

Suppose we have $f'(a) < \alpha < f'(b)$. Then let $g(x) = \alpha(x - a) - f(x)$ for x in $[a, b]$. Since $g(x)$ is continuous on the closed interval $[a, b]$, it follows from the extreme value theorem that $g(x)$ attains a maximum on $[a, b]$. Since

$$g'(a) = \alpha - f'(a) > 0 \quad \text{and} \quad g'(b) = \alpha - f'(b) < 0$$

it follows that $g(x)$ is increasing at $x = a$ and is decreasing at $x = b$. Then the maximum value for $g(x)$ cannot occur at either of these points but must occur at an interior point c where, according to Theorem 4.6, we have $g'(c)=0$. Then

$$0 = g'(c) = a - f'(c).$$

Note that this result implies that the function

$$H(x)= \begin{cases} 0 & \text{if } -1 < x < 0 \\ 1 & \text{if } 0 < x < 1 \end{cases}$$

is not the derivative of any differentiable function $f(x)$. For if $H(x)$ were equal to $f'(x)$ for some $f(x)$, then $f'(-1) < 1/2 < f'(1)$ but there is no point c such that $f'(c) = H(c) = 1/2$. Then the result implies that no such function $f(x)$ can exist.

PROBLEM 4.18

Use the mean value theorem to show that for $h > 0$ and $\alpha > 1$,

$$(1 + h)^\alpha > 1 + \alpha h \qquad (1)$$

SOLUTION 4.18

We shall first rewrite the mean value result in an alternative form. Let $h = b - a$ so that $b = a + h$. Then $f(b) - f(a) = f'(c)(b - a)$ for $a < c < b$, can be rewritten as

$$f(a + h) = f(a) + hf'(a + \lambda h) \tag{2}$$

for some λ, $0 < \lambda < 1$. Now we apply (2) to the function $f(x) = x^\alpha$ at the point $a = 1$, to obtain

$$(1 + h)^\alpha = 1 + h\alpha(1 + \lambda h)^{\alpha-1} \tag{3}$$

Since $1 + \lambda h > 1$ and $\alpha > 1$, (1) follows at once from (3).

PROBLEM 4.19

Prove Theorem 4.10, the so-called *Cauchy mean value theorem.*

SOLUTION 4.19

Let

$$\frac{f(b) - f(a)}{g(b) - g(a)} = \lambda \tag{1}$$

and define

$$F(x) = f(x) - f(a) - \lambda(g(x) - g(a)). \tag{2}$$

Then $F(a) = 0$, and because of the definition (1), $F(b) = 0$ as well. We are assuming that $f(x)$ and $g(x)$ are continuous on the closed interval and differentiable on the open interval (a, b) and hence the same is true of $F(x)$. Then Rolle's theorem applies to $F(x)$; i.e. $F'(c) = 0$ for some c, $a < c < b$. Then it follows from (2) that

$$f'(c) - \lambda g'(c) = 0. \tag{3}$$

Since $f'(x)$ and $g'(x)$ have been assumed not vanish simultaneously, it follows from (3) that neither $f'(c)$ nor $g'(c)$ is equal to zero. Then

$$\frac{f(b) - f(a)}{g(b) - g(a)} = \lambda = \frac{f'(c)}{g'(c)}.$$

This is the conclusion of Theorem 4.10. Note that this reduces to the usual mean value theorem if we choose $g(x) = x$.

PROBLEM 4.20

Prove Corollary 4.11, L'Hospital's rule.

SOLUTION 4.20

We assume that f(x) and g(x) are continuous on the closed interval, that they are differentiable on the open interval (a, b), and that f(a) = g(a) = 0. We suppose also that neither g(x) nor g′(x) vanishes on the open interval (a,b). Note that we do not suppose either f(x) or g(x) to be differentiable at x = a. Then we must show that if

$$\lim_{x \to a} \frac{f'(x)}{g'(x)} = L \tag{1}$$

then

$$\lim_{x \to a} \frac{f(x)}{g(x)} = L \tag{2}$$

It follows from (1) that for all $\epsilon > 0$ there exists a $\delta > 0$ such that

$$\left| \frac{f'(x)}{g'(x)} - L \right| < \epsilon \quad \text{if } a < x < a + \delta \tag{3}$$

We apply the Cauchy mean value theorem on the interval (a, x) to conclude that there exists a point c_x, $a < c_x < x$, such that

$$\frac{f(x)}{g(x)} = \frac{f'(c_x)}{g'(c_x)} \tag{4}$$

But c_x satisfies $a < c_x < x < a + \delta$, and hence it follows from (3) and (4) that

$$\left| \frac{f(x)}{g(x)} - L \right| = \left| \frac{f'(c_x)}{g'(c)} - L \right| < \epsilon.$$

Since this holds for all x such that $a < x < a + \delta$, (2) now follows from (5).

This proves L'Hospital's rule in the case that L and a are both real numbers. A similar argument proves the result in the case that L is $+\infty$ or $-\infty$, and these arguments can even be modified for the case where a is allowed to be infinite.

Indeterminate Limits

PROBLEM 4.21

Use L'Hospital's rule to evaluate the limit:

$$\lim_{x \to 0} \frac{1 - \cos x}{2\sqrt{x}}.$$

SOLUTION 4.21

The functions $f(x) = 1 - \cos x$ and $g(x) = 2\sqrt{x}$ are continuous on the closed interval [0,1], and are differentiable on the open interval (0, 1). In addition, both vanish at x = 0 so we are dealing with an indeterminate limit of the form 0/0. Although g(x) is not differentiable at x = 0, the hypotheses of Corollary 4.11 are satisfied. Thus

$$\lim_{x \to 0} \frac{1 - \cos x}{2\sqrt{x}} = \lim_{x \to 0} \frac{\sin x}{1/\sqrt{x}} = \lim_{x \to 0} \sqrt{x} \sin x = 0.$$

PROBLEM 4.22

Use L'Hospital's rule to evaluate the limit:

$$\lim_{x \to 0} \frac{\sin x - x}{x^3}.$$

SOLUTION 4.22

The hypotheses of Corollary 4.11 are satisfied for $f(x) = \sin x - x$ and $g(x) = x^3$. In particular, $f(0)=g(0)=0$ so that this is an indeterminate limit of the form 0/0. Then the corollary implies,

$$\lim_{x \to 0} \frac{\sin x - x}{x^3} = \lim_{x \to 0} \frac{\cos x - 1}{3x^2}.$$

This new limit is again indeterminate but the hypotheses of the corollary are again satisfied for new functions f(x) and g(x) given by, $f(x) = \cos x - 1$ and $g(x) = 3x^2$. This leads to a third indeterminate limit,

$$\lim_{x \to 0} \frac{\cos x - 1}{3x^3} = \lim_{x \to 0} \frac{-\sin x}{6x}$$

to which the corollary may once more be applied in order to obtain

$$\lim_{x \to 0} \frac{-\sin x}{6x} = \lim_{x \to 0} \frac{-\cos x}{6} = -\frac{1}{6}.$$

Then the original limit exists and equals $-1/6$.

PROBLEM 4.23

Use L'Hospital's rule to evaluate the limit:

$$\lim_{x \to 0} \frac{e^{-1/x^2}}{x^p}. \qquad (1)$$

Here p denotes a fixed, positive number.

SOLUTION 4.23

If we let

$$f(x) = \begin{cases} e^{-1/x^2} & x \neq 0 \\ 0 & x = 0 \end{cases}$$

and let $g(x) = x^p$, then f(x) and g(x) are continuous on the closed interval [0, 1], and are differentiable on the open interval. In addition, each tends to zero as x tends to zero. Then L'Hospital's rule applies to the limit (1). However,

$$f'(x) = \frac{2e^{-1/x^2}}{x^3} \quad \text{and } g'(x) = px^{p-1}$$

and hence

$$\operatorname*{Lim}_{x\to 0} \frac{e^{-1/x^2}}{x^p} = \frac{2}{p} \operatorname*{Lim}_{x\to 0} \frac{e^{-1/x^2}}{x^{p+2}}.$$

The new limit is still indeterminate and is, in fact, a worse limit than the one we started with. Consider instead, a limit which is equivalent to the limit (1),

$$\operatorname*{Lim}_{x\to 0} \frac{x^{-p}}{e^{1/x^2}}. \tag{2}$$

This limit involves the indeterminate form ∞/∞, but L'Hospital's rule applies to this limit as well, and leads to

$$\operatorname*{Lim}_{x\to 0} = \frac{x^{-p}}{e^{1/x^2}} = \operatorname*{Lim}_{x\to 0} \frac{-px^{-p-1}}{e^{1/x^2}(-2/x^3)}.$$

The expression in the last limit simplifies as follows,

$$\frac{-px^{-p-1}}{e^{1/x^2}(-2/x^3)} = \frac{-px^{-p+2}}{2e^{1/x^2}}$$

Then the limit (2) is equal to the limit,

$$\operatorname*{Lim}_{x\to 0} \frac{-px^{-p+2}}{2e^{1/x^2}}. \tag{3}$$

The limit (3) is also indeterminate of the form ∞/∞, but the exponent in the numerator has been reduced. It is evident that by repeated application of L'Hospital's rule, we shall eventually arrive at a limit having a *positive* power of x in the numerator. Then this limit will equal zero and it follows that for all positive numbers p, the limit (1) exists and equals zero.

PROBLEM 4.24

Use L'Hospital's rule to evaluate the limit:

$$\operatorname*{Lim}_{x\to 0} x^p \ln x \quad \text{for } p > 0. \tag{1}$$

SOLUTION 4.24

As it stands, the limit (1) is not of the form to which L'Hospital's rule may be applied. However, if we rewrite it in the form,

$$\operatorname*{Lim}_{x\to 0} \frac{\ln x}{x^{-p}} \tag{2}$$

then it is easily verified that we are dealing with an indeterminate limit of the form ∞/∞ and that L'Hospital's rule may be applied to obtain,

$$\lim_{x\to 0} \frac{\ln x}{x^{-p}} = \lim_{x\to 0} \frac{1/x}{-px^{-p-1}} = \lim_{x\to 0} \frac{x^{p+1}}{-p} = 0 . \tag{3}$$

Note that after applying L'Hospital's rule, we had to simplify the result before the limit could be evaluated. It follows from (3) that for all $p > 0$, the limit (1) exists and equals zero.

PROBLEM 4.25

Use L'Hospital's rule to evaluate the limits,

$$\text{(a)} \lim_{x\to\infty} \frac{\ln x}{x^p} \qquad \text{(b)} \lim_{x\to\infty} \frac{e^x}{x^p} \tag{1}$$

where p denotes a fixed positive number.

SOLUTION 4.25

Note that in each of these limits, x tends to infinity rather than to a finite limit point. However, L'Hospital's rule applies even in this case. In the first limit in (1), we have an indeterminate limit of the form ∞/∞ and L'Hospital's rule leads to

$$\lim_{x\to\infty} \frac{\ln x}{x^p} = \lim_{x\to\infty} \frac{1/x}{px^{p-1}} = \lim_{x\to\infty} \frac{1}{px^p} = 0 \tag{2}$$

In the second limit, repeated application of L'Hospital's rule leads to,

$$\lim_{x\to\infty} \frac{e^x}{x^p} = \lim_{x\to\infty} \frac{e^x}{px^{p-1}} = \cdots = \lim_{x\to\infty} \frac{e^x}{p!} = \infty; \tag{3}$$

i.e. each time L'Hospital's rule is applied the power of x in the denominator decreases by one while the exponential function in the numerator is not changed. After p applications of L'Hospital's rule, the degree of the denominator equals zero and the limit is no longer indeterminate.

We showed that the limit (a) exists and equals zero while the limit (b) is infinite. As a result, we say that $\ln x$ tends to infinity more *slowly* than any positive power of x, while e^x tends to infinity more *rapidly* than any positive power of x.

PROBLEM 4.26

Evaluate the indeterminate limits

$$\text{(a)} \lim_{x\to 0+} x^x \qquad \text{(b)} \lim_{x\to\pi/2} (\operatorname{Sec} x - \operatorname{Tan} x).$$

SOLUTION 4.26

L'Hospital's rule does not apply to either of these limits as they stand since neither is of the form f(x)/g(x). However, if we write,

$$y = x^x,$$

then

$$\ln y = x \ln x,$$

and by the result of Problem 4.24, we have $\text{Lim}_{x \to 0+} \ln y(x) = 0$. Then,

$$\lim_{x \to 0} y(x) = \lim_{x \to 0} e^{\ln y(x)} = e^{\text{Lim} \ln y(x)} = e^0 = 1.$$

This last step is a result of the fact that the exponential function is continuous at every x. In the case of the limit (b) we can use the definition of Sec x andTan x to write

$$\text{Sec } x - \text{Tan } x = \frac{1 - \text{Sin } x}{\text{Cos } x}.$$

Then it is evident that we are dealing with an indeterminate limit of the form 0/0. L'Hospital's rule may be applied, leading to,

$$\lim_{x \to \pi/2} \frac{1 - \text{Sin } x}{\text{Cos } x} = \lim_{x \to \pi/2} \frac{-\text{Cos } x}{-\text{Sin } x} = 0.$$

Taylor's Theorem

PROBLEM 4.27

Prove Theorem 4.12, Taylor's theorem.

SOLUTION 4.27

Let x and x_0 denote fixed points in the interval [a, b] and let I denote the interval whose endpoints are the points x and x_0. Now for t in I let

$$F(t) = f(x) - f(t) - (x - t)f'(t) - \cdots - \frac{(x - t)^n}{n!} f^{(n)}(t) \tag{1}$$

and

$$G(t) = F(t) - \left(\frac{x - t}{x - x_0}\right)^{n+1} F(x_0) \tag{2}$$

Then $G(x) = G(x_0) = 0$ and Rolle's theorem implies the existence of a point c in I such that $G'(c) = 0$. That is,

$$F'(c) + (n + 1)\frac{(x - c)^n}{(x - x_0)^{n+1}} F(x_0) = 0. \tag{3}$$

But it is easy to compute the derivative of F(t) from (1) to obtain,

$$F'(t) = -\frac{(x-t)^n}{n!} f^{(n+1)}(t).$$

Then

$$F(x_0) = -\frac{1}{n+1}\frac{(x-x_0)^{n+1}}{(x-c)^n} F'(c) = \frac{1}{n+1}\frac{(x-x_0)^{n+1}}{(x-c)^n}\frac{(x-c)^n}{n!} f^{(n+1)}(c)$$

i.e.,

$$F(x_0) = f^{(n+1)}(c)\frac{(x-x_0)^{n+1}}{(n+1)!} \tag{4}$$

Now it follows from (1) that (4) is precisely the result (4.4). Note that when $n = 0$, the result (4.4) reduces to the expression which appears in the mean value theorem,

$$f(x) - f(x_0) = f'(c)(x - x_0).$$

Thus Taylor's theorem is the generalization of the mean value theorem to derivatives of order higher than one.

PROBLEM 4.28

Prove the following theorem about local extreme points: Let a denote an interior point of an interval I and suppose that the function $f(x)$ together with its derivatives up to and including order n are all continuous in a neighborhood of $x = a$ for $n \geq 2$. Suppose

$$f'(a) = \ldots = f^{(n+1)}(a) = 0 \quad \text{and} \quad f^{(n)}(a) \neq 0.$$

Then

a) if n is even and $f^{(n)}(a) > 0$, $x = a$ is a local minimum for $f(x)$

b) if n is even and $f^{(n)}(a) < 0$, $x = a$ is a local maximum for $f(x)$

c) if n is odd, then $x = a$ is not a local extreme point for $f(x)$

SOLUTION 4.28

Expanding $f(x)$ in a Taylor's series about $x = a$ leads to

$$f(x) = f(a) + 0 + \cdots + 0 + f^{(n)}(c)\frac{(x-a)^n}{n!} \tag{1}$$

for some c lying between a and x. Since $f^{(n)}(x)$ is continuous and nonzero at $x = a$, it follows that there is a neighborhood N(a) throughout which $f^{(n)}(x)$ does not change sign. If x belongs to N(a), then c must also belong to N(a), and it follows that $f^{(n)}(a)$ and $f^{(n)}(c)$ have the same sign.

If n is even and $f^{(n)}(a) > 0$ then $f^{(n)}(c) > 0$ if x belongs to N(a). Then it follows from (1) that for all x in N(a),

$$f(x) = f(a) + f^{(n)}(c)\frac{(x-a)^n}{n!} \geq f(a);$$

i.e $x = a$ is a local minimum for $f(x)$.

If n is even and $f^{(n)}(a) < 0$, then $f^{(n)}(c) < 0$ if x belongs to N(a). In this case we have, for each x in N(a),

$$f(x) = f(a) + f^{(n)}(c)\frac{(x-a)^n}{n!} \leq f(a);$$

$x = a$ is a local maximum for $f(x)$.

If n is odd, then $(x - a)^n$ changes sign as x passes through the point $x = a$. Consequently, $f(x)-f(a)$ changes sign as x passes through $x = a$, and it is evident that $x = a$ is neither a local maximum nor local minimum for $f(x)$.

A function f(x) is said to be differentiable at the point c if the difference quotient

$$D_h f(c) = \frac{f(c+h) - f(c)}{h}$$

tends to a limit as h tends to zero. If f(x) is differentiable at x=c then it is continuous there.

In addition to the well known differentiation formulas and rules for differentiation, there are a number of important consequences of differentiability. These include:

Location of internal local extreme points
Rolle's theorem
Mean value theorem
Cauchy mean value theorem
Taylor's theorem

The mean value theorem has a surprising number of applications. Each of the following is a consequence of the mean value theorem.

$$|\text{Sin } x| < |x| \quad \text{for } x \neq 0$$

$$e^x > 1 + x \quad \text{for } x \neq 0$$

$$(1+h)^\alpha > 1 + \alpha h \quad \text{for } h > 0 \text{ and } \alpha > 1.$$

In addition, the Cauchy mean value theorem can be used to prove L'Hospital's rule which is useful for evaluating indeterminate limits of the form 0/0 and ∞/∞.

The Taylor theorem, which is an extension of the mean value theorem to derivatives of order higher than one, implies that a smooth function may be approximated locally by a polynomial. This result has many useful applications including the result of Problem 4.28 for the classification of interior extreme points.

SUPPLEMENTARY PROBLEMS

4.1 Let

$$f(x) = \begin{cases} 0 & \text{if } x < 0 \\ 1 & \text{if } x > 0 \end{cases}$$

Then use the definition of the derivative to show that $g(x) = xf(x)$ is continuous but not differentiable at $x = 0$.

4.2 Use the definition of the derivative to show that $h(x)=x^2f(x)$ is differentiable at $x = 0$.

4.3 For $f(x) = \sqrt{x}$ and x, y in [0, 10] does there exist a point c between x and y such that $|f(x) - f(y)| = |f'(c)||x - y|$? Does there exist a positive constant M such that $|f(x) - f(y)| < M|x - y|$ for all x, y in [0, 10]?

4.4 For f as in the previous problem, does there exist a positive constant K such that $|f(x) - f(y)| < K|x - y|$ for all x, y in [1, 10]?

4.5 Find all the extreme points for $f(x) = x^3 - 3x$ on the interval [–4, 4].

4.6 Find all the extreme points for $f(x) = |x^3 - 3x|$ on the interval [–4, 4].

4.7 Find all the extreme points on [–3, 3] for the function $f(x) = |x^2 - 1|$.

4.8 For $f(x) = |\text{Sin } x|$ does Rolle's theorem imply the existence of a point c between 0 and 2π such that $f'(c) = 0$?

4.9 Describe the set of points where $f(x) = e^x/x$ is decreasing.

4.10 Describe the set of points where $f(x) = xe^{-x}$ is decreasing.

4.11 Describe the set of points in $\mathbb{R}^+ = \{x > 0\}$ where $f(x) = \text{Sin}(1/x)$ is decreasing.

4.12 Let

$$f(x) = \begin{cases} 0 & \text{if } x = 0 \\ x\ln x & \text{if } x > 0 \end{cases}$$

Is f continuous on [0, 5]? Is f uniformly continuous on [0, 5]?
Is f differentiable on (0,5)? Is f differentiable on [0, 5]?

5

Integration

In this chapter we introduce the notion of the Riemann integral of a bounded function f on a closed, bounded interval I. Two equivalent definitions for the integral are given; one is stated in terms of upper and lower sums, while the other defines the integral as a limit of Riemann sums. The equivalence of these two definitions is asserted in Theorem 5.4. It is convenient to have the two definitions for the integral since certain facts about the integral follow most easily from one definition while other facts are more easily proved using the second definition. Once the integral has been defined, it then follows that in order for the integral of a given function f(x) over an interval I to exist, it is sufficient for f(x) to be either continuous or monotone on I. A condition which is both necessary and *sufficient to imply the integrability of f can be stated in terms of the more advanced concept of sets of measure zero.*

The integral has a number of fundamental properties including linearity, additivity, and positivity. In addition there are several inequalities that apply to integrals allowing us to estimate the magnitude of an integral without actually evaluating the integral. The Cauchy-Schwarz and Minkowski inequalities are two well known inequalities for integrals. There is a mean value theorem for integrals and this leads to a variation of Taylor's theorem that involves an integral form for the so-called remainder term.

The fundamental theorem of the calculus is the most powerful tool for evaluating integrals. The integration by parts formula and the change of variables theorem extend the applicability of the fundamental theorem. Finally, when it is not possible to evaluate an integral by means of the fundamental theorem, we can resort to approximation techniques referred to generally as numerical quadrature schemes. The trapezoidal rule and Simpson's rule are two widely used numerical integration schemes and Filon's approximation is a method that is specifically designed for numerical integration of oscillatory integrals.

PARTITIONS AND SUMS

Partitions

Throughout this chapter we will let I denote a closed bounded interval [a, b]. It will sometimes be convenient to denote the length of I by writing $|I|$; i.e., $|I| = b - a$. By a *partition* for I we mean a set of points in I

$$P = \{x_0, x_1, \cdots, x_n\} \quad \text{such that } a = x_0 < x_1 < \cdots < x_N = b.$$

The points of the partition P determine a set of nonoverlapping intervals whose union is I; i.e.,

$$I = [x_0, x_1] \cup [x_1, x_2] \cup \cdots \cup [x_{N-1}, x_N] = \bigcup_{k=1}^{N} I_k.$$

MESH SIZE

We define the mesh size for the partition P to be the number

$$\| P \| = \max [\, | I_k | = x_k - x_{k-1} : k = 1, 2, \ldots, N].$$

REFINEMENTS

The partition P′ is said to be a refinement of the partition P if each mesh point x_k of P is also a mesh point of the partition P′. If P′ is a refinement of P then $\| P' \| \leq \| P \|$. We shall let $\prod(I)$ denote the set of all possible partitions for the interval I.

Sums

Suppose the function f(x) is defined and bounded on the interval I and let m and M denote, respectively, the greatest lower bound and least upper bound for the bounded set $\{f(x) : x \in I\}$. Let $P = \{x_0, x_1, \ldots, x_N\}$ denote a partition for I. For k = 1, 2, …, N we may let m_k and M_k denote, respectively, the greatest lower bound and least upper bound for the bounded set of numbers $\{f(x) : x_{k-1} \leq x \leq x_k\}$. Then we can define various sums for the function f on the partition P.

LOWER SUM

The *lower sum* of f(x) based on the partition P of I equals

$$s[f : P] = \sum_{k=1}^{N} m_k \, | I_k |$$

UPPER SUM

The upper sum of f(x) based on the partition P of I equals

$$S[f : P] = \sum_{k=1}^{N} M_k \, | I_k |$$

In addition, if $\xi_1, \ldots, \xi_N$ denote points such that $x_{k-1} \le \xi_k \le x_k$ for $1 \le k \le N$, then we define:

RIEMANN SUM

The *Riemann sum* for f(x) based on the partition P and *evaluation points* ξ_k is equal to

$$RS[f : P; \xi] = \sum_{k=1}^{N} f(\xi_k) \mid I_k \mid$$

Note that the s[f : P] and S[f : P] are not Riemann sums if m_k and M_k are not in the range of f(x). We have the following properties of sums based on partitions:

Theorem 5.1

Theorem 5.1 Let f(x) be defined and bounded on I = [a, b]. Then

(a) for every P in $\prod(I)$ and any choice of evaluation points ξ_k,

$$m \mid I \mid \le s[f : P] \le RS[f : P; \xi] \le S[f : P] \le M \mid I \mid$$

(b) if P′ in P(I) is a refinement of P in P(I), then

$$s[f : P] \le s[f : P'] \quad \text{and} \quad S[f : P'] \le S[f : P]$$

(c) for all P and Q in $\prod(I)$, $s[f : P'] \le S[f : Q]$

DEFINITIONS OF THE INTEGRAL

Upper and Lower Integrals

There are two ways in which we can define the integral of a bounded function on a bounded interval. The first definition is based on the notion of upper and lower sums.

Let f(x) be defined and bounded on I = [a, b]. Then for any P in $\prod(I)$, the upper sum S[f : P] is an upper bound for the set of numbers $\{s[f : Q] : Q \in \prod(I)\}$ and the lower sum s[f : P] is a lower bound for the set $\{S[f : Q] : Q \in \prod(I)\}$. Then by the completeness axiom the set of lower sums has a least upper bound and by Theorem 1.3, the set of upper sums has a greatest lower bound. We define

LOWER INTEGRAL OF f(X) ON I

The lower integral of f(x) on I equals $s[f] = \text{Lub}\,\{s[f : Q] : Q \in \prod(I)\}$

UPPER INTEGRAL OF f(X) ON I

The upper integral of f(x) on I equals $S[f] = \text{Glb}\,\{S[f : Q] : Q \in \prod(I)\}$

Theorem 5.2

Theorem 5.2 Let f(x) be defined and bounded on I = [a, b]. Then f has both an upper integral and a lower integral on I and,

$$m|I| \leq s[f] \leq S[f] \leq M|I|.$$

Riemann Integral Defined in Terms of Upper and Lower Sums

Definition The function f(x), defined and bounded on I = [a, b], is Riemann integrable on I if s[f] = S[f]. Then the notation

$$\int_I f \quad \text{or} \quad \int_a^b f(x)dx$$

is used to denote the common value s[f] = S[f] which we refer to as the Riemann integral of f(x) over I.

We can also define the integral in terms of a limit. For f(x) defined on I, we say that the limit of RS[f : P; ξ] as ‖ P ‖ tends to zero exists and equals L if, for every $\in > 0$ there exists a $\delta > 0$ such that for any $P \in \prod(I)$ with $\| P \| < \delta$, and any choice of the evaluation points ξ, we have

$$| RS[f : P; \xi] - L | < \in.$$

Theorem 5.3

Theorem 5.3 Suppose f(x) is defined on I = [a, b]. If the limit of RS[f : P; ξ] as ‖ P ‖ tends to zero exists then the limiting value is unique. Moreover, if the limit exists, then f(x) must be bounded.

Limit Definition of Riemann Integral

Definition The function f(x), defined and bounded on I = [a, b] is Riemann integrable on I if the limit of RS[f : P; ξ] as ‖ P ‖ tends to zero exists and equals L. Then the notation

$$\int_I f \quad \text{or} \quad \int_a^b f(x)dx$$

is used to denote the limiting value L, which we refer to as the Riemann integral of f(x) over I.

Theorem 5.4

Theorem 5.4 Let f(x) be defined and bounded on I = [a, b]. Then the following statements are all equivalent:

(a) s[f] = S[f] = L

(b) the limit of RS[f : P; ξ] as ‖ P ‖ tends to zero exists and equals L

(c) for each $\in > 0$ there exists P in $\prod(I)$ such that $\sum_k (M_k - m_k) | I_k | < \in$

(d) for each $\in > 0$, there exists P in $\prod(I)$ such that $\sum_k \Omega_k | I_k | < \in$ where

$$\Omega_k = \text{Lub } \{f(x) - f(y) : x, y \text{ in } I_k\} \quad k = 1, 2, \ldots, N$$

It is convenient to have the two definitions of the integral since some properties of the integral follow more easily from one definition than from the other. The following results indicate that there is no shortage of Riemann integrable functions.

Conditions for Integrability Theorem 5.5

Theorem 5.5 Let f(x) be defined on I = [a, b]. Then f is Riemann integrable on I if either of the following conditions holds:

(a) f(x) is monotone on I

(b) f(x) is continuous on I

The conditions of monotonicity and continuity are each *sufficient* to imply that f(x) is integrable but neither is *necessary*. It is possible to state a condition which is both necessary and sufficient to imply f is integrable.

SETS OF MEASURE ZERO

A set W in $\mathbb{R}$ is said to have measure zero if, for all $\in > 0$, there is a family of intervals $I_n = \{(a_n, b_n): a_n < b_n\}$ $n \in \mathbb{N}$ such that W is contained in the union of the intervals I and

$$\sum_n (b_n - a_n) < \in$$

Any set of finitely many points is a set of measure zero. In fact, the set of integers $\mathbb{Z}$ is an *infinite* set of measure zero in $\mathbb{R}$; the set of rationals $\mathbb{Q}$ is another example of an infinite set of measure zero. Finally, any *subset* of a set of measure zero must also have measure zero. Then we have the following result, whose proof is beyond the scope of this text.

Theorem 5.6 Lebesgue's Theorem

Theorem 5.6 Suppose f(x) is defined and bounded on I = [a, b]. Then f is Riemann integrable on I if and only if the set of points of I where f is not continuous has measure zero.

Example 5.1 Integrable and Nonintegrable Functions

5.1 (a) Let I = [−1, 1]. Then f(x) is integrable on I if f(x) is a polynomial, or any of the functions Sin x, Cos x, e^x, because all of these functions are continuous on I. However, f(x) = 1/x is not integrable on I since this function is not bounded on I.

5.1 (b) Let

$$f(x) = \begin{cases} 0 & \text{if } 0 \le x < 1 \text{ or } 2 < x \le 3 \\ 2 & \text{if } 1 \le x \le 2 \end{cases}$$

Then f(x) is neither continuous nor monotone on I = [0, 3] but f is integrable on I since the set of discontinuities for f consists of just two points and so has measure zero. We can also prove directly that f is integrable by using Theorem 5.4. For n = 3, 4, ... let P_n denote the partition {0, 1 − 1/n, 1 + 1/n, 2 − 1/n, 2 + 1/n, 3}. Then as shown in Figure 5.1,

$I_1 = (0, 1 - 1/n)$, $I_2 = (1 - 1/n, 1 + 1/n)$,

$I_3 = (1 + 1/n, 2 - 1/n)$, $I_4 = (2 - 1/n, 2 + 1/n)$,

$I_5 = (2 + 1/n, 3)$

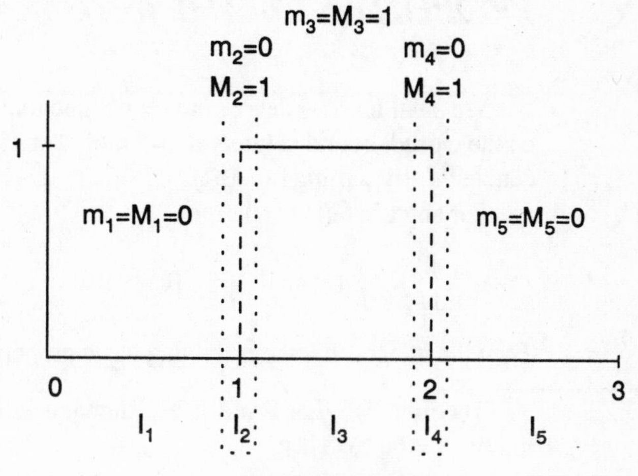

Figure 5.1
Example 5.1(b)

hence $m_1 = M_1 = 0$, $m_2 = 0$ $M_2 = 1$, $m_3 = M_3 = 1$, $m_4 = 0$ $M_4 = 1$, and $m_5 = M_5 = 0$. Thus

$$s[f, P_n] = 0\,|I_1| + 0\,|I_2| + 1|I_3| + 0|I_4| + 0|I_5| = 1 - 2/n$$

$$S[f, P_n] = 0\,|I_1| + 1\,|I_2| + 1|I_3| + 1\,|I_4| + 0|I_5| = 1 + 2/n$$

and $$S[f; P_n] - s[f; P_n] = |I_2| + |I_4| = \frac{4}{n}.$$

Then for any $\in > 0$, we have $S[f; P_n] - s[f; P_n] < \in$ for $n > \in/4$. This proves f is integrable on I the value of the integral equal to

$$\lim_n S[f; P_n] = \lim_n s[f; P_n] = 1.$$

5.1 (c) Let I = [0, 1] and consider the so called *Dirichlet function*,

$$f(x) = \begin{cases} 1 & \text{if x is rational} \\ 0 & \text{if x is irrational} \end{cases}$$

Let P denote any partition of I and note that each of the subintervals, I_k that is induced by the partition contains both rational and irrational numbers. Then for each k, it follows that $m_k = 0$ and $M_k = 1$. Then $s[f:P] = 0$ and $S[f:P] = 1$. Since this holds true for *any* partition P of I, it follows that $s[f] = 0$ and $S[f] = 1$ and the fact that the upper and lower integrals are unequal implies that f(x) is not integrable on I.

PROPERTIES OF THE INTEGRAL

We shall let $\mathbb{R}(I)$ denote the set of functions which are Riemann integrable on the closed, bounded interval $I = [a, b]$. Thus if f(x) is integrable on I we may denote this by writing $f \in \mathbb{R}(I)$.

For any f in $\mathbb{R}(I)$ we define

$$\int_I^b f = -\int_b^a f \quad \text{and} \quad \int_a^a f = 0$$

Then the integral has the following basic properties:

Theorem 5.7 Linearity, Additivity, and Positivity

Theorem 5.7 Let f and g be Riemann integrable on the closed bounded interval $I = [a, b]$. Then

i) $\int_I \alpha f + \beta g = \alpha \int_I f + \beta \int_I g$ for all α, β in $\mathbb{R}$

ii) if p lies between a and b then

$$\int_a^b f = \int_a^p f + \int_p^d f$$

More generally, if $I_1, \ldots, I_n$ are nonoverlapping intervals whose union equals I, then

$$\int_I f = \int_{I_1} f + \cdots + \int_{I_n} f$$

iii) if $f \geq 0$ on I, then $\int_I f \geq 0$

We refer to properties i), ii) and iii), respectively, by saying that the Riemann integral is *linear*, *additive*, and *positive*.

Theorem 5.8 Sets of Measure Zero

Theorem 5.8 Suppose $f, g \in \mathbb{R}(I)$ and the set of points $\{x \in I: f(x) \neq g(x)\}$ is a set of measure zero, then

$$\int_I f = \int_I g.$$

Theorem 5.9 Integrability and Composition of Functions

Theorem 5.9 Suppose f is integrable on I, f[I] is contained in the interval J, and the function g is continuous on J. Then the composition of functions, g(f(x)) is integrable on I.

COROLLARY 5.10

If f, g $\in \mathbb{R}(I)$ then $|f|$, f^n for $n \in \mathbb{N}$, and the product fg are also in $\mathbb{R}(I)$.

Inequalities Theorem 5.11

Theorem 5.11 Suppose f, g $\in$ R(I) with $|f(x)| < M$ for x in I. Then

i) $\left|\int_I f\right| \le \int_I |f| \le M|I|$

ii) $\left|\int_I fg\right| \le \left(\int_I f^2\right)^{1/2}\left(\int_I g^2\right)^{1/2}$

iii) $\left(\int_I (f+g)^2\right)^{1/2} \le \left(\int_I f^2\right)^{1/2} + \left(\int_I g^2\right)^{1/2}$

The results ii) and iii) are commonly referred to as the *Cauchy Schwarz*, and *Minkowski inequalities*, respectively.

Mean Value Theorem for Integrals

The following result is known as the mean value theorem for integrals.

Theorem 5.12

Theorem 5.12 Suppose that f is continuous on the closed bounded interval I, g is integrable on I and that g(x)> 0 for x in I. Then there exists a point c in I such that

$$\int_I fg = f(c)\int_I g$$

Finally, we have the following variation on Theorem 4.12.

Theorem 5.13 Taylor's Theorem

Theorem 5.13 Suppose f(x) is continuous, together with all its derivatives up to the order n + 1, on the interval I = [a, b]. Then for each x in I,

$$f(x) = f(a) + \frac{f'(a)}{1!}(x-a) + \cdots + \frac{f^{(n)}(a)}{n!}(x-a)^n + R_{n+1}$$

where

$$R_{n+1} = \frac{1}{n!}\int_a^x (x-t)^n f^{(n+1)}(t)\,dt.$$

EVALUATION OF INTEGRALS

Theorems 5.5 and 5.6 state conditions on f(x) sufficient to imply that the integral of f(x) over I exists. For purposes of evaluating the integral we have the following.

Theorem 5.14 Fundamental Theorem of Calculus

Theorem 5.14 If f(x) is Riemann integrable on I = [a, b] and if $F'(x) = f(x)$ for all x in I, then

$$\int_a^b f(x)dx = F(b) - F(a)$$

The function F(x) is called a *primitive* or *antiderivative* for f(x) on I if $F'(x) = f(x)$ for all x in I. Often it is not easy to discover F given f. The process is made easier by the following consequences of the fundamental theorem.

Theorem 5.15 Integration by parts

Theorem 5.15 If f and g are continuous with continuous first derivatives on I = [a, b], then

$$\int_a^b f'(x)g(x)dx = f(x)g(x)\Big|_a^b - \int_a^b g'(x)f(x)dx$$

Theorem 5.16 Change of variable

Theorem 5.16 Suppose φ is continuous with a continuous first derivative on I = [a, b] and that f is continuous on J = φ[I]. Then

$$\int_a^b f(\varphi(t))\varphi'(t)dt = \int_{\varphi(a)}^{\varphi(b)} f(x)dx$$

NUMERICAL QUADRATURE

In the event that no primitive can be found for a given f, we can use *numerical quadrature* to approximate the value of the integral of f on the interval I = [a, b]. For this purpose it is convenient to define a *uniform partition* P = $\{x_0, x_1, \ldots, x_N\}$ for I where $x_k = a + k(b - a)/N \quad k = 0, 1, \ldots, N$.

Trapezoidal Rule

On a uniform partition of I into N subintervals we can define the Riemann sum corresponding to evaluation points $\xi_k = (x_{k-1} + x_k)/2 \quad k = 1, \ldots, N$. This leads to the trapezoidal rule for approximating the integral of f on I,

$$\int_a^b f(x)dx \approx T_N = \frac{b-a}{2N}[f(x_0) + 2f(x_1) + \cdots + 2f(x_{N-1}) + f(x_N)]$$

The accuracy of the trapezoidal rule approximation to the integral is described by

Theorem 5.17

Theorem 5.17 Suppose that f is continuous on I = [a, b], together with its derivatives up to order 2. Then there exists a point c in I such that

$$\left|T_N - \int_I f\right| = \frac{(b-a)}{12N^2}|f''(c)| < CN^{-2}$$

where $C = M(b-a)^3/12$ and M denotes the maximum of the continuous function $|f''(x)|$ on the compact interval I.

It is evident from Theorem 5.17 that T_N becomes a better approximation for the integral of f(x) on I as the number, N, of subintervals in the partition of I is increased.

Method of Undetermined Coefficients

A different approach to numerical integration is provided by the method of undetermined coefficients. For example, if the constants A, B, and C are chosen so that that the integration formula

$$\int_{-h}^{h} f(x)dx = h[Af(-h) + Bf(0) + Cf(h)]$$

is correct for each of the three cases, $f(x) = 1$, x, and x^2, then we are led to

$$\int_{-h}^{h} f(x)dx = \frac{h}{3}[f(-h) + 4\,f(0) + f(h)].$$

If the interval [a, b] is divided into 2N subintervals of width $h = (b-a)/2N$, then we may apply this last result N times to obtain

Simpson's Rule

$$\int_{a}^{b} (f(x)dx \approx S_N = \frac{b-a}{3N}[f(x_0) + 4f(x_1) + 2f(x_2) + 4f(x_3) + \cdots$$
$$+ 2f(x_{N-2}) + 4f(x_{N-1}) + f(x_0)]$$

The accuracy of this approximation is described by

Theorem 5.18

Theorem 5.18 Suppose that f is continuous on I = [a, b], together with its derivatives up to order 4 and suppose that N is chosen to be even. Then there exists a point c in I such that

$$\left|S_N - \int_I f\right| = \frac{(b-a)}{180N^4}|f^{(4)}(c)| \leq CN^{-4}$$

where $C = M(b-a)^5/180$ and M denotes the maximum of the continuous function $|f^{(4)}(x)|$ on the compact interval I.

Filon's Approximation

Neither the trapezoidal rule nor Simpson's rule is very effective for approximating an integral with an oscillating integrand. However, we can seek coefficients A_m such that the formula

$$\int_{a}^{b} f(x)\text{Sin}\,kx\,dx \approx h\sum_{m=0}^{2N} A\,f(a + mh) \text{ with } h = \frac{b-a}{2N}$$

is an equality for $f(x) = x^p$ for $p = 0, 1, \ldots, 2N$. The approximation is called Filon's approximation when the coefficients A_m are given by,

$$A_0 = \gamma_1 \text{Cos ka} + \gamma_2 \text{Sin ka}$$

$$A_{2m-1} = 4\gamma_3 \text{Sin kx}_{2m-1} \quad m = 1, 2, \ldots, N$$

$$A_{2m} = 2\gamma_2 \text{Sin kx}_{2m} \quad m = 1, 2, \ldots, N$$

$$A_{2N} = -\gamma_1 \text{Cos kb} + \gamma_2 \text{Sin kb}$$

where

$$\gamma_1 = \frac{1}{kh} + \frac{\text{Sin}2kh}{2k^2h^2} - \frac{2\text{Sin}^2kh}{k^3h^3}$$

$$\gamma_2 = \frac{1 + \text{Cos}^2kh}{k^2h^2} - \frac{\text{Sin } 2kh}{k^3h^3}$$

$$\gamma_3 = \frac{\text{Sin kh}}{k^3h^3} - \frac{\text{Cos kh}}{k^2h^2}$$

Partitions and Sums

Filon's approximation is often more accurate than other numerical integration schemes for approximating integrals with oscillatory integrands.

Note that the trapezoidal rule is a Riemann sum corresponding to the evaluation points $x = (x_{k+1} + x_k)/2$. Since it is a Riemann sum, it follows that T_N tends to the value of the integral as the mesh size $h = (b - a)/N$ tends to zero. By contrast, Simpson's rule and Filon's approximation are based on the method of undetermined coefficients and are not Riemann sums. Nevertheless, these methods tend to the value of the integral as N tends to infinity.

SOLVED PROBLEMS

PROBLEM 5.1

Suppose $m \le f(x) \le M$ for all x in $I = [a, b]$. Then show that for every $P \in \prod(I)$, and every set of evaluation points $\{\xi_1, \ldots, \xi_N\}$,

$$m|I| \le s[f : P] < RS[f : P; \xi] < S[f : P] < M|I| \tag{1}$$

SOLUTION 5.1

Let $P = \{x_0, x_1, \ldots, x_N\}$ and let m_k, M_k denote, respectively, the minimum and maximum value for f(x) on $[x_{k-1}, x_k]$, $k = 1, \ldots, N$. Then for each k,

$$m \le m_k \le \xi_k \le M_k \le M,$$

and hence

$$m \sum_k |I_k| \le \sum_k m_k |I_k| \le \sum_k \xi_k |I_k| \le \sum_k M_k |I_k| \le M \sum_k |I_k|.$$

But this is precisely the result, (1).

PROBLEM 5.2

Suppose f(x) is defined and bounded on I = [a, b], let P = $\{x_0, \dots, x_N\}$ denote a partition of I, and let P′ denote a refinement of P. Then show that

$$s[f:P] \le s[f:P'] \quad \text{and} \quad S[f:P'] \le S[f:P]. \tag{1}$$

SOLUTION 5.2

Suppose first that the refinement P′ is obtained from P by adding a single point x between x_0 and x_1. Let M_1' and M_1'' denote the maximum of f(x) on $[x_0, \xi]$ and $[\xi, x_1]$, respectively. Then, clearly,

$$M_1' \le M_1 \quad \text{and} \quad M_1'' \le M_1,$$

and

$$M_1' \,|[x_0, \xi]| + M_1'' \,|[\xi, x_1]| \le M_1 \,|[x_0, \xi]| + M_1 \,|[\xi, x_1]| \le M_1 |\, I_1| \tag{2}$$

If we add $\sum_{k=2}^{N} M_k \,|\, I_k \,|$ to both sides of (2), we obtain $S[f:P'] \le S[f:P]$.

A similar argument leads to the result $s[f:P] < s[f:P']$. Repeating the argument several times then provides a proof in the case that more than one point is added to P to obtain P′.

PROBLEM 5.3

Suppose f(x) is defined and bounded on I = [a, b]. Then show that for arbitrary partitions P and P′ in $\prod(I)$,

$$s[f:P] \le S[f:P]'. \tag{1}$$

SOLUTION 5.3

Let Q denote the partition obtained by combining the points of P with those of P′. Then Q is a refinement of both P and of P′, and it follows by the results of the previous problems that

$$s[f:P] \le s[f:Q] \le S[f:Q] \le S[f:P'].$$

PROBLEM 5.4

Let I = [0, 1], and for n a positive integer, let $P_n = \{x_0, \dots, x_n\}$ denote the partition of I such that $x_k = a + k/n$, $k = 0, 1, \dots, n$. Then for $f(x) = x^2$, compute $s[f:P_n]$, $S[f:P_n]$ and show that

$$\operatorname*{Lub}_{n \in \mathbb{N}} s[f:P_n] = \operatorname*{Glb}_{n \in \mathbb{N}} S[f:P_n] \tag{1}$$

SOLUTION 5.4

The function $f(x) = x^2$ is an increasing function and hence, for k = 1, …, n the minimum value for f(x) on the interval $[x_{k-1}, x_k]$ is attained at the left end point of the interval. Similarly, the maximum value for f(x) on the interval occurs at the right end point of the interval. That is,

$$m_k = \left(\frac{k-1}{n}\right)^2 \quad \text{and} \quad M_k = \left(\frac{k}{n}\right)^2.$$

Since $|I_k| = 1/n$ for every k (the partition is *uniform*), we have

$$s[f:P_n] = \sum_{k=1}^{n}\left(\frac{k-1}{n}\right)^2\frac{1}{n} = (0^2 + 1^2 + \cdots + (n-1)^2)/n^3 \tag{2}$$

$$S[f:P_n] = \sum_{k=1}^{n}\left(\frac{k}{n}\right)^2\frac{1}{n} = (1^2 + \cdots + (n-1)^2 + n^2)/n^3 \tag{3}$$

In Problem 1.5, we proved that for any positive integer M,

$$1^2 + 2^2 + \cdots + M^2 = M(M+1)(2M+1)/6.$$

Using this result in (2) with $M = n-1$, and in (3) with $M = n$, leads to,

$$s[f:P_n] = (n-1)n(2n-1)/6n^3 = \left(1 - \frac{3}{2n} + \frac{1}{2n^2}\right)\frac{1}{3}$$

$$S[f:P_n] = n(n-1)(2n+1)/6n^3 = \left(1 + \frac{3}{2n} + \frac{1}{2n^2}\right)\frac{1}{3}$$

Then

$$\frac{1}{3} = \text{Lub}\,\{s[f:P_n]: n \in \mathbb{N}\} \le \text{Lub}\,\{s[f:P]: P \in \prod(I)\} = s[f] \tag{4}$$

and

$$S[f] = \text{Glb}\,\{S[f:P]: P \in \prod(I)\} \le \text{Glb}\,\{s[f:P_n]: n \in \mathbb{N}\} = \frac{1}{3}. \tag{5}$$

We have proved that $1/3 \le s[f] \le S[f] \le 1/3$; i.e, $s[f] = S[f] = 1/3$. Since each of the inequalities in (4) and (5) must be an equality, it follows that (1) holds.

DEFINITION OF THE INTEGRAL

PROBLEM 5.5

Suppose that f(x) is defined and bounded on $I = [a, b]$. Then show that

$s[f] = S[f]$ if and only if for all $\in > 0$ there exists a $P_\in$ in $\prod(I)$ such that $S[f:P_\in] - s[f:P_\in] < \in$.

SOLUTION 5.5

Suppose first that $s[f] = S[f]$ and let $\in > 0$ be given. Since

$$s[f] = \text{Lub}\{s[f:P]: P \in \prod(I)\} \quad \text{and} \quad S[f] = \text{Glb}\{S[f:P]: P \in \prod(I)\}$$

it follows from the definition of least upper bound and greatest lower bound that there exist partitions P′ and P″ for I such that

$$s[f] - \in/2 < s[f:P'] \quad \text{and} \quad S[f] + \in/2 > S[f:P'']. \tag{1}$$

Now let $P_\in$ denote the partition consisting of all the points in P′ combined with all the points of P″. Then $P_\in$ is a refinement of both of the partitions, P′ and P″. Then (1) together with the result of Problem 5.3 implies that

$$s[f] - \in/2 < s[f : P'] \le s[f : P_\in] \le S[f : P_\in] \le S[f : P''] < S[f] + \in/2.$$

But we have assumed that $s[f] = S[f]$ and it follows that $S[f : P_\in] - s[f : P_\in] < \in$.

Suppose now that for all $\in > 0$ there exists a $P_\in$ in $\prod(I)$ such that $S[f : P_\in] - s[f : P_\in] < \in$. Note that for all partitions $P \in \prod(I)$, we have

$$s[f : P] \le s[f] \quad \text{and} \quad S[f] \le S[f : P];$$

i.e.

$$S[f] - s[f] < S[f : P] - s[f : P]. \tag{2}$$

In particular, for $P = P_\in$, (2) implies

$$S[f] - s[f] \le S[f : P_\in] - s[f : P_\in] < \in. \tag{3}$$

Since (3) holds for all $\in > 0$, it follows that $S[f] \le s[f]$. But $S[f] \ge s[f]$ must also hold, and hence $S[f] = s[f]$.

PROBLEM 5.6

Suppose $f(x)$ is monotone on $I = [a, b]$. Then show that f is Riemann integrable on I.

SOLUTION 5.6

Suppose $f(x)$ is monotone increasing on I. We will show that for every $\in > 0$, there exists a partition $P_\in$ for I such that $S[f : P_\in] - s[f : P_\in] < \in$. Then by Theorem 5.4 it follows that $f(x)$ is integrable on I.

Let $\in > 0$ be given and choose a positive integer n such that

$$n \in > (b - a)(f(b) - f(a)). \tag{1}$$

Then let $P_\in = \{x_0, x_1, \ldots, x_n\}$ where $x_k = a + k(b - a)/n$ for $k = 0, 1, \ldots, n$. Then for each k, we have $|I_k| = (b - a)/n$.

Since $f(x)$ is monotone increasing, it follows that for each k,

$$m_k = \min\{f(x) : x \in I_k\} = f(x_{k-1})$$

and

$$M_k = \max\{f(x) : x \in I_k\} = f(x_k).$$

Then

$$S[f : P_\in] - s[f : P_\in] = \sum_{k=1}^{n} (M_k - m_k)\,|I_k| = \sum_{k=1}^{n} (f(x_k) - f(x_{k-1}))\,|I_k|$$

But $|I_k| = (b - a)/n$ for each k, and

$$\sum_{k=1}^{n} (f(x_k) - f(x_{k-1})) = f(x_n) - f(x_0) = f(b) - f(a).$$

Then, according to (1),

$$S[f : P_\in] - s[f : P_\in] = (f(b) - f(a))(b - a)/n < \in.$$

This proves the result.

PROBLEM 5.7

Suppose that f(x) is continuous on I = [a, b]. Then show that f(x) is Riemann integrable on I.

SOLUTION 5.7

Since f(x) is continuous on the compact interval, it follows that f(x) is uniformly continuous on I. Then for each $\in > 0$, there exists a $\delta > 0$ such that for any x, y in I with $|x - y| < \delta$, we have $|f(x) - f(y)| < \in/(b - a)$. Then with $\in > 0$ fixed, let $P_\in$ denote a uniform partition of I as in the last problem with n chosen so that

$$n\delta > b - a. \qquad (1)$$

Then for k = 1, 2, ..., n

$$|I_k| = x_k - x_{k-1} = (b - a)/n < \delta \qquad (2)$$

and

$$M_k - m_k = \max\{f(x) - f(y) : x, y \in I_k\} < \in/(b - a). \qquad (3)$$

Then it follows that,

$$S[f : P_\in] - s[f : P_\in] = \sum_{k=1}^{n} (M_k - m_k)|I_k| < n \frac{\in}{b - a} \frac{b - a}{n} = \in.$$

By Theorem 5.4, it follows that f(x) is integrable on I.

PROBLEM 5.8

Show that if f(x) is Riemann integrable on I = [a, b], then f must be bounded on I.

SOLUTION 5.8

Suppose f is integrable on I. Then for any choice of evaluation points, $\{\xi_k\}$, the limit of the Riemann sums, $RS[f : P; \xi]$, as $\|P\|$ tends to zero exists; denote the value of this limit by L. We shall show that if f is not bounded on I then a contradiction arises.

Fix $\in = 1$ and for an arbitrary $\delta > 0$, let P denote a partition of I with $\|P\| < \delta$. If f is not bounded on I, then there exist (at least) points x_{N-1}, x_N in P such that for any M > 0 there is a point ξ_N in $[x_{N-1}, x_N]$ with $f(\xi_N) > M$. In particular, choose $M = (L + 1)/\delta$. Let the remaining evaluation points $\{\xi_k : k \neq N\}$ be arbitrarily chosen. Then the Riemann sum $RS[f : P; \xi]$ satisfies

$$RS[f : P; \xi] = \sum_{k \neq N} f(\xi_k)|I_k| + f(\xi_N)|I_N| > \sum_{k \neq N} f(\xi_k)|I_k| + L + 1.$$

and

$$|RS[f : P; \xi] - L| > 1.$$

Evidently, the limit of the Riemann sums as $\|P\|$ tends to zero fails to exist, in contradiction to our assumption that f is integrable on I. It follows that f can not be unbounded.

PROBLEM 5.9

Show that the discontinuous function

$$f(x) = \begin{cases} -1 - x & \text{if } -1 \le x < 0 \\ 1 + x & \text{if } 0 \le x \le 1 \end{cases}$$

is integrable on $I = [-1, 1]$.

SOLUTION 5.9

The function f(x) has a finite jump discontinuity at x = 0; i.e

$$f(0+) = \lim_{h \to 0} f(x + h) = 1 > -1 = f(0-) = \lim_{h \to 0} f(x - h)\ h > 0.$$

Then for $\in > 0$, let $P' = \{x_0, x_1, \ldots, x_n\}$ and $P'' = \{x_{n+1}, \ldots, x_m\}$ denote partitions of $[-1, -\in/8]$ and of $[\in/8, 1]$, respectively, such that

$$-1 = x_0 < x_1 < \cdots < x_n = -\in/8 \quad \text{and} \quad \in/8 = x_{n+1} < \cdots < x_m = 1.$$

Let $P = \{x_0, x_1, \ldots, x_n, x_{n+1}, \ldots, x_m\}$ denote the partition of $[-1, 1]$ composed of the points of P′ and P″ taken together. Then,

$$S[f : P] = \sum_{k=1}^{n} M_k \mid I_k \mid + 1\frac{\in}{4} + \sum_{k=n+1}^{m} M_k \mid I_k \mid$$

$$s[f : P] = \sum_{k=1}^{n} m_k \mid I_k \mid + 1\frac{\in}{4} + \sum_{k=n+1}^{m} m_k \mid I_k \mid$$

and we obtain by subtraction,

$$S[f : P] - s[f : P] = \sum_{k \neq n} (M_k - m_k) \mid I_k \mid + \frac{\in}{2}. \tag{1}$$

Since f(x) is continuous on $[-1, -\in/8]$ and on $[\in/8, 1]$, the partitions P′ and P″ can be chosen so the sum on the right side of (1) is less than $\in/2$. Then the partition P is such that $S[f : P] - s[f : P]$ is less than $\in$ and it follows that f(x) is integrable on $I = [-1, 1]$.

This argument can be modified and repeated finitely often in order to prove that any function with a *finite* number of *finite* jump discontinuities on an interval I is integrable on I. Finally, we can apply mathematical induction to show that a function having an infinite number of finite jump discontinuities on an interval I is integrable on I, provided that the set of discontinuities is countable; i.e, the points of discontinuity can be placed in one to one correspondence with the integers. A countable set of discontinuities forms a set of measure zero.

PROBLEM 5.10

Consider the function, defined on [0, 1] by,

$$f(x) = \begin{cases} 1 & \text{if } x = 0 \\ 0 & \text{if } x \text{ is irrational} \\ 1/n & \text{if } x = m/n \text{ (lowest terms)} \in \mathbb{Q}. \end{cases}$$

Show that f is integrable on I = [0, 1] by showing:

(a) f(x) is discontinuous at each rational point x = m/n in I

(b) f(x) is continuous at each irrational x in I.

SOLUTION 5.10

The rational numbers are a set of measure zero in $\mathbb{R}$. Then the rational numbers in I are a subset of a set of measure zero; i.e. they are also a set of measure zero. If we show (a) and (b), we will have shown that the points of I where f is discontinuous is a set of measure zero in I. Then it will follow from Theorem 5.6 that f is integrable on I.

To prove (a), suppose that $x_1 = m/n$ is a point in I with positive integers m and n having no factors in common. Then $f(x_1) = 1/n$. Fix $\in$ such that $0 < \in < 1/n$. Then for every $\delta > 0$, there are irrational points y such that $|x - y| < \delta$. At each such point y we have $f(y) = 0$ and hence

$$|f(x_1) - f(y)| = 1/n > \in \quad \text{while } |x - y| < \delta.$$

Then f is not continuous at $x = x_1$.

To prove (b) suppose x_0 in I is an irrational point. If we fix an $\in > 0$, then there can be only *finitely many* positive integers n such that $n < 1/\in$. Then there are only finitely many rational numbers q = m/n in I such that $f(q) > \in$. We can choose $\delta > 0$ sufficiently small that none of these numbers lie in $N_\delta[x_0]$. Then it follows that

$$|f(x_0) - f(y)| = |f(y)| < \in \quad \text{for all } y \text{ in } N_\delta[x_0].$$

This proves that f is continuous at each irrational point in I. Then the set of discontinuities of f in I is just the set of rational points in I. Since this set has measure zero, f is integrable on I.

Properties of the

PROBLEM 5.11

Show that if $f \in \mathbb{R}(I)$ and $f \geq 0$ on I, then $\int_I f \geq 0$.

SOLUTION 5.11

Let $P = \{x_0, ..., x_n\}$ be a partition of I. Then f>0 on I implies that for each k,

$$m_k = \text{Glb}\,\{f(x) : x \in I_k\} \geq 0.$$

Then

$$s[f : P] \geq 0,$$

and since this holds for every partition $P \in \prod(I)$, it follows that

$$\int_I f = s[f] = \text{Lub}\,\{s[f:P]: P \in \prod(I)\} \geq 0.$$

Note that this result implies:

$$\text{for f, g in } \mathbb{R}(I) \text{ such that } f \geq g \text{ on I, we have } \int_I f \geq \int_I g. \tag{1}$$

PROBLEM 5.12

Give an example of functions f and g such that each is integrable on its domain of definition, but the composition of g with f is not integrable.

SOLUTION 5.12

Let f(x) be the function of Problem 5.10. Then f is defined on I = [0, 1] and f is integrable on I since the points of discontinuity are just the rational points of I, a set of measure zero. Let g(x) be defined on I as follows

$$g(x) = \begin{cases} 0 & \text{if } x = 0 \\ 1 & \text{if } 0 < x \leq 1. \end{cases}$$

Then

$$g[f(x)] = \begin{cases} 0 & \text{if x is irrational} \\ 1 & \text{if x is rational} \end{cases}$$

But this composition is precisely the nonintegrable Dirichlet function of Example 5.1(c).

PROBLEM 5.13

Suppose $f \in \mathbb{R}(I)$ satisfies $|f(x)| \leq M$ for x in I = [a, b]. Then

$$\left|\int_I f\right| \leq \int_I |f| \leq M\,|I|. \tag{1}$$

SOLUTION 5.13

Corollary 5.10 asserts that $|f|$ is integrable on I if $f \in \mathbb{R}(I)$. Since

$$-|f(x)| \leq f(x) < |f(x)| \quad \text{for all x,}$$

the result (1) in Problem 5.11 leads to $|\int_I f| \leq \int_I |f|$. Using this result again, together with $|f(x)| \leq M$, leads to (1).

PROBLEM 5.14

Suppose f is integrable on I = [a, b]. Then show that f^2 is integrable on I.

SOLUTION 5.14

If f is integrable on I, then f is bounded on I and we may denote by M the least upper bound for f on I. In addition, for any $\in > 0$ there exists a partition P of I into subintervals $I_1, \ldots, I_n$ such that

$$S[f;P] - s[f;P] = \sum_{k=1}^{n}(M_k - m_k)\,|I_k| < \frac{\in}{2M}$$

where m_k, M_k denote respectively the greatest lower bound and least upper bound for f on I_k, $k = 1, \ldots, n$. Then

$$S[f^2;P] - s[f^2;P] = \sum_{k=1}^{n}(M_k^2 - m_k^2)\,|I_k|$$

$$= \sum_{k=1}^{n}(M_k + m_k)(M_k - m_k)\,|I_k|$$

$$\leq 2M \sum_{k=1}^{n}(M_k - m_k)\,|I_k| < \in$$

and it follows from Theorem 5.4 that f^2 is integrable on I.

PROBLEM 5.15

Suppose that $f, g \in \mathbb{R}(I)$. Then show that

$$\left|\int_I fg\right| \leq \left(\int_I f^2\right)^{1/2}\left(\int_I f^2\right)^{1/2} \qquad (1)$$

SOLUTION 5.15

Corollary 5.10 asserts that if f and g are integrable then f^2, g^2 and fg are all in $\mathbb{R}(I)$. Then let

$$A = \int_I f, \quad B = \left|\int_I fg\right| \quad \text{and} \quad C = \int_I g.$$

Note that the quadratic function $Q(x) = Ax^2 - 2Bx + C$ is greater than or equal to zero for all x. To see this, write

$$Q(x) = x^2\int_I f^2 - 2x\left|\int_I fg\right| + \int_I g^2 =$$

$$\geq \int_I\left(x^2f^2 - 2x\,|fg| + g^2\right) = \int_I\left(x\,|f| - |g|\right)^2.$$

Then $(x\,|f| - |g|) \geq 0$ implies $Q(x) \geq 0$ by the positivity of the integral.

Now if $A = 0$, then B must also vanish, otherwise Q(x) would be negative for large positive x . This proves (1) in the case $A = 0$. If $A > 0$, then substituting $x = B/A$ into $Q(x) \geq 0$ leads to $B^2/A - 2B^2/A + C = -B^2/A + C \geq 0$. This is (1) in the case $A > 0$.

PROBLEM 5.16

Prove Theorem 5.12, the mean value theorem for integrals.

SOLUTION 5.16

Suppose f(x) is continuous on I and g(x) is integrable on I with $g(x) \geq 0$ for all x in I. Then f(x) is integrable by Theorem 5.5 and the product f(x)g(x) is also integrable by Corollary 5.10. Let M, m denote, respectively, the maximum and minimum values assumed by f(x) on I. Then

$$mg(x) \leq f(x)g(x) \leq Mg(x) \quad \text{for all x in I.} \tag{1}$$

and the result (1) of Problem 5.11 implies

$$m\int_I g < \int_I fg < M\int_I g \tag{2}$$

If $\int_I g = 0$, then for *any* c in I,

$$\int_I f(x)g(x)dx = f(c)\int_I g(x)dx. \tag{3}$$

If $\int_I g$ is not zero, then (2) implies that the value,

$$y = \int_I fg \Big/ \int_I g,$$

lies *between* m and M. But f(x) is continuous on I and hence by the intermediate value theorem, there exists some c in I where y = f(c). But y = f(c) is equivalent to (3) and the result is proved.

In the special case that g(x) = 1, I = [a, b], (3) reduces to

$$\text{for some c in I: } \int_I f(x)dx = f(c)\int_I dx = f(c)(b - a) \tag{4}$$

PROBLEM 5.17

Prove that if f is integrable on I = [a, b] and if $F'(x) = f(x)$ for all x in I, then

$$\int_I f(x)dx = F(b) - F(a). \tag{1}$$

SOLUTION 5.17

Let $P = \{x_0, x_1, \ldots, x_n\}$ denote a partition of I. Then the mean value theorem for derivatives implies that for k = 1, 2, ..., n, there exists ξ_k in $I_k = [x_{k-1}, x_k]$ such that,

$$F(x_k) - F(x_{k-1}) = F'(\xi_k)(x_k - x_{k-1}) = f(\xi_k)\,|I_k| \tag{2}$$

Then,

$$F(b) - F(a) = \sum_{k=1}^{n}(F(x_k) - F(x_k)) = \sum_{k=1}^{n} f(\xi_k)\,|I_k| = RS[f : P; \xi];$$

i.e., $$F(b) - F(a) = RS[f : P; \xi] \text{ for all } P \in \prod(I). \tag{3}$$

Since (3) holds for all partitions P of I, we can take the limit as $\| P \|$ tends to zero. Since f has been assumed to be integrable on I, the limit exists and we obtain (1).

PROBLEM 5.18

Suppose f, g and f′, g′ are all continuous on I = [a, b]. Then show that

$$\int_I f'(x)g(x)dx = f(b)g(b) - f(a)g(a) - \int_I f(x)g'(x)dx \qquad (1)$$

SOLUTION 5.18

The hypotheses on f, g imply that F(x) = f(x)g(x) is continuous with a continuous first derivative, F′ = f′g + fg′. Then by the fundamental theorem of calculus,

$$\int_I F' = F(b) - F(a);$$

i.e.,

$$\int_I (f'g + fg') = f(b)g(b) - f(a)g(a)\,.$$

But this is (1).

PROBLEM 5.19

Suppose that f(x) is continuous, together with its derivatives up through the order n+1 on the interval I = [a, b]. Then show that for each x in I,

$$f(x) = f(a) + \frac{f'(a)}{1!}(x - a) + \cdots + \frac{f(a)}{n!}(x - a)^n + R_{(n+1)} \qquad (1)$$

where

$$R_{(n+1)} = \frac{1}{n!}\int_a^x (x - t)^n f^{(n+1)}(t)dt \qquad (2)$$

SOLUTION 5.19

The fundamental theorem implies that

$$f(x) = f(a) + \int_a^x f'(t)dt. \qquad (3)$$

Integration by parts may be applied to show that

$$\int_a^x f'(t)dt = -f'(t)(x-t)\Big]_{t=a}^{t=x} + \int_a^x f''(t)(x - t)dt$$

$$= f'(a)(x - a) + \int_a^x f''(t)(x - t)dt\,.$$

If we substitute this into (3), we obtain,

$$f(x) = f(a) + f'(a)(x - a) + \int_a^x f''(t)(x - t)dt. \qquad (4)$$

This is (1) for the case n = 1. If we apply integration by parts to the integral that appears in (4), we obtain

$$f(x) = f(a) + f'(a)(x - a) + \frac{f''(a)}{2!}(x - a)^2 + \int_a^x f^3(t)(x - t)^2 dt \qquad (5)$$

which is just (1) for the case n = 2. Continuing in this way, we see that for general n we have (1).

Note that we may apply the mean value theorem for integrals to the expression (2) to conclude that for some c, a < c < x, we have

$$R_{n+1} = \frac{1}{n!}\int_a^x (x-t)^n f^{(n+1)}(t)dt = \frac{f^{(n+1)}(c)}{n!}\int_a^x (x-t)^n dt.$$

But,

$$\int_a^x (x-t)^n dt = \frac{(x-a)^{n+1}}{n+1}$$

and hence

$$R_{n+1} = \frac{f^{(n+1)}(c)}{(n+1)!}(x-a)^{n+1}.$$

This is the formula obtained in Theorem 4.12 for the remainder term.

Evaluation of Integrals

PROBLEM 5.20

Find values for the constants A, B, and C such that the equation

$$\int_{-h}^{h} f(x)dx = h[Af(-h) + Bf(0) + Cf(h)] \tag{1}$$

holds for $f(x) = 1$, x and x^2. Use the result to derive Simpson's rule.

SOLUTION 5.20

We can easily compute,

$$\int_{-h}^{h} 1\,dx = 2h, \quad \int_{-h}^{h} x\,dx = 0, \quad \text{and} \quad \int_{-h}^{h} x^2 dx = 2/3.$$

Then applying (1) in each of the three cases, $f(x) = 1$, x, and x^2, we find

$$\begin{aligned} A + B + C &= 2 \\ -A + \quad\; + C &= 0 \\ A \quad\quad\; + C &= 4/3. \end{aligned}$$

We solve this system to find, $A = C = 1/3$, and $B = 4/3$. Then (1) becomes

$$\int_{-h}^{h} f(x)dx = \frac{h}{3}[f(-h) + 4f(0) + f(h)]. \tag{2}$$

To approximate an integral on the interval $I = [a, b]$, let $h = (b - a)/2N$ for N a positive integer. Then if $x_k = a + kh$ for $k = 0, 1, \ldots, 2N$ we can write

$$\int_I f(x)dx = \sum_{k=1}^{N}\int_{x_{k-1}}^{x_{k+1}} f(x)\,dx \tag{3}$$

and

$$\int_{x_{k-1}}^{x_{k+1}} f(x)\,dx = \int_{x_k-h}^{x_k+h} f(x)\,dx \approx \frac{h}{3}[f(x_k - h) + 4f(x_k) + f(x_k + h)] \tag{4}$$

Then, substituting (4) into (3) produces the Simpson's rule formula.

PROBLEM 5.21

Use the Filon approximation to compute, for p = 8, 16, 64, and 128,

$$\int_0^{2\pi} \sqrt{x}\, \text{Sin } px\, dx = J(p), \quad \text{for } N = 4, 8, 16, 64, 128, 256, \text{and } 4096.$$

Compare the result with the Simpson's rule approximation with the same choices for N.

SOLUTION 5.21

If we let S_N and F_N denote, respectively, the Simpson's rule and Filon approximation to the integral J(p) for a partition of N points, then we compute for the case p = 8,

N	S_N	F_N
4	$-6.8\ 10^{-10}$	−.313
8	$-6.8\ 10^{-10}$	−.313
16	$-1.87\ 10^{-10}$	−.313
64	−.2859	−.2856
128	−.2856	−.2856
256	−.2856	−.2856
4096	−.2856	−.2856

Repeating the computation for p = 16, we find

N	S_N	F_N
4	$-1.36\ 10^{-9}$	−.156
8	$-1.36\ 10^{-9}$	−.156
16	$-5.35\ 10^{-10}$	−.156
64	−.152	−.1479
128	−.147	−.1468
256	−.1468	−.1468
4096	−.1468	−.1468

So far, the Filon approximation appears to be slightly better than the Simpson's rule, particularly when N is not large. Now repeat the computation, this time with two very highly oscillatory integrands, p = 64 and p = 128,

	p = 64		p = 128	
N	S_N	F_N	S_N	F_N
4	$-5.45\ 10^{-9}$	-.0391	$-1.09\ 10^{-8}$	-.0195
8	$-5.44\ 10^{-9}$	-.0391	$-1.08\ 10^{-8}$	-.0195
16	$-2.14\ 10^{-9}$	-.0391	$-4.28\ 10^{-9}$	-.0195
64	$-4.54\ 10^{-9}$	-.0391	$-9.08\ 10^{-9}$	-.0195
128	$2.14\ 10^{-9}$	-.0391	$-9.96\ 10^{-9}$	-.0195
256	-.0396	-.0380	$4.23\ 10^{-9}$	-.0195
4096	-.0379	-.0379	-.01915	-.01915

Evidently, when the integrand is highly oscillatory, the Filon approximation gets very close to the correct value of the integral with as few as 4 points, while the Simpson's rule requires many more points before it begins to converge on the correct result.

Let $I = [a, b]$ be a closed bounded interval and let P denote any partition for I. If f(x) is bounded on I then we can define:

Lower Sum for f based on P: $s[f : P] = \sum_k m_k |I_k| \quad m_k = \min\{f(x) : x \in I_k\}$

Upper Sum for f based on P: $S[f : P] = \sum_k M_k |I_k| \quad M_k = \max\{f(x): x \in I_k\}$

Riemann Sum for f based on P: $RS[f : P; \xi] = \sum_k f(\xi_k) \,|\, I_k|, \quad \xi_k \in I_k$

Then the set of all lower sums is bounded above; any upper sum serves as an upper bound. The least upper bound for the set of all lower sums is called the Lower Integral for f on I. The set of all upper sums is bounded below; any lower sum serves as a lower bound for the set. The greatest lower bound is called the Upper integral for f on I. For any bounded function, the upper integral can be no less than the lower integral; if the upper integral is equal to the lower integral, then we say f is integrable on I, and the common value of the upper and lower integrals is called the Integral of f over I.

Equivalently, if the Riemann sums $RS[f : P; \xi]$ tend to a limit L as the mesh size $\|P\|$ of the partition P tends to zero, then f is integrable on I and the value of the integral is equal to L.

Every function that is continuous on I is integrable on I. In fact, if f is monotone on I it is integrable on I. Each of these conditions is sufficient but not necessary for integrability. A condition which is sufficient and necessary for integrability on I is that the points of I where f is discontinuous is a set of measure zero.

The Riemann integral is linear, positive and additive, and two functions that are equal except at finitely many points of I have equal integrals over I. If f and g are each integrable on I, then so are the functions $|f|$, fg and f^n for all n in N. If *f is continuous and g is integrable and nonnegative on I, then for some c in I, we have*

$$\int_I fg = f(c)\int_I g.$$

This is called the Mean Value theorem for integrals.

If f is integrable on I = [a, b], and F is any primitive for f, then

$$\int_I f = F(b) - F(a).$$

This is the Fundamental Theorem of Calculus, and it provides the most powerful tool for evaluating integrals. However, not every function which is integrable has a primitive which can be expressed in terms of a finite number of elementary functions. When this is the case, there are various numerical methods for computing an approximation to the value of the integral of f over I.

SUPPLEMENTARY PROBLEMS

5.1 Let f(x) be defined by

$$f(x) = \begin{cases} 1 & \text{if } 0 \le x \le 1 \\ -2 & \text{if } 1 < x \le 4 \end{cases}$$

Prove f is integrable on I = [0, 4] by showing that for all $\in > 0$ there exists a partition P for I such that $S[f; P] - s[f; P] < \in$. Do this in the following steps:

(a) Define a partition P and the resulting subintervals I_k. Compute the lengths $|I_k|$ for the subintervals.

(b) Find m_k and M_k for each subinterval I_k

(c) Compute the upper sum and the lower sum for the partition

(d) Show that the difference $S[f; P] - s[f; P]$ can be made as small as you like

5.2 Repeat the procedure of the previous problem for f(x) defined on I by

$$f(x) = \begin{cases} 1 & \text{if } x = 1, 2, 3 \\ 0 & \text{otherwise} \end{cases}$$

Compute the integral of f over I.

5.3 Let $F(x) = \int_0^x \text{Sin}(e^t + t)\,dt$. Prove that F(x) is continuous at every x.

5.4 Show that F(x) is differentiable at each x and compute F′(x).

5.5 Let $F(x) = \int_0^x f(t)dt$ for f as in the first supplemenary problem.
Is F(x) continuous on I? Is F(x) differentiable on I?

5.6 Repeat the previous problem for the f(x) from the second problem.

6

Improper Integrals and Infinite Series

The Riemann integral of the function $f = f(x)$ over the interval $I = [a, b]$, defined in the previous chapter, has meaning only if $|I| = b - a$ is finite and the function f is bounded on I. If either one, or both of these assumptions is not satisfied, then the integral is not a proper Riemann integral. However, the "improper" integral may still have meaning. We shall specify conditions under which an integral on an unbounded interval of integration can be assigned a finite value and other conditions under which a finite value can be assigned to an integral whose integrand f becomes unbounded at a point in $I = [a, b]$. In such cases we say that the improper integral is convergent. *If it is not possible to assign a finite value to the improper integral then we say the integral is* divergent. *We shall discuss certain applications of improper integrals including the* Gamma function *and the* Laplace *and* Fourier transforms.

Improper integrals on unbounded intervals have certain points of similarity to the topic of constant term infinite series. We shall give a precise definition of an infinite series *and we shall see that every such series is either convergent or divergent. Many of the tests for determining the convergence or divergence of an infinite series are closely parallel to the tests that are applied to improper integrals on unbounded intervals. The material on infinite series of numbers lays the foundation for a later chapter on infinite series of functions.*

IMPROPER INTEGRALS

Integrals of Unbounded Functions

If f(x) is Riemann integrable on I = [a, b] then f is bounded on I. If f is not bounded on I then the integral of f over I is said to be improper.

CONVERGENCE AND DIVERGENCE

Definition Suppose that f(x) is defined on the interval [a, b) but f(x) tends to infinity as x tends to b. Suppose also that for each $\in > 0$, f(x) is Riemann integrable on the interval $[a, b - \in]$. Finally, suppose

$$\lim_{\in \to 0} \int_a^{b-\in} f(x)dx = L.$$

If L is finite then the improper integral of f(x) over [a, b) is said to be convergent with value equal to L. If L is infinite or if the limit fails to exist, then the improper integral of f(x) over [a, b) is said to be divergent.

OTHER IMPROPER INTEGRALS

A similar definition applies when f(x) is defined on (a, b] and f tends to infinity as x tends to a. If c is an interior point of the compact interval I = [a, b], and if f(x) is defined on [a, c) and on (c, b] but f tends to infinity as x tends to c then the improper integral

$$\int_a^b f(x)\,dx = \int_a^c f(x)\,dx + \int_c^b f(x)\,dx$$

is convergent if and only if *both* the improper integrals on the right side of the equation are convergent. If either or both of the integrals on the right diverges then the improper integral over [a, b] diverges.

Example 6.1 Improper Integrals of Unbounded Functions

6.1 (a) For $0 < p < 1$, the integral

$$\int_0^1 x^{-p}\,dx$$

is improper because $f(x) = x^{-p}$ tends to infinity as x tends to zero. For each $\in > 0$, we have

$$\int_\in^1 x^{-p}\,dx = \left.\frac{x^{1-p}}{1-p}\right|_{x=\in}^{x=1} = \frac{1-\in^{1-p}}{1-p}$$

Now $1 - p > 0$ if $0 < p < 1$, and hence,

$$\lim_{e \to 0} \frac{1-\in^{1-p}}{1-p} = \frac{1}{1-p}.$$

Since the limit exists and is finite, the improper integral is convergent. Note that if $p > 1$, then $1 - p < 0$, and in this case $\in^{1-p}$ tends to infinity as $\in$ tends to zero. Then the improper integral is divergent if $p > 1$.

6.1 (b) The integral

$$\int_0^1 \frac{1}{x}\,dx$$

is similarly improper and for each $\in > 0$,

$$\int_{\in}^{1} \frac{1}{x}\,dx = \ln x \Big|_{x=\in}^{x=1} = 0 - \ln\in .$$

Since $\ln\in$ tends to $-\infty$ as $\in$ tends to zero, this improper integral is divergent.

Improper Integrals on Unbounded Intervals

The Riemann integral is defined for bounded intervals I = [a, b]. If the interval I is not bounded then the integral of f over I is said to be improper.

Definition Suppose f(x) is Riemann integrable on the bounded interval [a, N] for every N > a. Suppose also that

$$\lim_{N\to\infty} \int_a^N f(x)\,dx = L.$$

If L is finite, then the *improper integral* of f over [a, ∞) is *convergent* with value equal to L. If L is infinite or if the limit fails to exist, we say that the improper integral of f over [a, ∞) is *divergent*.

OTHER IMPROPER INTEGRALS

A similar definition applies for the improper integral of f(x) over (−∞, b]. The improper integral

$$\int_{-\infty}^{\infty} f(x)\,dx = \int_{-\infty}^{c} f(x)\,dx + \int_{c}^{\infty} f(x)dx$$

is convergent if and only if both the improper integrals on the right side of the equation are convergent. If either or both of the integrals on the right diverges then the improper integral over (−∞, ∞) diverges.

Suppose that f(x) tends to infinity as x tends to a. Then the improper integral

$$\int_{a}^{\infty} (x)\,dx = \int_{a}^{b} f(x)\,dx + \int_{b}^{\infty} f(x)dx, \quad a < b < \infty,$$

is convergent if both of the improper integrals on the right side of the equation are convergent. If either or both of the integrals on the right diverges then the improper integral of f(x) over (a, ∞) diverges. The improper integral of f(x) over (−∞, a) is treated in a similar fashion.

Example 6.2 Improper Integrals on Unbounded Intervals

6.2 (a) For p a real number, consider the integral

$$\int_{1}^{\infty} x^{-p}\,dx = I(x^{-p}).$$

For p ≠ 1 and for each N > 1,

$$\int_{1}^{N} x^{-p}\,dx = \frac{x^{1-p}}{1-p}\Big|_{x=1}^{x=N} = \frac{1}{1-p}(N^{1-p} - 1)$$

If p > 1, then 1 − p < 0 and N^{1-p} tends to zero as N tends to infinity. In this case the improper integral $I(x^{-p})$ is convergent and has value 1/(p − 1).

If p < 1, then 1 − p > 0 and N^{1-p} tends to infinity as N tends to infinity. In this case the improper integral $I(x^{-p})$ is divergent.

In the case p = 1 we have

$$\int_1^N x^{-p}\,dx = \ln x \Big|_{x=1}^{x=N} = \ln N.$$

Since $\ln N$ tends to infinity as N tends to infinity, the improper integral $I(x^{-1})$ is divergent.

6.2 (b) Consider the improper integral

$$\int_0^\infty e^{-px}dx \text{ for } p > 0.$$

For each $N > 0$,

$$\int_0^N e^{-px}\,dx = \frac{e^{-px}}{-p}\Big|_{x=1}^{x=N} = \frac{-1}{p}(e^{-pN}-1).$$

For $p > 0$, e^{-pN} tends to zero as N tends to infinity. Then the improper integral is convergent and has value equal to $1/p$.

TESTS FOR CONVERGENCE

The following tests can be used to decide the convergence or divergence of an improper integral with an integrand for which no elementary primitive can be found.

Theorem 6.1 Comparison Test

Theorem 6.1 Suppose functions f and g are defined and nonnegative on the interval $[a, b)$ where $b > a$ (and $b = \infty$ is allowed). Suppose also that

(i) f and g are Riemann integrable on $[a, \lambda]$ for all λ, $a < \lambda < b$

(ii) $f(x) \le g(x)$ for all x, $a \le x \le b$.

Let $\quad I(f) = \int_a^b f \quad$ and $\quad I(g) = \int_a^b g.$

Then I(f) is convergent if I(g) is convergent, and I(g) is divergent if I(f) is divergent. If condition (ii) is replaced by

(iii) $\lim_{x\to b} f(x)/g(x) = L \quad$ for $0 < L < \infty$

then either I(f) and I(g) both converge or else they both diverge. If $L = 0$, then the convergence of I(g) implies the convergence of I(f).If $L = \infty$, then the divergence of I(f) implies the divergence of I(g).

Theorem 6.2

Theorem 6.2 Suppose g is continuously differentiable with $g'(x) < 0$ and g(x) tends to zero as x tends to infinity. Suppose also that f is continuous and that

$$F(x) = \int_a^x f(t)dt$$

is bounded for all $x > a$. Then the improper integral $\int_a^\infty f(x)g(x)dx$ is convergent.

FUNCTIONS DEFINED FROM IMPROPER INTEGRALS

A number of important functions in mathematics are defined by improper integrals. One of the most important of these is the *Gamma function*.

Definition For each $x > 0$ let the *Gamma function*, $\Gamma(x)$ be defined by

$$\Gamma(x) = \int_0^\infty e^{-t}t^{x-1}dx .$$

Theorem 6.3

Theorem 6.3 The gamma function has the following properties:

(a) For each $x > 0$, the improper integral defining $\Gamma(x)$ is convergent

(b) For each $x > 0$, $\Gamma(x + 1) = x\Gamma(x)$; in particular, for $n \in \mathbb{N}$ $\Gamma(n + 1) = n!$

(c) For each positive integer n, $\Gamma(x)$ can be defined for $x < 0$ by

$$\Gamma(x) = \frac{1}{x}\Gamma(x+1) \quad \text{if } -n < x < -n + 1$$

(d) $\Gamma(x)$ tends to infinity as x tends to zero through positive values

(e) $\Gamma(1/2) = \sqrt{\pi}$

INTEGRAL TRANSFORMS

Integral transforms, which are of considerable use in the solution of differential equations, are often defined by improper integrals.

LAPLACE TRANSFORM

Definition Let $f(t)$ be defined for $t > 0$. Then the Laplace Transform of f is given by

$$F(s) = \int_0^\infty e^{-st} f(t)\, dx\,.$$

The set of all s for which the improper integral F(s) is convergent is the domain of a function which we denote by F(s) and refer to as the Laplace transform of the function f(t).

Theorem 6.4

Theorem 6.4 Suppose f(t) is Riemann integrable on [0, b] for each b>0 and there exist constants M and k such that

$$|f(t)| \le Me^{kt} \quad \text{for all } t \ge 0.$$

Then the improper integral F(s) exists for each $s > k$.

FOURIER TRANSFORM

Definition Let f(x) be defined for all real values x. Then the *Fourier Transform* of f(x) is given by

$$F(\alpha) = \int_{-\infty}^{\infty} f(x) e^{-ix\alpha} dx$$

As in the case of the Laplace transform, the set of all real α for which the improper integral $F(\alpha)$ is convergent is the domain of a function we call the Fourier transform of f(x). We denote this transform by $F(\alpha)$.

Theorem 6.5

Theorem 6.5 Suppose f(x) is such that the improper integral

$$\int_{-\infty}^{\infty} |f(x)|\, dx$$

is convergent. Then the improper integral for $F(\alpha)$ is convergent for all real α.

INFINITE SERIES

Let $\{a_n\}$ denote a sequence of real numbers and define from this sequence a new sequence,

$$S_k = \sum_{n=1}^{k} a_n, \quad k \in \mathbb{N}.$$

Then the expression

$$\sum_{n=1}^{\infty} a_n$$

will be used to denote the limit of the sequence $\{S_k\}$ as k tends to infinity.

Sequence of Terms, Sequence of Partial Sums

We refer to the infinite sum as the *infinite series* having $\{a_n\}$ as its *sequence of terms*. The sequence $\{S_k\}$ is referred to as the *sequence of partial sums*.

CONVERGENCE

The infinite series is said to be *convergent* with sum equal to S if the sequence of partial sums is convergent with limit equal to S. If the sequence of partial sums does not converge then the infinite series is said to be *divergent*.

Example 6.3 Geometric Series

Consider the infinite series for which the series of terms is equal to $a_n = Ar^n$ for n = 0, 1, ... Then for any N > 0

$$S_N = A + Ar + \cdots + Ar^N$$

and

$$rS_N = Ar + Ar^2 + \cdots + Ar^N + Ar^{N+1}.$$

Subtraction leads to

$$(1 - r)S_N = A(1 - r^{N+1})$$

i.e.

$$S_N = A\frac{1 - r^{N+1}}{1 - r}$$

Evidently the sequence S_N converges to the limit $A/(1 - r)$ if $|r| < 1$ and if $|r| \geq 1$, the sequence diverges. Then

$$\sum_{n=0}^{\infty} Ar = \begin{cases} A/(1 - r) & \text{if } |r| < 1 \\ \text{diverges} & \text{if } |r| \geq 1 \end{cases}$$

This infinite series is called the geometric series. It is one of the few examples of an infinite series for which it is possible to express S_N explicitly in terms of N.

TESTS FOR CONVERGENCE

The convergence or divergence of series for which no explicit formula for S_N is available must be decided by indirect tests for convergence.

Theorem 6.6
Nth term test

Theorem 6.6 If the infinite series $\sum a_n$ is convergent, the sequence of terms $\{a_n\}$ must converge to zero as n tends to infinity.

Positive Term Series

An infinite series whose terms a_n are all positive is called a *positive term* series. We list now some convergence tests that apply to positive terms series.

Theorem 6.7

Theorem 6.7 If $\{a_n\}$ denotes a sequence of positive numbers then the infinite series $\sum a_n$ is convergent if and only if the sequence of partial sums, $\{S_N\}$, is bounded. In this case the sum of the series is equal to the least upper bound for the sequence of partial sums.

Theorem 6.8
Comparison Test

Theorem 6.8 If $\{a_n\}$ and $\{b_n\}$ are two sequences of positive numbers such that

(i) for some integer N, $a_n \le b_n$ for all $n \ge N$

then the convergence of $\sum b_n$ implies the convergence of the series $\sum a_n$. Also, the divergence of the series $\sum a_n$ implies the divergence of $\sum b_n$. If condition (i) is replaced by

(ii) $\lim_{n\to\infty} a_n/b_n = L$ for $0 < L < \infty$,

then the two series are either both convergent or both divergent. Test (i) is called the *direct comparison test*; test (ii) is referred to as the *limit comparison test.*

A convergence test which emphasizes the similarity between improper integrals and infinite series is the following.

Theorem 6.9
Integral Test

Theorem 6.9 Let $\{a_n\}$ be a sequence of positive numbers. If there exists a positive, monotone function $f = f(x)$ such that $f(n) = a_n$ for $n \in \mathbb{N}$, then either

$$\int_1^\infty f(x)dx \quad \text{and} \quad \sum a_n$$

are both convergent or they are both divergent.

Theorem 6.10
Ratio Test and Root Test

Theorem 6.10 Let $\{a_n\}$ denote a sequence of positive numbers and suppose

i) $\lim_{n\to\infty} a_{n+1}/a_n = L$

Then

$$\sum a_n \text{ is } \begin{cases} \text{convergent} & \text{if } L < 1 \\ \text{divergent} & \text{if } L > 1 \end{cases}$$

If $L = 1$ the test provides no information about the convergence of the series. The same conclusions apply if i) is replaced by

ii) $\lim_{n\to\infty} (a_n)^{1/n} = L$

Absolute and Conditional Convergence

We may also consider series which are not positive term series.

Definition The series $\sum a_n$ is said to be *absolutely convergent* if the positive term series $\sum | a_n |$ is convergent. If $\sum a_n$ converges but $\sum | a_n |$ is divergent, then $\sum a_n$ is said to be *conditionally convergent.*

Theorem 6.11

Theorem 6.11 If the series, $\sum a_n$ is absolutely convergent, then it is convergent.

Theorem 6.12 Alternating Series Test

Theorem 6.12 Let $\{a_n\}$ denote a decreasing sequence of positive numbers, tending to zero as n tends to infinity. Then the alternating series,

$$\sum_{n=1}^{n} (-1)^n a_n$$

converges to a finite sum, S, and for each N, we have $| S - S_N | \leq a_{N+1}$.

Theorem 6.12 Alternating Series Test

Now we consider the question of whether altering the order of the terms in an infinite series can affect the value of the sum or destroy the convergence of the series.

Definition The infinite series $\sum a_n$ is a *rearrangement* of the series $\sum b_n$ if there is a one to one mapping, μ, of $\mathbb{N}$ onto $\mathbb{N}$ such that for each n in $\mathbb{N}$, $a_n = b_{\mu(n)}$; i.e. each a_n is equal to one and only one $b_{\mu(n)}$.

Theorem 6.13

Theorem 6.13 If $\sum a_n$ is absolutely convergent with sum equal to S then every rearrangement of this series is absolutely convergent with sum equal to S.

Theorem 6.14

Theorem 6.14 If $\sum a_n$ is conditionally convergent then for any real number T, there is a rearrangement of $\sum a_n$ that is conditionally convergent with sum equal to T. There are also divergent rearrangements for $\sum a_n$.

See Problem 6.24 for an example of rearrangement of a conditionally convergent series can affect the sum.

Computations with Series

We describe now the rules for performing arithmetic operations with convergent series to form other convergent series.

Theorem 6.15 Adding and Subtracting Series

Theorem 6.15 Suppose $\sum a_n$ and $\sum b_n$ are convergent series with sums equal to A and B respectively. Then for any real numbers α and β, the series $\sum c_n$ whose sequence of terms is equal to $c_n = \alpha a_n + \beta b_n$ for every n is also convergent with sum equal to $\alpha A + \beta B$. If the series $\sum a_n$ and $\sum b_n$ are absolutely convergent then the series $\sum c_n$ is also absolutely convergent.

PRODUCT OF TWO SERIES

If $\sum a_n$ and $\sum b_n$ are convergent with sums equal to A and B respectively, then the *product* of $\sum a_n$ with $\sum b_n$ is the series $\sum c_n$ with the sequence of terms given by

$$c_n = \sum_{k=0}^{n} a_k b_{n-k} \quad \text{for every n.}$$

Theorem 6.16 Product of Two Series

Theorem 6.16 If $\sum a_n$ and $\sum b_n$ are *absolutely* convergent with sums equal to A and B, respectively, then the product series $\sum c_n$ is absolutely convergent with sum equal to AB.

If the series $\sum a_n$ and $\sum b_n$ are convergent but not absolutely convergent then the convergence of $\sum c_n$ is not guaranteed. However, if the series $\sum a_n$ and $\sum b_n$ are convergent and $\sum c_n$ is also convergent, then the sum of $\sum c_n$ is equal to AB. See Problem 6.25 for an example of this behavior.

SOLVED PROBLEMS

Improper Integrals

PROBLEM 6.1

Determine the convergence of

$$\int_{-1}^{1} 1/x \, dx.$$

SOLUTION 6.1

The integral is improper because the integrand grows without bound as x tends to zero. Since x = 0 is an interior point of the interval of integration, we write the integral as the sum of two improper integrals

$$\int_{-1}^{1} 1/x \, dx = \int_{-1}^{0} 1/x \, dx + \int_{0}^{1} 1/x \, dx \tag{1}$$

Then the improper integral on the left in (1) is convergent if and only if each of the improper integrals on the right side of the equation is convergent. In Example 6.1 we showed that the integral of 1/x on (0, 1) is divergent and hence the integral over (–1, 1) is also divergent. The fact that for each $\in > 0$ we have

$$\int_{-1}^{-\in} 1/x \, dx + \int_{\in}^{1} 1/x \, dx = 0$$

does not imply that the improper integral in (1) is convergent with value equal to zero. The definition requires the integrals on the right side of equation (1) to converge independently.

PROBLEM 6.2

Determine the convergence of:

$$\int_1^\infty \frac{1}{x(x-1)}\,dx\,.$$

SOLUTION 6.2

The integrand becomes unbounded at x = 1 and the interval of integration is unbounded as well. Then we write

$$\int_1^\infty \frac{1}{x(x-1)}\,dx = \int_1^2 \frac{1}{x(x-1)}\,dx + \int_2^\infty \frac{1}{x(x-1)}\,dx. \tag{1}$$

The first integral on the right in (1) is improper because the integrand is not bounded and the second integral is improper because of an unbounded interval of integration. Partial fractions implies that

$$\frac{1}{x(x-1)} = \frac{1}{x-1} - \frac{1}{x} \tag{2}$$

and a simple change of variable shows

$$\int_1^2 \frac{1}{x-1}\,dx = \int_0^1 \frac{1}{x}\,dx = \text{divergent.}$$

It follows without checking any of the other integrals on the right in (1), that the improper integral on the left in (1) is divergent.

PROBLEM 6.3

Determine the convergence of:

$$\int_0^\infty \text{Sin}\,x\,dx\,.$$

SOLUTION 6.3

The integrand is bounded but the interval of integration is unbounded. Then the improper integral is convergent if and only if the following limit exists

$$\underset{N\to\infty}{\text{Lim}} \int_0^N \text{Sin}\,x\,dx = \underset{N\to\infty}{\text{Lim}}\,(1 - \text{Cos}\,N) = 1 - \underset{N\to\infty}{\text{Lim}}\,\text{Cos}\,N$$

But Cos N tends to no limit as N tends to infinity and hence the improper integral is divergent.

PROBLEM 6.4

Suppose that

$$\int_1^\infty f(x)dx \text{ is convergent.} \tag{1}$$

Is it true or false that f(x) must tend to zero as x tends to infinity.

SOLUTION 6.4

It is false. Consider the monotone function f(x) defined by

$$f(x) = \begin{cases} n & \text{if } n < x < n + n^{-3} \; n \in \mathbb{N} \\ 0 & \text{otherwise} \end{cases} \qquad (2)$$

The graph of this function is a sequence of square "pulses" located at x = n for n = 1, 2, ... The pulse at x = n is n units high and n^{-3} units in width. Thus the area under the pulse at x = n is equal to n^{-2} and the total area under the graph of f(x) is given by the following infinite sum,

$$\int_1^\infty f(x)dx = \sum_{n=1}^{\infty} n \cdot n^{-3} = \sum_{n=1}^{\infty} n^{-2}. \qquad (3)$$

The infinite series in (3) can be shown to converge by means of the integral test of Theorem 6.9. Then the function f defined in (2) is integrable on $[1, \infty)$ but f does not tend to zero as x tends to infinity. Other examples of functions which are integrable on $[1, \infty)$ and which do not tend to zero as x tends to infinity include,

$$f(x) = \text{Sin}\, x^2,\; g(x) = e^{-(x^2 \text{Sin}\, x)^2}, \quad \text{and} \quad h(x) = x^2 e^{-(x^4 \text{Sin}\, x)^2}.$$

Convergence Tests

PROBLEM 6.5

Determine whether the following improper integral converges,

$$\int_0^1 \frac{dx}{\sqrt{x}\,\text{Ln}\, x} \qquad (1)$$

SOLUTION 6.5

L'Hospital's rule shows that for each p > 0,

$$\lim_{x \to 0} x^p \,\text{Ln}\, x = 0\,. \qquad (2)$$

Then the integrand becomes unbounded at x = 0 as well as at x = 1. For any b, 0 < b < 1, we can break the integral into two improper integrals,

$$\int_0^1 \frac{dx}{\sqrt{x}\,\text{Ln}\, x} = \int_0^b \frac{dx}{\sqrt{x}\,\text{Ln}\, x} + \int_b^1 \frac{dx}{\sqrt{x}\,\text{Ln}\, x} \qquad (1)$$

The integral in (1) converges if and only if both integrals on the right side of (3) are convergent. Letting

$$f(x) = \frac{1}{\sqrt{x}\,\text{Ln}\, x} \quad \text{and} \quad g(x) = \frac{1}{x^{1/2-\epsilon}}$$

it follows from (2) that for $\epsilon > 0$,

$$\lim_{x \to 0} f(x)/g(x) = 0$$

Then since g(x) is improperly integrable on [0, b], Theorem 6.1 implies that f(x) is also integrable on [0, b]. However, on the interval [b, 1] we have

$$\lim_{x \to 1} f(x)/h(x) = 1 \quad \text{for} \quad h(x) = 1/(x-1)\,.$$

Since the improper integral of h(x) over [b, 1] is divergent, it follows from Theorem 6.1 that the integral of f(x) over [b, 1] also diverges. Then the integral (1) diverges.

PROBLEM 6.6

Show that the Gamma function integral,

$$\Gamma(x) = \int_0^\infty e^{-t} t^{x-1} dt$$

is convergent for x > 0.

SOLUTION 6.6

The interval of integration is unbounded and, for x<1, the integrand becomes unbounded at t = 0. Therefore, we write

$$\Gamma(x) = \int_0^1 e^{-t} t^{x-1} dt + \int_1^\infty e^{-t} t^{x-1} dt \tag{1}$$

Now

$$e^{-t} t^{x-1} \le t^{x-1} \text{ for } t \ge 0$$

and we showed in Example 6.1 that,

$$\int_0^1 t^{x-1} dt \text{ is convergent for } x > 0.$$

Then by the direct comparison test of Theorem 6.1, the first integral on the right in (1) is convergent. Next, use L'Hospital's rule to observe that,

$$\lim_{t \to \infty} \frac{e^{-t} t^{x-1}}{t^{-2}} = \lim_{t \to \infty} \frac{t^{x-1}}{e^t} = 0.$$

Then, since Example 6.2 showed that,

$$\int_1^\infty t^{-2} dt \quad \text{is convergent},$$

it follows from the limit comparison test of Theorem 6.1 that the second integral on the right side of (1) is convergent for x > 0. This proves the result.

PROBLEM 6.7

Discuss the convergence of the improper integral,

$$\int_0^\infty \frac{\sin x}{x^2} dx. \tag{1}$$

SOLUTION 6.7

We cannot apply Theorem 6.1 directly in this case since the integrand is not nonnegative. However, it follows from Theorem 5.11 that for each $N > 1$,

$$\left| \int_1^N \frac{\text{Sin } x}{x^2} dx \right| \le \int_1^N \frac{|\text{Sin } x|}{x^2} dx \le \int_1^N \frac{1}{x^2} dx \tag{2}$$

In Example 6.2 we showed that the last integral in (2) tends to the value 1 as N tends to infinity and it follows that the integral in (1) is convergent.

An improper integral with the property that the integral of the *absolute value* of the integrand is convergent is said to be absolutely convergent. Thus an improper integral that is absolutely convergent is also convergent. We have proved that the improper integral (1) is absolutely convergent.

PROBLEM 6.8

Determine whether the following improper integral is absolutely convergent.

$$\int_0^\infty \frac{\sin x}{x} dx. \tag{1}$$

SOLUTION 6.8

Note that the integrand does not become unbounded at $x = 0$ so this integral is improper only because of the unboundedness of the interval of integration. It is clear that

$$\left| \frac{\text{Sin } x}{x} \right| \ge \frac{|\text{Sin } x|}{(n+1)\pi} \quad \text{for } n\pi < x < (n+1)\pi \text{ and } n \in \mathbb{N}$$

Then

$$\int_{n\pi}^{(n+1)\pi} \left| \frac{\text{Sin } x}{x} \right| dx \ge \frac{1}{(n+1)\pi} \int_{n\pi}^{(n+1)\pi} |\text{Sin } x|\, dx = \frac{2}{\pi} \frac{1}{n+1}$$

and hence, for each L, $N\pi \le L \le (N+1)\pi$,

$$\int_0^L \left| \frac{\text{Sin } x}{x} \right| dx > \frac{2}{\pi} \sum_{n=0}^{N-1} \frac{1}{n+1} \tag{2}$$

The sum on the right in (2) diverges as N tends to infinity and it follows that the improper integral (1) is not absolutely convergent.

On the other hand, $g(x) = 1/x$ is continuously differentiable and decreases to zero as x tends to infinity. In addition, $f(x) = \text{Sin } x$ is continuous and

$$\int_0^b \text{Sin } x dx = 1 - \text{Cos } b$$

is bounded for every b > 0. Then by Theorem 6.2, the integral in (1) is convergent. An integral which is convergent but is not *absolutely convergent* is said to be conditionally convergent. The integral in (1) is conditionally convergent.

PROBLEM 6.9

Determine whether the following integral is convergent,

$$\int_0^{\infty} \text{Sin } x^2\, dx. \tag{1}$$

SOLUTION 6.9

In view of Problem 6.3, it may seem at first that the integral in (1) is divergent. However, the change of variable,

$$x = \sqrt{t} \quad dx = \frac{dt}{2\sqrt{t}}$$

leads to the result that for each N > 0,

$$\int_0^N \text{Sin } x^2\, dx = \int_0^N \frac{1}{2\sqrt{t}} \text{ Sin } t\, dt\,. \quad (2)$$

Now Theorem 6.2 with f(t) = Sin t and $g(t) = 1/2\sqrt{t}$ implies that

$$\int_0^N \frac{1}{2\sqrt{t}} \text{ Sin } t\, dt \text{ is convergent.}$$

Then we can let N tend to infinity in (2) to conclude that the improper integral in (1) is convergent. Of course, this integral is conditionally convergent not absolutely convergent. Note that $f(x) = \text{Sin } x^2$ is an example of a function that is integrable on $[0, \infty)$ but does not tend to zero as x tends to infinity.

Gamma Function, Laplace and Fourier Transforms

PROBLEM 6.10

Show that for x > 0, $\Gamma(x + 1) = x\Gamma(x)$.

SOLUTION 6.10

It follows from the definition that for x > 0,

$$\Gamma(x+1) = \int_0^{\infty} t^x e^{-t} dt \tag{1}$$

For each N > 0, we integrate by parts, using

$$u = t^x,\ du = xt^{x-1}dt \quad \text{and} \quad dv = e^{-t}\, dt,\ v = -e^{-t}$$

to obtain

$$\int_0^N t^x e^{-t} dt = -e^{-t} t^x \Big]_{x=0}^{x=N} + \int_0^N x t^{x-1} e^{-t} dt.$$

L'Hospital's rule implies

$$\lim_{N\to\infty} N^x e^{-N} = 0,$$

and hence

$$\Gamma(x+1) = \lim_{N\to\infty}\int_0^N t^x e^{-t}dt = \lim_{N\to\infty}\int_0^N x t^{x-1} e^{-t}dt = x\Gamma(x).$$

PROBLEM 6.11

Show that $\Gamma(x)$ tends to ∞ as $x > 0$ tends to zero

SOLUTION 6.11

For $x > 0$,

$$\Gamma(x) = \int_0^\infty t^{x-1}e^{-t}dt > \int_0^1 t^{x-1}e^{-t}dt$$

$$\geq \int_0^1 t^{x-1}e^{-t}dt = e^{-1}\int_0^1 t^{x-1}dt.$$

For $x > 0$, we have from Example 6.1,

$$\int_0^\infty t^{x-1}dt = 1/x$$

and

$$\Gamma(x) > 1/(ex) \to \infty \quad \text{as} \quad x \to 0.$$

PROBLEM 6.12

Suppose $f(t)$ is Riemann integrable on $[0, b]$ for every $b > 0$ and that there exist constants M and k such that

$$|f(t)| \leq Me^{kt} \quad \text{for all } t > 0. \text{ (1)}$$

Show that for $s > k$, the Laplace transform integral $F(s)$ is convergent.

SOLUTION 6.12

We have

$$F(s) = \int_0^\infty f(t)e^{-st}dt$$

and for each $N > 0$,

$$\left|\int_0^N f(t)e^{-st}dt\right| \leq \int_0^N |f(t)|e^{-st}dt$$

$$\leq \int_0^N Me^{kt}e^{-st}dt = \frac{M}{s-k}(1 - e^{-(s-k)N})$$

Then for $s>k$,

$$F(s) = \lim_{N\to\infty}\int_0^N f(t)e^{-st}dt \quad \text{exists.}$$

A function that satisfies (1) for some M and k is said to be of exponential type.

PROBLEM 6.13

Compute the Laplace transform of $f(t) = 1$.

SOLUTION 6.13

The function $f(t) = 1$ is clearly of exponential type (i.e. choose $M = 1, k = 0$). Then

$$F(s) = \lim_{N\to\infty} \int_0^N 1\, e^{-st}dt = \lim_{N\to\infty} \frac{(1 - e^{-Ns})}{s} = \frac{1}{s} \quad \text{for } s > 0.$$

PROBLEM 6.14

Compute the Fourier transform of the function $f(x) = e^{-x^2}$.

SOLUTION 6.14

Note that,

$$\int_{-\infty}^{\infty} e^{-x^2}dx = 2\int_0^{\infty} e^{-x^2}dx \qquad (1)$$

and

$$\int_1^{\infty} e^{-x^2}dx \leq \int_1^{\infty} e^{-x}dx. \qquad (2)$$

Since the last integral in (2) is convergent, it follows from (1) and (2) that

$$\int_{-\infty}^{\infty} |e^{-x^2}|\, dx \quad \text{is convergent.}$$

Then by the definition of the Fourier transform,

$$F(\alpha) = \int_{-\infty}^{\infty} e^{-x^2}e^{ix\alpha}dx \quad \text{for } \alpha \in \mathbb{R}.$$

Note that

$$-(x^2 + ix\alpha - \alpha^2/4 + \alpha^2/4) = -(x + i\alpha) - \alpha^2/4$$

and hence

$$F(\alpha) = e^{-\alpha^2/4}\int_{-\infty}^{\infty} e^{-(x+i\alpha)^2}dx = e^{-\alpha^2/4}\int_{-\infty}^{\infty} e^{-z^2}dz$$

Note that

$$\Gamma(1/2) = \int_0^{\infty} e^{-t}t^{-1/2}dt = 2\int_0^{\infty} e^{-z^2}dz,$$

where we have used the change of variable, $z = \sqrt{t}$. Then

$$F(\alpha) = \frac{1}{2}\Gamma(1/2)\, e^{-\alpha^2/4}.$$

PROBLEM 6.15

Show that $\Gamma(1/2) = 1/2\sqrt{\pi}$.

SOLUTION 6.15

There are various ways of evaluating the improper integral,

$$I = \int_0^{\infty} e^{-z^2} dz;$$

this method is due to Robert Weinstock and appeared in the AMS Monthly in January 1990. Let

$$f(x) = \int_0^1 \frac{e^{-x(1+t^2)}}{1+t^2}\, dt, \tag{1}$$

and,

$$I(x) = \int_0^x e^{-z^2} dz. \tag{2}$$

Then f(x) and I(x) are defined and continuously differentiable for all $x \geq 0$. In particular,

$$f(0) = \int_0^1 \frac{1}{1+t^2}\, dt = \text{ArcTan}(1) = \pi/4$$

and since

$$0 < f(x) < e^{-x} \int_0^1 \frac{e^{-xt^2}}{1+t^2}\, dt < e^{-x}(\pi/4)$$

we have $f(\infty) = 0$. Clearly, $I(0) = 0$ and $I(\infty)$ is the value we are trying to compute.

Differentiating (1) with respect to x leads to,

$$f'(x) = -\int_0^1 e^{-x(1+t^2)} dt = -e^{-x} \int_0^1 e^{-xt^2} dt$$

i.e.

$$f'(x) = -\frac{e^{-x}}{\sqrt{x}} \int_0^{\sqrt{x}} e^{-z^2}\, dz = -\frac{e^{-x}}{\sqrt{x}}\, I(\sqrt{x}) \tag{3}$$

Differentiating (2) with respect to x leads to

$$I'(x) = e^{-x^2}. \tag{4}$$

These two differentiations anticipate Leibniz' rule for differentiating an integral with respect to a parameter. Leibniz' rule is discussed in chapter 8.

Now integrate both sides of (3) with respect to x from zero to L; this leads to

$$f(L) - f(0) = -\int_0^L \frac{e^{-x}}{\sqrt{x}} \int_0^{\sqrt{x}} e^{-z^2} dz\, dx$$

$$= -2\int_0^{L^2} e^{-z^2} I(z)\, dz = -2\int_0^{L^2} I'(z)I(z)\, dz$$

$$= -2\int_0^{L^2} I'(z)I(z)\, dz = I(0)^2 - I(L^2)^2.$$

Letting L tend to infinity, and recalling that $f(\infty) = I(0) = 0$, we have

$$I(\infty)^2 = f(0) = \pi/4;$$

i.e.

$$\Gamma(1/2) = I(\infty) = 1/2\sqrt{\pi}.$$

Infinite Series

PROBLEM 6.16

Show that if the series $\sum a_n$ is convergent, then the sequence of terms must converge to zero.

SOLUTION 6.16

If $\sum a_n$ is convergent with sum equal to S, then the sequence $\{S_N\}$ of partial sums converges to S. Then for each n, $a_n = S_n - S_{n-1}$ and

$$\lim_{n\to\infty} a_n = \lim_{n\to\infty} (S_n - S_{n-1}) = S - S = 0.$$

Positive Term Infinite Series

PROBLEM 6.17

Show that the harmonic series

$$\sum_{n=1}^{\infty} \frac{1}{n} \quad \text{is divergent.} \tag{1}$$

SOLUTION 6.17

By definition, an infinite series is convergent if its sequence of partial sums is convergent. By Theorem 2.9, the sequence of partial sums is convergent if and only if the sequence of partial sums is a Cauchy sequence. To see that the sequence of partial sums of the series in (1) is *not* a Cauchy sequence, note that for any $N > 1$

$$S_{2N} - S_N = \frac{1}{N+1} + \frac{1}{N+2} + \cdots + \frac{1}{N+N} > N\frac{1}{2N} = 1/2$$

Then the difference $|S_{2N} - S_N|$ does not tend to zero as N tends to infinity and the sequence $\{S_N\}$ is not a Cauchy sequence.

PROBLEM 6.18

Show that the *p-series*

$$\sum_{n=1}^{\infty} \frac{1}{n^p}$$

is convergent if $p > 1$, and is divergent for $p \leq 1$.

SOLUTION 6.18

In Example 6.2, we showed that the integral

$$\int_1^{\infty} x^{-p} dx$$

converges if $p > 1$ and diverges if $p \leq 1$. The non-increasing function $f(x) = x^{-p}$ satisfies,

$$f(n) = n^{-p} = a_n \quad n = 1, 2, \ldots$$

Then Theorem 6.9 implies that the series has the same convergence properties as the integral.

PROBLEM 6.19

Show that the positive terms series $\sum a_n$ is convergent if and only if the sequence of partial sums is bounded.

SOLUTION 6.19

Let the sequence of partial sums be denoted by $\{S_N\}$. Then for each N, $S_{N+1} = S_N + a_{N+1}$ and since $a_n \geq 0$ for each N, the sequence of partial sums is non decreasing. Then by Corollary 2.2, the sequence of partial sums is convergent if and only if it is bounded.

PROBLEM 6.20

Test the following series for convergence:

$$\sum_{n=1}^{\infty} \left(\frac{1}{n} - \frac{1}{n+p} \right), p > 0.$$

SOLUTION 6.20

Note that for $p > 0$ and every $n \in \mathbb{N}$

$$a_n = \frac{1}{n} - \frac{1}{n+p} = \frac{p}{n(n+p)} < \frac{p}{n^2} = b_n$$

It follows from the result of Problem 6.18 that $\sum b_n$ is a convergent series. Thus, by either the direct or limit comparison test, $\sum a_n$ also converges.

PROBLEM 6.21

Test the following series for convergence:

(a) $\sum_{n=1}^{\infty} \frac{n!}{2^n}$ (b) $\sum_{n=1}^{\infty} \left(\frac{1}{\text{Log}(n+1)}\right)^n$

SOLUTION 6.21

6.21 (a) We form the ratio of consecutive terms for this series

$$\frac{a_{n+1}}{a_n} = \frac{(n+1)!}{2^{n+1}} \cdot \frac{2^n}{n!} = \frac{n+1}{2}$$

Since the ratio tends to infinity as n tends to infinity, we use part i) of Theorem 6.10 to conclude that the series is divergent.

6.21 (b) For this series, the root test is more convenient. We compute

$$\sqrt[n]{b_n} = \frac{1}{\text{Log}(n+1)}$$

Since this expression tends to zero with increasing n, we conclude by the root test, part ii) of Theorem 6.10, that the series is convergent.

Absolute and Conditional Convergence

PROBLEM 6.22

Prove that if $a_n > a_{n+1}$ for every n and a_n tends to zero as n tends to infinity, then the alternating series

$$\sum_{n=1}^{\infty} (-1)^{n+1} a_n = a_1 - a_2 + \ldots$$

is convergent.

SOLUTION 6.22

If we let S_m denote the mth partial sum of this alternating series, then

$$S_{2m+2} = S_{2m} + a_{2m+1} - a_{2m+2}$$

Since $a_{2m+2} < a_{2m+1}$, it follows that for each m, $S_{2m+2} > S_{2m}$. Similarly, we can show $S_{2m+1} < S_{2m-1}$ for every m. Thus, the sequence $\{S_{2m}\}$ of even order partial sums is an increasing sequence while the sequence $\{S_{2n-1}\}$ of odd order partial sums is decreasing.

Now observe that if $2m < 2n-1$ then

$$S_{2n-1} - S_{2m} = (a_{2m+1} - a_{2m+2}) + \cdots + (a_{2n-3} - a_{2n-2}) + a_{2n-1} > 0.$$

On the other hand, if $2m > 2n-1$, then

$$S_{2m} - S_{2n-1} = -(a_{2n} - a_{2n+1}) - \cdots - (a_{2m-2} - a_{2m-1}) - a_{2m} < 0.$$

In either case we have, $S_{2m} < S_{2n-1}$. This shows that the increasing sequence of even order partial sums is bounded above by every odd order partial sum. It also shows that the decreasing sequence of odd order partial sums is bounded below by every even order partial sum. It follows from Theorem 1.7, the monotone sequence theorem, that each of these sequences converges. To see that they must have the same limit, write

$$\lim_{n\to\infty} S_{2n+1} - \lim_{n\to\infty} S_{2n} = \lim_{n\to\infty}(S_{2n+1} - S_{2n}) = \lim_{n\to\infty} a_{2n+1} = 0 .$$

This proves the result. Note that the sum of the alternating series equals the limit S of the sequence of partial sums and that

$$0 < S_{2n-1} - S < S_{2n-1} - S_{2n} = a_{2n}.$$

Similarly,

$$0 < S - S_{2n} < S_{2n+1} - S_{2n} = a_{2n+1}.$$

It follows that for each n, the sum S of the alternating series satisfies

$$|S - S_n| < a_{n+1}.$$

Thus the error made in approximating S with S_n is less than a_{n+1}.

PROBLEM 6.23

Test the convergence of the alternating series

$$\sum_{n=2}^{\infty}(-1)^n \frac{\operatorname{Log} n}{n}$$

SOLUTION 6.23

The series begins with n = 2 since $a_n = (-1)^n \operatorname{Log} n/n$ vanishes for n = 1. The series is alternating in sign and a tends to zero as n tends to infinity (apply L'Hospital's rule). However, it is also necessary that the sequence of terms is strictly decreasing. To see that this is, in fact, the case note that

$$\text{if } f(x) = \frac{\operatorname{Log} x}{x} \quad \text{then} \quad f'(x) = \frac{1 - \operatorname{Log} x}{x^2}$$

hence $f'(x) < 0$ for $x > e$; that is, $a_{n+1} < a_n$ for $n > 3$. Since the terms are strictly decreasing after the first two, the series is convergent by the alternating series test. It is easily seen by comparison with the divergent harmonic series that the series is not absolutely convergent. Thus, this series is conditionally convergent.

PROBLEM 6.24

Show that the alternating harmonic series,

$$\sum_{n=2}^{\infty}(-1)^{n+1} \frac{1}{n}$$

(a) converges conditionally to the sum S
(b) has a rearrangement that converges conditionally to S/2.

SOLUTION 6.24

6.24 (a) Since $a_n = 1/n > 1/n + 1 = a_{n+1}$ and a_n tends to zero as n tends to infinity, it follows by the alternating series test that the alternating harmonic series converges. However, it is clear from the result of Problem 6.17 that the series is not absolutely convergent. Let the sum of the alternating series be denoted by S. Then according to the result of the previous problem, the error made in approximating S by the sum of the first n terms of the alternating series is less than $1/n + 1$.

6.24 (b) We have that

$$\sum a_n = 1 - \frac{1}{2} + \frac{1}{3} - \frac{1}{4} + \frac{1}{5} - \frac{1}{6} + \cdots \tag{1}$$

and

$$2\sum a_n = 2 - 1 + \frac{2}{3} - \frac{1}{2} + \frac{2}{5} - \frac{1}{3} + \frac{2}{7} - \frac{1}{4} + \frac{2}{9} - \frac{1}{5} \cdots \tag{2}$$

Here we reduced all fractions to lowest terms. Now we rearrange the terms in the sum on the right in (2) so that we have

$$\begin{aligned} 2\sum a_{n'} &= (2-1) - \frac{1}{2} + \left(\frac{2}{3} - \frac{1}{3}\right) - \frac{1}{4} + \left(\frac{2}{5} - \frac{1}{5}\right) - \frac{1}{6} + \cdots \\ &= 1 - \frac{1}{2} + \frac{1}{3} - \frac{1}{4} + \frac{1}{5} - \frac{1}{6} + \cdots \end{aligned} \tag{3}$$

We have denoted the rearranged sum by $\sum a_{n'}$. Since the last sum on the right in (3) adds to S by definition, we have shown that

$$\sum a_{n'} = S/2;$$

i.e., this rearrangement of the terms in the alternating harmonic series changes the sum from S to S/2.

PROBLEM 6.25

Compute the product of the conditionally convergent series

$$\sum_{n=1}^{\infty} (-1)^{n+1}/\sqrt{n} \tag{1}$$

with itself and show that the result is a divergent series.

SOLUTION 6.25

If we let

$$a_n = (-1)^{n+1}/\sqrt{n} \quad \text{for } n = 1, 2, \ldots$$

then the terms c_n of the product series are

$$c_n = \sum_{k=1}^{n} a_{n-k+1}a_k \quad \text{for } n = 1, 2, \ldots$$

Consider now the subsequence $\{c_{2m}\}$ of the sequence of terms for the product series. We have

$$c_{2m} = \sum_{k=1}^{2m} (-1)^{2m+1-k}(-1)^{k+1}/\sqrt{2m+1-k}\,\sqrt{k}$$

$$= \sum_{k=1}^{2m} 1/\sqrt{2m+1-k}\,\sqrt{k}$$

We will show that $1/\sqrt{2m+1-k}\,\sqrt{k} \geq 1/2m + 1$, which will imply that

$$c_{2m} > (2m + 1)/2m + 1 = 1 \quad \text{for all m;}$$

i.e., c_{2m} does not tend to zero as m tends to infinity, so the product series can not be convergent. It remains only to show that for all m

$$1/\sqrt{2m+1-k}\,\sqrt{k} \geq 1/2m + 1. \tag{2}$$

But

$$(2m + 1 - k)^2 > 0 \quad \text{for all m and k,}$$

hence

$$4m^2 - 4m(k - 1) + (k - 1)^2 \geq 0 \quad \text{for all m and k.}$$

This implies

$$4m^2 + 4m + 1 \geq 4mk - k^2 + 2k \geq 2mk - k^2 + k = (2m + 1 - k)k$$

which leads immediately to (2). Note that this example does not contradict Theorem 6.16 since the series involved are not absolutely convergent.

The Riemann integral of the function f over the interval I is defined if and only if the interval I has finite length and the function f is bounded on I. If either or both of these conditions is violated, the integral is not properly defined. However, when one or both conditions fails the integral may still exist as an improper integral. For example, the integral

$$\int_a^{\infty} f(x)dx$$

is not a proper Riemann integral but is a convergent improper integral if

$$\lim_{N\to\infty} \int_a^{N} f(x)dx$$

exists and has a finite value. An improper integral of an unbounded function on a bounded domain has a similar definition as a limit of proper integrals. An improper integral for which the limit fails to exist or is infinite is said to be divergent.

When the convergence of an improper integral can not be decided by a direct application of the definition in terms of a limit, it is sometimes possible to apply a convergence test such as the comparison test, Theorem 6.1.

Functions defined in terms of improper integrals occur frequently. For example, the gamma function

$$\Gamma(x) = \int_0^\infty e^{-t}t^{x-1}dx \quad \text{for } x > 0$$

has the important property that $\Gamma(n + 1) = n!$ *for* $n \in \mathbb{N}$; *i.e. it extends the factorial function to noninteger values. In addition, many integral transforms are defined in terms of improper integrals. The transforms of Fourier and Laplace are two of the most important of these.*

The concept of an infinite series of constants is closely related to that of an improper integral on an unbounded interval. An infinite series is said to converge if the sequence of partial sums converges. This is analogous to the definition of convergence for an improper integral on an unbounded interval.

Convergence or divergence of an infinite series is most often decided by appealing to one of many convergence tests. Several of the convergence tests are restricted to positive term series. These include the comparison test, integral test, ratio test and root test.

The series $\sum a_n$ *is said to be absolutely convergent if the positive term series* $\sum |a_n|$ *is convergent. If* $\sum a_n$ *converges but* $\sum |a_n|$ *diverges, then the series is said to be conditionally convergent. An alternating series in which successive terms alternate in sign is a particular type of series that is not a positive term series. The alternating series test provides a set of sufficient conditions for the convergence of such a series.*

For an absolutely convergent series, the convergence and the value of the sum are unaffected by rearrangement of the order of the terms of the series. On the other hand, if the series is conditionally convergent the terms of the series may be rearranged to produce any desired sum.

Adding or subtracting convergent series produces another convergent series with the expected sum. However, multiplying two series can produce unex pected results unless both series are absolutely convergent.

SUPPLEMENTARY PROBLEMS

Determine whether the following integrals converge or diverge.

6.1 $\int_0^\infty e^{-x}\mathrm{Sin}(1/x)dx$

6.2 $\int_0^1 x^{-3/2}\mathrm{Sin}(1/x)dx$

6.3 $\int_1^\infty \mathrm{Cos}^2(1/x)dx$

6.4 $\int_0^\infty x^{-1}\mathrm{Sin}(x^2 - x)dx$

6.5 $\int_0^\infty x^2\mathrm{Sin}x\, dx$

Determine whether the following series converge or diverge and tell whether the convergence is conditional or absolute.

6.6 $\sum \frac{1}{\sqrt[n]{n+1}}$

6.7 $\sum e^n\mathrm{Sin}\, 2^{-n}$

6.8 $\sum \frac{1}{e^n - n}$

6.9 $\sum \frac{(-2)^n}{3^n - 10^6}$

6.10 $\sum \frac{(-1)^n}{1 + \ln n}$

7

Infinite Series of Functions

In previous chapters we have considered sequences and infinite series of numbers. When convergent, these sequences and series converge to limits that are real numbers. In this chapter we will build on these concepts in order to consider the more complicated situation in which the terms of the series and sequences are functions of a single real variable. A convergent sequence of functions has for its limit a new function we call the limit of the sequence and a convergent infinite series of functions defines a new function by it sum. Functions defined as the limit of function sequences may arise in connection with solving certain problems by means of iteration procedures and solutions to differential equations are frequently defined by infinite series.

In dealing with sequences and infinite series we are obliged to consider the notion of convergence. For infinite series of numbers, convergence and divergence are mutually exclusive possibilities. Convergence is the single alternative to divergence. For infinite series of functions, however, there is more than just a single type of convergence. Here we will define two modes of convergence for sequences and series of functions:

Pointwise convergence
Uniform convergence

In addition to discussing convergence for function sequences and series, we will be interested in the more delicate matter of determining which properties of the terms are passed to the limit or the sum. The following questions are fundamental for convergent sequences and series of functions:

1. *If each of the terms of a convergent sequence or series is a continuous function, is the limit or sum also continuous*
2. *If a convergent sequence or series is integrated term by term is the resulting sequence or series still convergent and is the new limit equal to the antiderivative of the original limit*
3. *If a convergent sequence or series is differentiated term by term, is convergence preserved and if so, is the new limit equal to the derivative of the original limit*

We shall establish conditions under which each of these questions has an affirmative answer. In particular we shall see that the answer to each of these questions depends on whether the convergence is pointwise or uniform.

We conclude this chapter with a discussion of power series. Power series are function series of the form

$$\sum_{n=0}^{\infty} c_n (x-a)^n$$

where x denotes the function variable and c1, c2... and a denote constants. All such series converge on an interval I = {| c – a | < R } where the nonnegative number R, referred to as the radius of convergence, depends on the coefficients cn. Power series have the important property that a power series having positive radius of convergence and sum equal to f(x) may be differentiated or integrated term by term to produce a new power series which converges on the same interval I to the derivative or antiderivative of f(x). Only functions that have derivatives of all orders at each point of I can have a power series expansion converging on I.

SEQUENCES AND SERIES OF FUNCTIONS

SUM OF A SERIES

Let $u_1(x), u_2(x), \ldots = \{u_n(x)\}$ denote a family of functions having a common interval of definition I in $\mathbb{R}^1$. Then we can formally consider an infinite series of functions formed from the family $u_n(x)$:

$$\sum_{n=1}^{\infty} u_n(x) \qquad (7.1)$$

For each x in I this is an infinite series of numbers to which the results of Chapter 6 apply. The set of all x in I where convergence occurs is the domain for a new function, u(x), which we will refer to as the *sum of the series* (7.1).

SEQUENCE OF PARTIAL SUMS

In Chapter 6 we defined convergence for infinite series of numbers in terms of the convergence of the sequence of partial sums. Similarly, for the series (7.1) we define the associated *sequence of partial sums* by

$$S_N(x) = \sum_{n=1}^{N} u_n(x) \quad N = 1, 2, \ldots \qquad (7.2)$$

Then we say that the infinite series (7.1) is convergent on the set D in I if the sequence $S_N(x)$ is convergent on D.

Function Sequences

The sequence of partial sums (7.2) is an example of a function sequence. Function sequences are of interest in their own right apart from any association with infinite series.

POINTWISE CONVERGENCE

The function sequence $\{f_n\}$ is said to be *pointwise convergent* on the set D in I if, for each x in D the number sequence $\{f_n(x)\}$ is convergent. If $\{f_n\}$ is pointwise convergent on D, the limiting values define a function f(x) on D as follows:

$$f(x) = \lim_{n\to\infty} f_n(x) \quad \text{for all x in D.}$$

We indicate this by writing $f_n \to f$ on D. We can state the definition of pointwise convergence more precisely as follows:

Pointwise Convergence The sequence of functions $\{f_n\}$ defined on I converges pointwise to f on $D \subset I$ if for each $\in > 0$ and each x in D there is a positive integer $N = N(\in, x)$ such that $|f_n(x) - f(x)| < \in$ when $n > N$.

The infinite series (7.1) is pointwise convergent to the sum u(x) on D if and only if the associated sequence of partial sums converges pointwise to u on D.

THE CAUCHY CRITERION

We can state an alternative characterization for pointwise convergence that does not require knowledge of the limit function.

Theorem 7.1

Theorem 7.1 The sequence of functions $\{f_n\}$ defined on I converges pointwise on $D \subset I$ if and only if for each $\in > 0$ and each x in D there is positive integer $N = N(\in, x)$ such that $|f_n(x) - f_m(x)| < \in$ when $m, n > N$.

Example 7.1 Function Sequences

7.1 (a) For I = [0, 1] let $f_n(x) = x/n$, n = 1, 2, ... Then for each x in I

$$\lim_{n\to\infty} f_n(x) = 0; \quad \text{i.e., } f \to 0 \text{ on I}$$

7.1 (b) For I = [0, 1] let $f_n(x) = x^n$, n = 1, 2, ... Then

$$\lim_{n\to\infty} f(x) = 0 \quad \text{for } 0 \le x < 1,$$

$$f_n(1) = 1 \quad \text{for all n.}$$

Thus

$$f_n \to f(x) = \begin{cases} 0 & \text{if } 0 < x < 1, \\ 1 & \text{if } x = 1 \end{cases}$$

Note that if we had said I = [0, 2] then since the sequence $\{f_n\}$ diverges on (1, 2] we would have $f_n \to f$ on D = [0, 1] in I.

7.1 (c) For $I = [0, 2]$ let

$$f_n(x) = \begin{cases} nx & \text{for } 0 \le x \le \frac{1}{n} \\ 2 - nx & \text{for } \frac{1}{n} \le x \le \frac{2}{n} \\ 0 & \text{for } \frac{2}{n} \le x \le 2 \end{cases}$$

For any $x > 0$ we have $x > 2/n$ for all integers n such that $n > 2/x$. Then $f_n(x) = 0$ for $n > 2/x$ and since $f_n(0) = 0$ for all n it follows that

$$f_n \to 0 \quad \text{on } I.$$

A typical f_n in this sequence is pictured in Figure 7.1. Note that for every n there exist points x in I such that $f_n(x) > 1/2$. If we denote the set of all such x by A_n, then for each n A_n is an interval of length $1/n$ and as n tends to infinity the length of A tends to zero.

UNIFORM CONVERGENCE

Pointwise convergence is only one possible mode of convergence for sequences and series. A second type of convergence, differing from pointwise convergence in subtle but important ways, is the following:

Uniform Convergence The sequence of functions $\{f_n\}$ defined on I converges uniformly to f on $D \subset I$ if for each $\in > 0$ there is a positive integer $N = N(\in)$ such that for all x in D, $|f_n(x) - f(x)| < \in$ when $n > N$.

The infinite series (7.1) is uniformly convergent to the sum u(x) on D if and only if the associated sequence of partial sums converges uniformly to u on D.

Note that in the definition of pointwise convergence the integer N depends on $\in$ and on x. For a uniformly convergent sequence or series, N depends *only* on $\in$. If $\{f_n\}$ converges uniformly to f on I then $n > N(\in)$ implies that $|f_n(x) - f(x)| < \in$ for all x in I whereas if $\{f_n\}$ converges pointwise to f on I then $n > N(\in, x)$ implies only that $|f_n(x) - f(x)| < \in$ for this particular x and does not preclude the possibility that $|f_n(y) - f(y)| > \in$ for y in I, $y \neq x$.

Theorem 7.2

Theorem 7.2 If the sequence of functions $\{f_n\}$ converges uniformly to f on D then $\{f_n\}$ converges pointwise to f on D.

The converse to Theorem 7.2 is false. Sequences that converge pointwise need not converge uniformly. See Problem 7.3 for an example.

EQUIVALENT CHARACTERIZATIONS OF UNIFORM CONVERGENCE

It will be convenient to have alternate but equivalent characterizations for uniform convergence.

Theorem 7.3

Theorem 7.3 The sequence of functions $\{f_n\}$ converges uniformly to f on D if and only if

$$\lim_{n\to\infty}\left(\operatorname{Lub}_{x\in D}|f_n(x) - f(x)|\right) = 0$$

Theorem 7.4 Cauchy Criterion

Theorem 7.4 The sequence of functions $\{f_n\}$ converges uniformly on D if and only if for each $\epsilon > 0$ there is positive integer $N = N(\epsilon)$ such that for all x in D we have $|f_n(x) - f_m(x)| < \epsilon$ when m, n > N.

THE WEIERSTRASS M-TEST

Each of the statements about the convergence of function sequences can be translated to a statement about convergence of infinite series of functions by applying it to the sequence of partial sums. The following condition applies directly to the series itself and is sufficient to imply uniform convergence of an infinite series of functions.

Theorem 7.5

Theorem 7.5 If the functions $u_n(x)$, defined on I, satisfy

$$\operatorname{Lub}_{x\in I}|u_n(x)| \leq M_n \quad n = 1, 2, \ldots$$

for positive constants M_n such that the infinite series $\sum M_n$ is convergent, then the infinite series of functions is uniformly convergent on I. Additionally the series of functions is then absolutely convergent on I.

Example 7.2

7.2 (a) The infinite series

$$\sum_{n=1}^{\infty}\frac{1}{n^2}\operatorname{Sin} nx$$

is uniformly convergent on $I = \mathbb{R}^1$ since

$$|n^{-2}\operatorname{Sin} nx| < n^{-2} = M_n \quad \text{for all real x and every } n \in \mathbb{N}$$

and

$$\sum_{n=1}^{\infty}\frac{1}{n^2} \quad \text{converges}$$

7.2 (b) The infinite series

$$\sum_{n=1}^{\infty}(-1)^n\frac{x^n}{n}$$

does not converge absolutely on I = [0, 1] since at x = 1 this is the alternating harmonic series which converges conditionally. Thus there can be no sequence of constants M_n as called for in Theorem 7.5. Nevertheless, this series does converge uniformly on I. For each x in I the series is an alternating series to which

Theorem 6.12 can be applied to obtain that for each N the difference between the Nth partial sum $S_N(x)$ and the sum S(x) satisfies

$$|S(x) - S_N(x)| \leq \frac{x^{n+1}}{n+1} \leq \frac{1}{n+1} \quad \text{for all x in I.}$$

Then for any $\in > 0$ we can choose $N(\in) > 1/\in$ to ensure that

$$\text{for all x in I} \quad |S(x) - S_n(x)| < \in \quad \text{for } n > N(\in).$$

This example shows that the Weierstrass M-test is sufficient but not necessary for uniform convergence.

Continuity of the Limit Function

If $f_n \to f$ on I then $f_n \in \mathbb{C}(I)$ for all n does not necessarily imply $f \in \mathbb{C}(I)$. The functions f_n in Example 7.1(b) are continuous functions converging to a limit f that is not continuous.

Theorem 7.6

Theorem 7.6 Suppose f_n is continuous on I for each n and that the function sequence $\{f_n\}$ converges uniformly on I to the limit f. Then f is continuous on I. That is, for each a in I, the following interchange of order of limits is valid

$$\lim_{n\to\infty} \lim_{x\to a} f(x) = \lim_{x\to a} \lim_{n\to\infty} f(x) = \lim_{x\to a} f(x)$$

This result is sometimes stated: The uniform limit of a sequence of continuous functions is continuous.

Corollary 7.7 Suppose u_n is continuous on I for each n and that the infinite series whose terms are $u_n(x)$ converges uniformly on I to the sum u(x). Then u(x) is continuous on I. Thus for each a in I, the following limit may be passed under the summation sign

$$\lim_{x\to a} \sum_{n=1}^{\infty} u_n(x) = \sum_{n=1}^{\infty} \lim_{x\to a} u_n(x)$$

Integration of Sequences and Series

Suppose the sequence $\{f_n\}$ converges to the limit f on an interval I. Then we may be interested in knowing:

if each of the functions f_n is integrable on I does it follow that f is integrable on I and if so then is it true that

$$\int_I f_n \to \int_I f$$

The next theorem asserts that uniform convergence is sufficient to imply an affirmative answer to both questions.

Theorem 7.8

Theorem 7.8 Suppose that for each n, f_n is integrable on the closed bounded interval I and that the function sequence $\{f_n\}$ converges uniformly on I to the limit f. Then f is integrable on I and

$$\lim_{n\to\infty}\int_I f_n = \int_I \lim_{n\to\infty} f = \int_I f.$$

Corollary 7.9 Suppose that for each n, u_n is integrable on the closed bounded interval I and that the associated infinite series converges uniformly on I to the sum u(x). Then u is integrable on I and

$$\sum_{n=1}^{\infty}\int_I u_n = \int_I \sum_{n=1}^{\infty} u_n = \int_I u.$$

Differentiation of Sequences and Series

It is easy to construct examples of function sequences such that $f_n \to f$ on I but the sequence of derivatives f_n' does not converge to f′. See Problem 7.5 for an example. Conditions sufficient to imply the convergence of the differentiated sequence can be stated as follows.

Theorem 7.10

Theorem 7.10 Suppose that for each n, f_n is continuously differentiable on the closed bounded interval I and that the sequence of derivatives f_n' converges uniformly on I to the limit g. Suppose further that for some a in I, the sequence of numbers $f_n(a)$ is convergent. Then $\{f_n\}$ converges uniformly on I to a differentiable limit function f whose derivative equals g. That is

$$\lim_{n\to\infty}\left(\frac{d}{dx} f_n(x)\right) = \frac{d}{dx}\left(\lim_{n\to\infty} f(x)\right)$$

Corollary 7.11 Suppose that for each n, u_n is continuously differentiable on the closed bounded interval I and that

$$\sum_{n=1}^{\infty} u_n(x) \text{ is convergent to the sum } u(x)$$

$$\sum_{n=1}^{\infty} u_n'(x) \text{ converges uniformly on I to the sum } v(x).$$

Then it follows that u′(x) = v(x). That is,

$$\frac{d}{dx}\sum_{n=1}^{\infty} u_n(x) = \sum_{n=1}^{\infty} u_n'(x).$$

POWER SERIES

Perhaps the most familiar example of an infinite series of functions is the *power series*. This is a series of the form

$$\sum_{n=0}^{\infty} c_n(x-a)^n \tag{7.3}$$

We refer to this as a power series in x, about the point $x = a$, with coefficients c_n. Here a denotes a fixed constant.

Interval of Convergence Theorem 7.12

Theorem 7.12 The power series (7.3) converges on an interval of the form $|x - a| < R$. There are three possibilities for R:

(a) $R = 0$ In this case the series converges only at $x = a$

(b) $R = \infty$ In this case the series converges for all real x

(c) $0 < R < \infty$ In this case the series converges uniformly on the open interval $|x - a| < R$

RADIUS OF CONVERGENCE

The number R in Theorem 7.12 is called the *radius of convergence* for the series (7.3). In cases (b) and (c) we say the series has a positive radius of convergence.

Theorem 7.13

The radius of convergence R for the series (7.3) is equal to

$$R = \lim_{n\to\infty} \left| \frac{c_n}{c_{n+1}} \right|$$

if the limit exists or equals $+\infty$.

Operations on Power Series

One of the more remarkable properties of power series is that if the radius of convergence is positive then the series may be differentiated or integrated term by term and the series obtained will have the same interval of convergence.

Theorem 7.14

Suppose the power series (7.3) has positive radius of convergence R and that the series then converges to f(x) on $I = \{|x - a| < R\}$. Then the following conclusions must hold:

1. f(x) is continuous on the open interval I
2. For each x in I

$$\int_a^x f = \sum_{n=0}^{\infty} \frac{c_n}{n+1}(x-a)^{n+1}$$

3. For each x in I

$$f'(x) = \sum_{n=1}^{\infty} n\, c_n(x-a)^{n-1}$$

$$f''(x) = \sum_{n=1}^{\infty} n(n-1)c_n(x-a)^{n-2}$$

etc.

4. For each n,

$$c_n = \frac{f^{(n)}(a)}{n!}.$$

Thus the series (7.3) is the Taylor series for f expanded about $x = a$.

Uniqueness of Power Series

If two power series converge to the same function then they must have the same coefficients. More precisely we have

Theorem 7.15

Theorem 7.15 Suppose that for two power series convergent on $I = \{|x - a| < R\}$ we have

$$\sum_{n=0}^{\infty} d_n(x-a)^n = \sum_{n=0}^{\infty} c_n(x-a)^n \quad \text{for all x in I.}$$

Then it follows that $d_n = c_n$ for all n.

Theorems 7.14 and 7.15 together imply that if f(x) has a convergent power series expansion about $x = a$ and if that series has a positive radius of convergence then f(x) must have derivatives of all orders at $x = a$. Thus if f(x) fails to have derivatives of all orders at some point $x = b$ then f(x) cannot have a power series expansion that converges on any interval I containing b. Having derivatives of all orders is a necessary condition for having a power series but it is not sufficient.

SOLVED PROBLEMS

Pointwise and Uniform Convergence

PROBLEM 7.1

Show that the sequence $f_n(x) = n^2xe^{-nx}$ converges pointwise to the limit zero on $I = [0, L]$ for every real number L.

SOLUTION 7.1

For each fixed $x > 0$, (see Problem 4.25),

$$n^2xe^{-nx} = \frac{xe^{-nx}}{1/n^2} \quad \to 0 \text{ as n tends to infinity}$$

In addition $f_n(0) = 0$ for all n. Thus $f_n \to 0$ on [0, L] for all $L > 0$. Note that $f_n(1/n) = ne^{-1} \to \infty$ as n tends to infinity.

Thus whatever the choice of $\in > 0$, there can be no $N = N(\in)$ such that

$$\text{for all } x \text{ in } [0, L] \quad |f_n(x) - 0| < \in \quad \text{when } n > N(\in).$$

Thus the sequence is not uniformly convergent on [0, L].

PROBLEM 7.2

For m = 1, 2, ... let

$$f_m(x) = \lim_{n \to \infty} (\text{Cos } m!\pi x)^{2n}$$

Show that on any interval I,

$$f_m \to f(x) = \begin{cases} 0 & \text{if } x \text{ is irrational} \\ 1 & \text{if } x \text{ is rational} \end{cases} \tag{1}$$

SOLUTION 7.2

For a fixed positive integer m, if x is such that m!x is an integer then Cos $m!\pi x = \pm 1$ and

$$f_m(x) = \lim_{n \to \infty} (\pm 1)^{2n} = 1.$$

For every other value of x $|\text{Cos } m!px| < 1$ so that $f_m(x) = 0$. In particular, if x is irrational, m!x is never an integer for any m and $f_m(x) = 0$ for every m. Then for an irrational value of x it follows that $f(x) = \text{Lim } f_m(x) = 0$. On the other hand if $x = p/q$ for integers p and q, then m!x is an integer for $m > q$ and thus $f(x) = 1$.

Note that for any closed bounded interval I, Theorem 5.6 implies that f_m is integrable on I for each m but the limit function f in (1) is not integrable on I.

PROBLEM 7.3

Find the limit on I = [0, 1] of the sequence $f(x) = nx(1 - x^2)^n$.

SOLUTION 7.3

We have $f_n(0) = f_n(1) = 0$ for every n and for $0 < x < 1$,

$$f_n(x) = \frac{x(1 - x^2)^n}{1/n} \to \frac{0}{0} \qquad \text{as } n \to \infty$$

Then by L'Hospital's rule, we find $f_n(x) \to 0$ as $n \to$. Thus $f_n \to 0$ on I. The convergence is only pointwise and not uniform on I since for each n we have

$$f_n(x_n) = \sqrt{2n}\left(\frac{2n}{2n+1}\right)^{3/2} \to \infty \text{ as } n \to \infty$$

for

$$x_n = (2n + 1)^{-1/2}.$$

Note also that while

$$\int_0^1 \lim_n f_n = 0$$

we have

$$\int_0^1 f_n(x)dx = \frac{n}{2n+2} \to \frac{1}{2} \text{ as } n \to \infty$$

Thus

$$\int_0^1 \lim_n f_n \neq \lim_n \int_0^1 f_n$$

even though both limits exist.

PROBLEM 7.4

Discuss the convergence of the infinite series

$$\sum_{n=1}^{\infty} x^n$$

SOLUTION 7.4

For $|x| \geq 1$ the nth term of this series does not tend to zero with increasing n and thus the series diverges for these values of x by Theorem 6.6. It follows from the ratio test, Theorem 6.10, that the series converges for $|x| < 1$. In fact, the series is a geometric series with ratio equal to x. Thus for $|x| < 1$ we have

$$S_N(x) = \sum_{n=0}^{N} x^n = \frac{1 - x^{N+1}}{1 - x} \to \frac{1}{1 - x} \text{ as N tends to infinity.}$$

The convergence of the sequence of partial sums is pointwise on the open interval $(-1, 1)$ and is uniform on the closed interval $[-1 + \rho, 1 - \rho]$ for every ρ, $0 < \rho < 1$. To see this, note that for every $\in > 0$ there exists an integer M such that

$$\frac{(1 - \rho)^{M+1}}{\rho} < \in$$

Then for all x, $|x| \leq 1 - \rho$

$$\left| S_N(x) - \frac{1}{1 - x} \right| = \left| \frac{x^{N+1}}{1 - x} \right| \leq \frac{(1 - \rho)^{M+1}}{\rho} < \in \text{ for } N > M.$$

PROBLEM 7.5

Discuss the convergence of the infinite series

$$\text{Sin } x + 2^{-2} \text{ Sin } 2^4 x + 3^{-2} \text{Sin } 3^4 x + \cdots$$

SOLUTION 7.5

For any real value x,

$$|u_n(x)| \le |n^{-2}\text{Sin } n^4x| \le n^{-2}$$

Thus

$$\left|\sum_{n=1}^{\infty} u_n(x)\right| \le \sum_{n=1}^{\infty} |u_n(x)| \le \sum_{n=1}^{\infty} n^{-2} < \infty$$

and the convergence is clearly uniform on any closed bounded interval I. Note further that

$$\sum_{n=1}^{\infty} u_n'(x) = \sum_{n=1}^{\infty} n^2\text{Cos } n^4x$$

and this series is divergent since the nth term fails to go to zero. Thus even for a series that converges uniformly, the differentiated series may fail to converge.

PROBLEM 7.6

Discuss the convergence of the series

$$\sum_{n=0}^{\infty} e^{-nx}$$

SOLUTION 7.6

The series diverges for $x \le 0$ since in this case the nth terms does not tend to zero. For $x > 0$ this is a geometric series with ratio equal to e^{-x} hence

$$S_N(x) = \sum_{n=0}^{N} e^{-nx} = \frac{1 - e^{-(N+1)x}}{1 - e^{-x}} \to \frac{1}{1 - e^{-x}} \quad \text{as } N \to \infty$$

For any $L > 0$, the convergence is pointwise on the interval $(0, L]$ and it is uniform on any interval of the form $[\in, L]$, $0 < \in < L$. In fact, the change of variable $z = e^{-x}$ reduces this to the series in Problem 7.4 with $0 < z(x) < 1$ for all $x > 0$. Note that the differentiated series

$$\sum_{n=0}^{\infty} u_n(x) = \sum_{n=0}^{\infty} -ne^{-nx}$$

converges for every $x > 0$ by Theorem 6.9. The convergence is uniform on any interval of the form $[\in, L]$, $0 < \in < L$.

PROBLEM 7.7

Show that the sequence of functions $\{f_n\}$ defined on I converges pointwise on I if and only if for each $\in > 0$ and each x in I there is positive integer $N = N(\in, x)$ such that $|f_n(x) - f_m(x)| < \in$ when $m, n > N$.

SOLUTION 7.7

Suppose $f_n \to f$ on I. Then for each x in I and any $\epsilon > 0$ there exists an integer $N = N(\epsilon, x)$ such that $|f_n(x) - f(x)| < \epsilon/2$ when $n > N$. Then

$$|f_n(x) - f_m(x)| < |f_n(x) - f(x)| + |f(x) - f_m(x)| \le \epsilon/2 + \epsilon/2 = \epsilon$$

for m, n > N. Thus convergence implies the Cauchy criterion. Now suppose the Cauchy criterion holds; i.e., for every x, $\{f_n(x)\}$ is a Cauchy sequence. Then by Theorem 2.9 it follows that $\{f_n\}$ is pointwise convergent. We have proved that the Cauchy characterization of pointwise convergence is equivalent to the definition.

PROBLEM 7.8

Suppose the sequence of functions $u_n(x)$ are such that for n = 1, 2, ...

$$|u_n(x)| \le M_n \quad \text{for all x in I.}$$

Show that $\sum u_n(x)$ converges uniformly on I if $\sum M_n$ converges.

SOLUTION 7.8

If the series $\sum M_n$ converges then for every $\epsilon > 0$ there is a positive integer N such that

$$\sum_{n=j}^{k} M_n < \epsilon \quad \text{provided that } j, k > N.$$

But then

$$|S_j(x) - S_k(x)| = \left|\sum_{n=j}^{k} u_n(x)\right| \le \sum_{n=j}^{k} M_n < \epsilon \quad \text{for all x in I}$$

and uniform convergence of the series follows from Theorem 7.4 applied to the sequence of partial sums.

General Properties of Sequences and Series

PROBLEM 7.9

Suppose $f_n \in C(I)$ for each n and that $\{f_n\}$ converges uniformly on I to the limit f. Then show that f must be continuous on I.

SOLUTION 7.9

Let a denote an arbitrary point of I. Then for x in I,

$$|f(x) - f(a)| \le |f(x) - f_n(x)| + |f_n(x) - f_n(a)| + |f_n(a) - f(a)|.$$

For every $\epsilon > 0$ there exists a $\delta > 0$ and an integer N such that

$$|f_n(x) - f_n(a)| < \epsilon/3 \quad \text{if } |x - a| < \delta \text{ since } f_n \in C(I) \text{ for all n}$$

and by the uniform convergence on I of the sequence $\{f_n\}$,

$$|f(x) - f_n(x)| < \in/3, \; |f_n(a) - f(a)| < \in/3 \quad \text{for } n > N.$$

It follows that f is continuous on I.

PROBLEM 7.10

It can be shown that the Fourier series

$$\sum_{n=1}^{\infty} (2n - 1)^{-1} \operatorname{Sin}(2n - 1)x \tag{1}$$

converges to a sum that equals $-\pi/4$ when $-\pi < x < 0$ and equals $\pi/4$ for $0 < x < \pi$. Show that the convergence cannot be uniform on any interval that contains $x = 0$.

SOLUTION 7.10

The terms $u_n(x) = (2n - 1)^{-1}\operatorname{Sin}(2n - 1)x$ of the infinite series (1) are continuous for all x. Then every partial sum of this series is an everywhere continuous function and if the sequence of partial sums converges uniformly on an interval I, then the limit function (which is also the sum of the infinite series) must be continuous by Theorem 7.6. But the sum of this series is discontinuous on any interval I containing the point $x = 0$. Thus the convergence cannot be uniform on such an interval.

PROBLEM 7.11

Suppose the sequence of functions $f_n(x)$ are integrable on the closed bounded interval I for each n and that $\{f_n\}$ converges uniformly on I to limit function f. Then show that f must be integrable on I and that

$$\operatorname{Lim} \int_I f_n = \int_I \operatorname{Lim} f_n = \int_I f \tag{1}$$

SOLUTION 7.11

We show first that f is integrable on I. For $\in > 0$ there exists an N such that for all x in I

$$|f_N(x) - f(x)| < \frac{\in}{3|I|}$$

For $P = \{x_0, x_1, \ldots, x_n\}$ any partition of I let

$$M_{k,N} = \operatorname{Lub} \{f_N(x): x_{k-1} \le x \le x_k\}$$

$$M_k = \operatorname{Lub} \{f(x): x_{k-1} \le x \le x_k\}$$

and note that

$$|M_k - M_{k,N}| < \frac{\in}{3|I|} \quad \text{for each } k.$$

Then

$$|S[f, P] - S[f_N, P]| \leq \sum_{k=1}^{n} |M_k - M_{k,N}| |I_k|, \frac{\in}{3|I|} \sum_{k=1}^{n} |I_k| = \in/3$$

Similarly, we can show that

$$|s[f, P] - s[f_N, P]| \leq \in/3.$$

Finally, since f_N is integrable on I, there is a partition of I (which we may call P since P is at this point arbitrary) such that

$$|S[f_N, P] - s[f_N, P]| \leq \in/3.$$

Then

$$|S[f, P] - s[f, P]| < |S[f, P] - S[f_N, P]| + |S[f_N, P] - s[f_N, P]|$$
$$+ |s[f, P] - s[f_N, P]| < \in$$

and it follows from Theorem 5.4 that f is integrable on I.

To prove (1) note that the uniform convergence of $\{f_n\}$ to f implies that for every $\in > 0$ there is an integer $N = N(\in)$ such that

$$\operatorname*{Lub}_{x \in I} |f_n(x) - f(x)| < \frac{\in}{|I|} \quad \text{for } n > N.$$

Then

$$\left|\int_I f_n - \int_I f\right| \leq \int_I |f_n - f| \leq \frac{\in}{|I|} = \in \text{ for } n > N(\in)$$

But this implies (1) and proves Theorem 7.8.

PROBLEM 7.12

Give an example that illustrates that uniform convergence of $\{f_n\}$ to f on I is sufficient but not necessary in order to have

$$\text{Lim} \int_I f_n = \int_I \text{Lim } f_n = \int_I f \qquad (1)$$

SOLUTION 7.12

Let $\{f_n\}$ denote the sequence of continuous functions in Example 7.1(c). Then f_n converges pointwise but not uniformly on I = [0, 2] to the limit f = 0. In addition, it is easy to see from Figure 7.1 that

$$\int_I f_n = \frac{1}{n} \quad n = 1, 2, \ldots$$

and thus

$$\text{Lim} \int_I f_n = 0 = \int_I \text{Lim } f_n$$

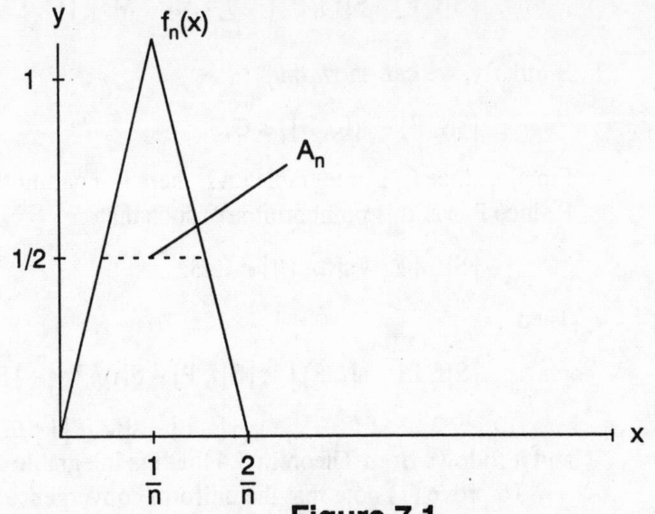

Figure 7.1
A Function Sequence Tending Pointwise to Zero

Then we have (1) in spite of the fact that the sequence is pointwise convergent but not uniformly convergent. On the other hand it is incorrect to expect that (1) necessarily follows from pointwise convergence. Each of the sequences of functions

$$g_n(x) = n\,f_n(x) \quad \text{and} \quad h_n(x) = n^2 f_n(x) \quad n = 1, 2, \ldots$$

converges pointwise to the limit 0 on I = [0, 2] but

$$\int_I g_n = 1 \quad \text{and} \quad \int_I h_n = n \quad \text{for } n = 1, 2, \ldots$$

Therefore

$$\text{Lim} \int_I g_n = 1 \neq 0 = \int_I \text{Lim}\ g_n$$

$$\int_I h_n \to \infty \neq 0 = \int_I \text{Lim}\ h_n$$

As these examples illustrate, pointwise convergence of a sequence $\{f_n\}$ on I implies nothing about the convergence of the integral of $\{f_n\}$ on I.

PROBLEM 7.13

Discuss the series obtained from term by term integration of the series

$$\sum_{n=0}^{\infty} x^n \tag{1}$$

SOLUTION 7.13

We showed in Problem 7.4 that for each $\in > 0$, the series (1) converges uniformly on $I = \{|x| < 1 - \in\}$ to the limit

$$f(x) = \frac{1}{1-x}.$$

Then for any t in I we have

$$\int_{-t}^{t} \sum_{n=0}^{\infty} x^n dx = \sum_{n=0}^{\infty} \frac{t^{n+1}}{n+1} - \frac{(-t)^{n+1}}{n+1} = 2\sum_{n=0}^{\infty} \frac{t^{2n+1}}{2n+1} \qquad (2)$$

$$\int_{-t}^{t} \frac{1}{1-x} dx = \ln \frac{1+t}{1-t} = F(t)$$

The series (2) is easily shown to be uniformly convergent on I and it follows then from Corollary 7.9 that it converges to F; that is

$$\ln \frac{1+t}{1-t} = 2\sum_{n=0}^{\infty} \frac{t^{2n+1}}{2n+1} \quad \text{for t in I.}$$

PROBLEM 7.14

We can show that the function

$$f(x) = \begin{cases} 0 & \text{if } -\pi \le x \le 0 \\ \text{Sin } x & \text{if } 0 \le x \le \pi \end{cases}$$

has as its Fourier series

$$\frac{1}{\pi} + \frac{1}{2}\text{Sin } x - \frac{2}{\pi}\sum_{n=1}^{\infty} \frac{1}{4n-1} \text{Cos } 2nx \qquad (1)$$

Discuss the series obtained by term by term integration of this series.

SOLUTION 7.14

It is easily shown, using the Weierstrass M-test that the series in (1) is uniformly convergent on $I = [-\pi, \pi]$. Integration of this series from $-\pi$ to x, $x < \pi$, leads to

$$\frac{x}{\pi} + \frac{1}{2} - \frac{1}{2}\text{Cos } x - \frac{1}{\pi}\sum_{n=1}^{\infty} \frac{1}{4n^2-1}\frac{1}{n}\text{Sin } 2nx \qquad (2)$$

and

$$\int_{-\pi}^{x} f(t)dt = F(x) = \begin{cases} 0 & \text{if } -\pi \le x \le 0 \\ 1 - \text{Cos } x & \text{if } 0 \le x \le \pi \end{cases}$$

It follows from Corollary 7.9 that the series (2) converges to F(x) on I.

PROBLEM 7.15

Suppose for each n that f_n is continuously differentiable on the closed bounded interval I. Suppose further that the sequence of derivatives $\{f_n'\}$ converges uniformly on I to a limit g. Finally suppose that for some a in I, the number sequence $f_n(a)$ converges to a limt A. Then show that the sequence $\{f_n\}$ must converge uniformly to a limit f and that $f' = g$.

SOLUTION 7.15

We have assumed that for each n, $f_n' \in \mathbb{C}(I)$. Then f_n' is integrable on I and it follows from the fundamental theorem of calculus that

$$f_n(x) = f_n(a) + \int_a^x f_n' \tag{1}$$

By hypothesis for each $\epsilon > 0$ there exists a positive integer $N = N(\epsilon)$ such that for all m, n > N

$$|f_n(a) - f_m(a)| < \epsilon/2 \quad \text{and for all x in I, } |f_n'(x) - f_m'(x)| < \frac{\epsilon}{2(b-a)}$$

Then it follows from (1) that

$$\text{for all x in I,} \quad |f_n(x) - f_m(x)| < \frac{\epsilon}{2} + \frac{\epsilon}{2(b-a)}(b-a) = \epsilon \quad \text{if } m, n > N$$

which implies that $\{f_n\}$ converges uniformly on I. Let the limit of this sequence be denoted by f(x). Letting n tend to infinity in (1) we find

$$f(x) = A + \int_a^x g$$

Then $f' = g$ and since f is the uniform limit of continuous functions, f is continuous on I so that $f(a) = A$.

PROBLEM 7.16

Suppose f(x, y) is continuous on a domain Ω in $\mathbb{R}^2$. Suppose further that for $M > 0$ and for $C > 0$ not depending on x

$$|f(x, y)| < M \quad \text{for } (x, y) \text{ in } \Omega \tag{1}$$

$$|f(x, y_1) - f(x, y_2)| < C|y_1 - y_2| \quad \text{for } (x, y_1), (x, y_2) \text{ in } \Omega \tag{2}$$

Finally suppose h, k are such that $Mh < k$ and $R = \{(x, y): |x - a| < h, |y - b| \le k\}$ is contained in Ω. Then show that there exists in R a unique solution $y = y(x)$ for the initial value problem

$$y'(x) = f(x, y) \quad y(a) = b \tag{3}$$

SOLUTION 7.16

If y(x) satisfies (3) then

$$y(x) = b + \int_a^x f(t, y(t))\, dt \tag{4}$$

This motivates us to recursively define the following sequence of functions

$$y_0(x) = b$$
$$y_n(x) = b + \int_a^x f(t, y_{n-1}(t))\, dt \quad n = 1, 2, \ldots \tag{5}$$

The series

$$Y(x) = y_0(x) + (y_1(x) - y_0(x)) + (y_2(x) - y_1(x)) + \cdots \tag{6}$$

is a telescoping series whose partial sums $Y_N(x)$ satisfy $Y_N(x) = y_N(x)$; i.e., the partial sums of the series for $Y(x)$ are the terms of the sequence $\{y_n\}$. Then the sequence $\{y_n\}$ converges if and only if the series for $Y(x)$ converges. But

$$|y_1(x) - y_0(x)| = \left|\int_a^x f(t, y_0(t))\, dt\right| < M|x - a| < Mh \tag{7}$$

$$|y_2(x) - y_1(x)| = \left|\int_a^x (f(t, y_1(t) - f(t, y_0(t)))\, dt\right|$$

$$\leq \int_a^x C|y_1(t) - y_0(t)|\, dt \leq \int_a^x CM|t - a|\, dt$$

$$\leq CM\frac{|x - a|^2}{2} \leq \frac{CM\, h^2}{2} \tag{8}$$

Note that we used (7) to obtain (8). Proceeding in this way we find in general that

$$|y_n(x) - y_{n-1}(x)| \leq \frac{MC^{n-1}h^n}{n!} \tag{9}$$

Then (7), (8) and (9) imply that the terms of the series (6) are dominated by the terms of the following positive term series of numbers

$$b + Mh + M\frac{Ch^2}{2!} + \cdots + M\frac{C^{n-1}h^n}{n!} + \cdots \tag{10}$$

The series (10) is easily shown to converge and it follows then from the Weierstrass M-test that the series (6) converges uniformly on $|x - a| < h$ to $Y(x)$.

It remains to show that $Y(x)$ solves the initial value problem (3). But since $y_n(x)$ converges uniformly to $Y(x)$, it follows that for any $\in > 0$ there is an N such that for all x in $I = \{|x - a| < h\}$

$$|y_n(x) - Y(x)1 < \frac{\in}{C} \quad \text{for} \quad n > N$$

Then for all x in I

$$|f(x, y_n(x)) - f(x, Y(x))| < C|y_n(x) - Y(x)| = \in \quad \text{for } n > N$$

and it follows that $f(x, y_n(x))$ converges uniformly to $f(x, Y(x))$ on I. Now Theorem 7.8 shows that

$$\operatorname*{Lim}_n y_n(x) = Y(x) = \operatorname*{Lim}_n \left(b + \int_a^x f(t, y_{n-1}(t))\, dt\right)$$

$$= b + \int_a^x \lim_n (f(t, y_n(t)))\,dt$$

i.e.,
$$Y(x) = b + \int_a^x f(t, Y(t))\,dt \tag{11}$$

Since $\{y_n\}$ is a sequence of continuous functions converging uniformly to Y on I it follows from Theorem 7.6 that Y(x) is continuous on I. Then f(t, Y(t)) is a continuous function of t on I and thus by the fundamental theorem of calculus we have

$$Y'(x) = f(x, Y(x)) \quad x \text{ in } I.$$

Finally, to show that Y(x) is unique, suppose there exists a second solution Z(x) for (3). Then

$$|Y(x) - Z(x)| = \left|\int_a^x (f(t, Y(t)) - f(t, Z(t)))dt\right|$$

and using (2), we have

$$|Y(x) - Z(x)| \le \int_a^x C\,|Y(t) - Z(t)|\,dt \tag{12}$$

Since Y(x) and Z(x) are to lie in R for $|x - a| < h$, it must be that

$$|Y(x) - Z(x)| \le 2k \quad \text{for } |x - a| < h.$$

Substituting this into (12) leads to

$$|Y(x) - Z(x)| \le 2k\,C\,|x - a|$$

Putting this result into the right side of (12), we get

$$|Y(x) - Z(x)| \le \int_a^x 2k\,C^2\,|x - a|\,dt = 2k\,C\,\frac{|x - a|^2}{2!}$$

Repeating this substitution n times leads to

$$|Y(x) - Z(x)| \le 2kC^n\frac{|x - a|^n}{n!} \le 2k\,\frac{(Ch)^n}{n!}$$

Since the expression on the right tends to zero as n tends to infinity, we can conclude that $|Y(x) - Z(x)|$ is smaller than every positive number; i.e., $Y(x) = Z(x)$ proving the solution is unique.

Power Series

PROBLEM 7.17

Suppose the power series

$$\sum_{n=0}^{\infty} c_n(x - a)^n \tag{1}$$

converges for some $x_0 \neq a$. Then show that the series converges absolutely for all x such that $|x - a| < |x_0 - a|$.

SOLUTION 7.17

If

$$\sum_{n=0}^{\infty} c_n(x_0 - a)^n$$

converges, then $c_n(x_0 - a)^n$ tends to zero as n tends to infinity. But this implies that $\{c_n(x_0 - a)^n\}$ is a bounded sequence, say

$$|c_n(x_0 - a)^n| \leq A \quad \text{for } n = 0, 1, \ldots$$

Now

$$|c_n(x - a)^n| = |c_n(x_0 - a)^n| \left(\frac{|x - a|}{|x_0 - a|}\right) \leq Ar^n$$

where

$$r = \frac{|x - a|}{|x_0 - a|}$$

Then the series (1) is absolutely convergent for $r < 1$ by comparison with the geometric series

$$\sum_{n=0}^{\infty} Ar^n$$

PROBLEM 7.18

Suppose the number sequence $\{c_n\}$ is such that the following limit exists

$$\lim_{n\to\infty} \frac{c_n}{c_{n+1}} = R$$

Then show that the power series $\sum c_n(x - a)^n$ converges on the set $|x - a| < R$ and diverges on the set $|x - a| > R$. We refer to R as the *radius of convergence.*

SOLUTION 7.18

If we apply the ratio test to the power series then

$$\lim_{n} \left| \frac{c_{n+1}(x - a)^{n+1}}{c_n(x - a)^n} \right| = \lim_{n} \left| \frac{c_{n+1}}{c_n} \right| |x - a| \to \frac{1}{R}|x - a|$$

and it follows that:

the series converges for $\frac{1}{R}|x - a| < 1$

the series diverges for $\frac{1}{R}|x - a| > 1$

Note that if $R = 0$ then the series converges only at $x = a$ and if $R = \infty$ then the series converges for all real values x. Note also that the ratio test implies neither convergence nor divergence at the points $x = a \pm R$.

PROBLEM 7.19

If the power series $\sum c_n(x - a)^n$ has radius of convergence equal to R then show that the series converges uniformly on the closed interval $|x - a| \le R - \in$ for every $\in > 0$.

SOLUTION 7.19

Let $b = a + R - \in$ and let $M_n = c_n(b - a)^n$. The power series is convergent at $x = b$ by the result of the previous problem which is to say the series of constants $\sum c_n$ is convergent. In addition,

$$|u_n(x)| = |c_n(x - a)^n| \le M_n \quad \text{for } x \text{ in } I = \{|x - a| < R - \in\}$$

Then the Weierstrass M-test implies the desired result.

PROBLEM 7.20

Suppose that

$$\operatorname*{Lim}_{n} \left| \frac{c_n}{c_{n+1}} \right| = R$$

so the power series

$$\sum_{n=0}^{\infty} c_n(x - a)^n$$

has radius of convergence equal to R. Then show that the series

$$\sum_{n=1}^{\infty} n\, c_n(x - a)^{n-1} \quad \text{and} \quad \sum_{n=0}^{\infty} c_n \frac{(x - a)^{n-1}}{n + 1} \tag{2}$$

have the same radius of convergence.

SOLUTION 7.20

Applying the ratio test to the first of the two series in (2) leads to

$$\left| \frac{(n + 1)\, c_{n+1}(x - a)^n}{n\, c_n(x - a)^{n-1}} \right| = \frac{n + 1}{n} \left| \frac{c_{n+1}}{c_n} \right| |x - a| \to \frac{1}{R} |x - a|$$

Thus the same radius of convergence results. Forming the ratio of consecutive terms in the second series yields

$$\frac{n}{n + 1} \left| \frac{c_{n+1}}{c_n} \right| |x - a| \to \frac{1}{R} |x - a|$$

Evidently since

$$\frac{n}{n + 1} \quad \text{and} \quad \frac{n + 1}{n}$$

each tend to 1 as n tends to infinity, the two series in (2) have the same radius of convergence as the series in (1). Note, however, that this discussion does not imply that the series in (2) have the same behavior as the series in (1) at the endpoints $x = a \pm R$.

PROBLEM 7.21

Suppose that the power series

$$\sum_{n=0}^{\infty} c_n(x-a)^n \tag{1}$$

has positive radius of convergence R and let f(x) denote the sum of the series for x in $I = \{|x-a| < R\}$. Then show that f(x) has derivatives of every order on I and that

$$c_n = \frac{f^{(n)}(a)}{n!} \qquad n = 0, 1, \ldots \tag{2}$$

SOLUTION 7.21

According to the result of the previous problem the series in (1) can be differentiated term by term to produce a new power series having the same radius of convergence. Corollary 7.11 implies that the sum of the derived series is f′(x). Proceeding in this way we find that for each $k = 1, 2, \ldots$

$$f^{(k)}(x) = \sum_{n=k}^{\infty} n(n-1)\cdots(n-k+1)c_n(x-a)^{n-k} \tag{3}$$

That is, for each $k = 1, 2, \ldots$ the kth derivative of the series (1) converges uniformly on I to the kth derivative of f(x). At $x = a$, the series (3) has a single nonzero term corresponding to $n = k$. Then

$$f^{(k)}(a) = k!c_k$$

This is (2). The series

$$\sum_{n=0}^{\infty} \frac{f^{(n)}(k)}{n!}(x-a) \tag{4}$$

is called the *Taylor series* for f(x) about the point $x = a$. Note that what we have shown is that if f(x) has a convergent power series expansion of the form (1) then the coefficients must be given by (2). We have not shown that any function having derivatives of all orders at $x = a$ has a convergent power series representation of the form (4).

PROBLEM 7.22

Show that

$$f(x) = \begin{cases} e^{-1/x^2} & x \neq 0 \\ 0 & x = 0 \end{cases}$$

has derivatives of all orders at x = 0 but f(x) does not equal the Taylor series expansion for f about x = 0.

SOLUTION 7.22

We compute, via the chain rule,

$$f'(x) = \frac{2}{x^3} e^{-1/x^2} \quad \text{for } x \neq 0$$

It follows from L'Hospital's rule (see Problem 4.23) that f'(x) tends to zero as x tends to zero. By repeated application of L'Hospital's rule we can show that

$$\text{for } k = 1, 2, \ldots\ f^{(k)}(x) \to 0 \quad \text{as } x \to 0.$$

This result together with the result from the previous problem implies that all of the coefficients in the Taylor series for f(x) at x = 0 are equal to zero. But f(x) is not identically zero in a neighborhood of x = 0 and so f(x) is not equal to the sum of its Taylor series about the point x = 0.

PROBLEM 7.23

Derive the following power series expansion for lnx about the point x = 1

$$\ln x = x - 1 - \frac{x-1}{2} + \frac{(x-1)^2}{3} - \frac{(x-1)^2}{4} + \cdots \quad 0 < x < 2$$

SOLUTION 7.23

From Problem 7.4 we have that

$$\frac{1}{1-t} = 1 + t + t^2 + t^3 + \cdots \text{ for } |t| < 1$$

Letting t = x − 1, we obtain

$$\frac{1}{1-(x-1)} = \frac{1}{x} = 1 + (x-1) + (x-1)^2 + (x-1)^3 + \cdots \text{ for } |x-1| < 1$$

Now

$$\int_0^z \frac{1}{x}\, dx = \ln z \text{ for } |z-1| < 1; \text{ i.e., for } 0 < z < 2$$

Term by term integration from 1 to z of the power series for 1/x leads to a series which converges uniformly on the interval $|z-1| < 1$ to the antiderivative lnz for 1/x. That is

$$\ln z = (z-1) - \frac{(z-1)}{2} + \frac{(z-1)^2}{3} - \frac{(z-1)^3}{4} + \cdots \quad 0 < z < 2$$

PROBLEM 7.24

Derive the following power series expansion for Arctan x

$$\text{Arctan } x = x - \frac{x^3}{3} + \frac{x^5}{5} - \frac{x^7}{7} + \cdots \quad |x| < 1 \tag{1}$$

SOLUTION 7.24

Substituting $t = -x^2$ into

$$\frac{1}{1-t} = 1 + t + t^2 + t^3 + \cdots \quad \text{for } |t| < 1$$

leads to

$$\frac{1}{1+x^2} = 1 - x^2 + x^4 - x^6 + \ldots \text{ for } |x^2| < 1 \tag{2}$$

We have

$$\int_0^z \frac{1}{1+x^2}\, dx = \text{Arctan } z$$

and thus by the result of Problem 7.20 term by term integration from 0 to z of the series in (2) leads to the result (1).

PROBLEM 7.25

Find a power series expansion about $x = 0$ for the solution of the initial value problem

$$y'(x) = y(x), \quad y(0) = 1 \qquad (1)$$

SOLUTION 7.25

If we suppose that

$$y(x) = \sum_{n=0}^{\infty} c_n x^n \quad \text{for } |x| < R \tag{2}$$

then

$$y'(x) = \sum_{n=1}^{\infty} n\, c_n x^{n-1} \quad \text{for } |x| < R$$

This last result can be rewritten in the form

$$y'(x) = \sum_{n=0}^{\infty} (n+1)\, c_{n+1} x^n \tag{3}$$

Substituting (3) and (2) into (1) leads to

$$\sum_{n=0}^{\infty} (n+1) c_{n+1} x^n = \sum_{n=0}^{\infty} c_n x^n$$

and then Theorem 7.15 implies that

$$(n+1)c_{n+1} = c_n \quad \text{for } n = 0, 1, \ldots \tag{4}$$

Combining (2) with $y(0) = 1$ produces $c_0 = 1$ and then we use (4) recursively to obtain

$$c_1 = \frac{c_0}{1} = 1$$

$$c_2 = \frac{c_1}{2} = \frac{1}{2}$$

$$c_3 = \frac{c_2}{3} = \frac{1}{3!}$$

etc.

$$c_{n+1} = \frac{c_n}{n+1} = \frac{1}{(n+1)!}$$

Thus

$$y(x) = \sum_{n=0}^{\infty} \frac{x^n}{n!}$$

and we can easily show that this power series has an infinite radius of convergence (i.e., $R = \infty$). Moreover, the theory of ordinary differential equations tells us that the unique solution of the initial value problem (1) is $y(x) = e^x$. It follows that

$$e^x = \sum_{n=0}^{\infty} \frac{x^n}{n!}$$

Let $\{u_n\}$ denote a family of functions having common domain of definition I in R. Then $\{u_n\}$ is called a function sequence. This function sequence is said to be pointwise convergent on I if, for each x in I, $\{u_n(x)\}$ is a Cauchy sequence of numbers. The sequence is said to converge to the limit $u = u(x)$ if for each $\in > 0$ and all x in I there is an $N = N(\in, x)$ such that

$$|u_n(x) - u(x)| < \in \quad \text{if } n > N.$$

Similarly the infinite series $\sum u_n(x)$ converges pointwise to the sum U(x) if the sequence of partial sums is pointwise convergent to the limit U(x).

The statememts above relate to pointwise convergence. Also of interest is the notion of uniform convergence. Uniform convergence can be characterized by the following equivalent statements about $\{u_n\}$:

1. *$\{u_n\}$ converges uniformly on I to the limit u*
2. *For all $\in > 0$ there is an $N = N(\in)$ such that $|u_n(x) - u(x)| < \in$ for all x in I*

3. *The number sequence* $M_n = Lub_{x \in I} | u_n(x) - u(x) |$ *converges to zero.*

Uniform convergence is stronger than pointwise convergence in the sense that every uniformly convergent sequence is pointwise convergent but the converse is false. An infinite series converges uniformly on I if the associated sequence of partial sums is a uniformly convergent function sequence.

The Weierstrass M-test is a test that can be applied directly to an infinite series to determine whether it is uniformly convergent. An infinite series $\sum u_n$ *converges uniformly on I if there exists a positive term infinite series of numbers* $\sum M_n$ *such that*

(a) *for all x in I* $| u_n(x) | < M_n$ *for every n*
(b) $\sum M_n$ *converges*

These conditions are sufficient but not necessary for uniform convergence of an infinite series. Note that the M-test determines only whether the series is uniformly convergent. It does not identify the sum of the series in the event of convergence.

Uniform convergence has a number of important consequences. Stating these consequences for uniformly convergent infinite series, we have:

Suppose $\sum u_n(x)$ *converges uniformly on I to the sum U(x).*

1. *If each* $u_n(x)$ *is continuous on I then U(x) must be continuous on I*
2. *If each* $u_n(x)$ *is integrable on I then the series may be integrated, term by term: i.e.,*

$$\sum_{n=1}^{\infty} \int_a^x u_n = \int_a^x U \quad \text{for a and x in I}$$

3. *If each* $u_n(x)$ *is differentiable on I and if the differentiated series* $\sum u_n'(x)$ *is uniformly convergent on I then the series may be differentiated, term by term; i.e.,*

$$\sum_{n=1}^{\infty} u_n'(x) = U'(x)$$

Similar results can be stated for uniformly convergent sequences.

Power series are a type of function series that is important for many applications. These series have the form

$$\sum_{n=1}^{\infty} c_n(x-a)^n$$

Power series converge on an interval of the form $I = \{| x - a | < R\}$ *where R depends on the coefficients* c_n. *In particular*

$$R = \operatorname{Lim}_n \left| \frac{c_n}{c_{n+1}} \right|$$

if this limit exists. If $R > 0$, then for every $\in > 0$, the convergence is uniform on the set $|x - a| \le R - \in$.

Power series have the important property that if R is positive and the sum of the series on I is denoted by f then:

1. *f(x) is continuous on the open interval I*
2. *For each x in I*

$$\int_a^x f = \sum_{n=0}^{\infty} \frac{c_n}{n+1}(x-a)^{n+1}$$

3. *For each x in I*

$$f'(x) = \sum_{n=1}^{\infty} n\, c_n (x-a)^{n-1}$$

$$f''(x) = \sum_{n=2}^{\infty} n(n-1)c_n (x-a)^{n-2}$$

etc.

4. *For each n,*

$$c_n = \frac{f^{(n)}(a)}{n!}.$$

Note that the integrated and differentiated series in statements 2 and 3 all have the same radius of convergence as the original power series for f.

The following is a short list of some power series representations for common functions. The interval of convergence is listed for each series:

$$e^x = \sum_{n=0}^{\infty} \frac{x^n}{n!} = 1 + x + \frac{x^2}{2} + \frac{x^3}{6} + \cdots \quad |x| < \infty$$

$$\text{Sin } x = \sum_{n=1}^{\infty} (-1)^{n+1} \frac{x^{2n-1}}{(2n-1)!} = x - \frac{x^3}{6} + \frac{x^5}{125} - \cdots \quad |x| < \infty$$

$$\text{Cos } x = \sum_{n=0}^{\infty} (-1)^n \frac{x^{2n}}{(2n)!} = 1 - \frac{x^2}{2} + \frac{x^4}{24} - \cdots \quad |x| < \infty$$

$$\ln x = \sum_{n=0}^{\infty} (-1)^n \frac{(x-1)^n}{n} = (x-1) - \frac{(x-1)^2}{2} + \frac{(x-1)^3}{3} - \cdots \quad 0 < x < 2$$

$$\text{Arctan } x = \sum_{n=1}^{\infty} (-1)^{n+1} \frac{x^{2n-1}}{2n-1} = x - \frac{x^3}{3} + \frac{x^5}{5} - \frac{x^7}{7} + \cdots \quad |x| < 1$$

$$\frac{1}{1-x} = \sum_{n=0}^{\infty} x^n = 1 + x + x^2 + x^3 + \cdots \quad |x| < 1$$

8

Partial Differentiation

In the previous chapters we have considered the notions of continuity, differentiation and integration for functions of a single variable. In this chapter we begin consideration of functions of more than one variable by considering the notion of differentiation for real valued functions of several variables. In the next chapter we will consider integration for such functions.

The concept of a partial derivative of a function of several variables is defined and some of the properties are developed. Examples are given to show that even at a point where a function of several variables has partial derivatives with respect to each of its independent variables, the function need not be continuous at that point. This is quite different from the case for functions of one variable.

A function of several variables for which the partial derivatives all exist and are continuous *in a neighborhood of a point is said to be $\mathbb{C}^1$ near that point. If a function is $\mathbb{C}^1$ in a neighborhood of a point then it must be continuous at the point. We also define differentiability, a degree of regularity that lies between continuity and $\mathbb{C}^1$. If a function is* differentiable *at a point it must be continuous there but it need not be $\mathbb{C}^1$.*

We state multivariable versions of the mean value theorem, Taylor's theorem and the chain rule. Implicit differentiation of functions of several variables is considered and Jacobian determinants are defined for convenience in expressing these results. We state versions of an implicit function theorem providing conditions sufficient to imply that a function is implicitly well defined by a set of equations. A several variable version of Theorem 3.14, an inverse function theorem is discussed as well

Several applications of partial differentiation are described. These include: Leibniz' rule for differentiating an integral depending on a parameter, several geometric applications and the location and classification of extreme points of a function of several variables.

REAL VALUED FUNCTIONS OF SEVERAL REAL VARIABLES

DEPENDENT AND INDEPENDENT VARIABLES

In this chapter we will consider real valued functions of several real variables. For example $z = f(x, y)$, $w = F(x, y, z)$, and $y = G(x_1, x_2, \ldots, x_n)$ denote functions of two, three, and n variables, respectively. We speak of (x, y), (x, y, z), and $(x_1, x_2, \ldots, x_n)$ as the *independent variables* or *arguments* for the functions f, F, and G respectively. Similarly, z, w and y are referred to as the *dependent variables* or of these functions.

DOMAIN AND RANGE

A real valued function of two real variables can be defined as a rule that assigns to each point (x, y) in a subset D of the plane, a unique real number $f(x, y)$. The set D in the plane is called the *domain* of the function f, and the set of values $z = f(x, y)$ generated as (x, y) varies over D is called the *range* of the function. More generally, we can define a real valued function of n real variables to be a rule that assigns to each n-tuple $(x_1, x_2, \ldots, x_n)$ in a subset D of n-space, a unique real number $G(x_1, x_2, \ldots, x_n)$. Then the set of n-tuples D is the domain of the function G, and as $(x_1, x_2, \ldots, x_n)$ ranges over D, the set of values that is assumed by G is the range.

FUNCTIONS OF N REAL VARIABLES

We can give a more precise definition of a real valued function of n real variables as a set of ordered pairs (X, y) where the first member of the pair $\underline{X} = (x_1, x_2, \ldots, x_n)$ is an ordered n-tuple and the second member y is a real number. The set of ordered pairs is a function if no two distinct pairs have the same first member—i.e., it never happens that pairs $(\underline{X}, y_1)$ and $(\underline{X}, y_1)$ occur with $y_1 \neq y_2$. Then the set of all the n-tuples $\underline{X}$ occuring in the collection of ordered pairs forms the domain of the function and the set of all second members y forms the range.

DISTANCE BETWEEN TWO POINTS IN R

We will use the notation $\mathbb{R}^n$ for n-space, the set of all ordered n-tuples of real numbers. Then $\mathbb{R}^2$ indicates the real plane and $\mathbb{R}^3$ is real three dimensional space. The distance $d(\underline{X}, \underline{Y})$ between two points $\underline{X}$ and $\underline{Y}$ in $\mathbb{R}^n$ is defined to be

$$d(\underline{X}, \underline{Y}) = \sqrt{(x_1 - y_1)^2 + \cdots + (x_n - y_n)^2}$$

NEIGHBORHOODS IN $\mathbb{R}^n$

For $\in > 0$ we define an $\in$-neighborhood of $\underline{P}$ in $\mathbb{R}^2$ to be the set of points $\underline{X}$ in $\mathbb{R}^n$ such that $d(\underline{X}, \underline{P}) < \in$. We will use the notation $N_{\in}(\underline{P})$ to denote this set of points and we will denote by $N_\epsilon'(\underline{P})$ the deleted neighborhood consisting of $N_\epsilon(\underline{P})$ with the point $\underline{P}$ deleted. In two dimensions, $N_\epsilon'(\underline{P})$ is a disc centered at $\underline{P}$ having radius ϵ; in three dimensions it is a ball.

INTERIOR POINTS AND BOUNDARY POINTS

A point $\underline{P}$ belonging to a set U in $\mathbb{R}^n$ is said to be an interior point of U if there is a neighborhood $N_\epsilon(\underline{P})$ contained within U. The point $\underline{P}$ in U is said to be a boundary point of U if every neighborhood $N_\epsilon(\underline{P})$ contains points of U and points that are not in U.

POINT SETS IN $\mathbb{R}^n$

A set U in $\mathbb{R}^n$ is said to be *open* if all its points are interior points. A set G in $\mathbb{R}^n$ is said to be *closed* if its complement is open. The *complement* of G is the point set consisting of all points in $\mathbb{R}^n$ which do not belong to G. A set is said to be *bounded* if it can be contained within some ball of finite radius and a *compact* set is a set that is both closed and bounded.

LIMITS IN $\mathbb{R}^N$

Let $y = G(x_1, x_2, \ldots, x_n) = G(\underline{X})$ be a real valued function of n real variables. Then we say that G *tends to the limit* L as $\underline{X} = (x_1, x_2, \ldots, x_n)$ tends to the point $\underline{P} = (p_1, p_2, \ldots, p_n)$ in the domain of G if and only if for every $\epsilon > 0$ there is a $\delta > 0$ such that

$$|G(\underline{X}) - L| < \epsilon \quad \text{for every } \underline{X} \text{ in the deleted neighborhood } N_\delta'(P).$$

Theorem 8.1 Limits

Theorem 8.1 If $y = G(x_1, x_2, \ldots, x_n)$ tends to distinct values L_1 and L_2 as $\underline{X} = (x_1, x_2, \ldots, x_n)$ approaches the point P along two different paths, then the limit as $\underline{X}$ approaches $\underline{P}$ fails to exist.

Example 8.1 Limits

8.1 (a) Consider the following function of two variables:

$$F(x, y) = \frac{3x^3 - 2y^3}{x^2 + y^2} \qquad (x, y) \neq (0, 0).$$

To show that F tends to 0 as $\underline{X} = (x, y)$ tends to $\underline{0} = (0, 0)$, we have to show that for every $\epsilon > 0$ there is a $\delta > 0$ such that $|F(x, y)| < \epsilon$ for all $\underline{X}$ in $N_\delta'(\underline{0})$; i.e., for (x, y) such that $0 < x^2 + y^2 < \delta^2$. But

$$|3x^3 - 2y^3| \leq 3|x|x^2 + 2|y|y^2$$

and

$$|x| \leq (x^2 + y^2)^{1/2}, |y| \leq (x^2 + y^2)^{1/2}$$

Hence

$$|3x^3 - 2y^3| \leq 3(x^2 + y^2)^{3/2}$$

and

$$|F(x, y)| \leq 3(x^2 + y^2)^{1/2} \quad \text{for } 0 < x^2 + y^2 < \delta^2.$$

It is now clear that $|F(x, y)| < \in$ for all (x, y) in $N_{\delta'}(0, 0)$ if we choose $\delta = \in/3$. Thus the limit of F(x, y) as (x, y) tends to (0, 0) exists and equals 0.

8.1 (b) Consider the limit as (x, y) tends to (0, 0) for the function

$$G(x, y) = \frac{x^2 - y^2}{x^2 + y^2} \quad (x, y) \neq (0, 0)$$

This limit does not exist. In order to show this, note that as (x, y) tends to (0, 0) along the x-axis (where y = 0) we have

$$G(x, 0) = \frac{x^2 - 0}{x^2 + 0} \quad \text{for all } x \neq 0.$$

This indicates that the limit, if it exists, is equal to 1. On the other hand, note that as (x, y) tends to the origin along the y-axis (where x = 0) we have

$$G(0, y) = \frac{0 - y^2}{0 + y^2} \quad \text{for all } x \neq 0.$$

Finally note that along the line y = x we have G(x, x) = 0 for all x. Then G(x, y) approaches different values as (x, y) approaches the origin along different paths and we conclude via Theorem 8.1 that the limit fails to exist.

CONTINUITY

Let $y = G(\underline{X})$ denote a real valued function of n real variables. Then we say that G is continuous at the point $\underline{P}$ in the domain D of G if and only if for each $\in > 0$ there exists a $\delta > 0$ such that $|G(\underline{X}) - G(\underline{P})| < \in$ for every $\underline{X}$ in $N_\delta(\underline{P})$.

We can show that this definition is equivalent to the condition that

$$\lim_{\underline{X} \to \underline{P}} G(\underline{X}) = G(\underline{P})$$

Here if $\underline{P}$ is an interior point of D then $\underline{X}$ may approach $\underline{P}$ from any direction. But if $\underline{P}$ is a boundary point of D then $\underline{X}$ may only approach $\underline{P}$ along paths that lie within D.

Example 8.2
Continuous Functions

8.2 (a) According to Example 8.1(a), the function

$$F(x, y) = \begin{cases} \dfrac{3x^2 - 2y^2}{x^2 + y^2} & (x, y) \neq (0, 0 \\ 0 & (x, y) = (0, 0) \end{cases}$$

is continuous at the origin. That is, F(x, y) tends to a limit as (x, y) tends to (0, 0) and this limiting value equals the function value at (0, 0). It is not hard to show that F(x, y) is also continuous at every other point of $\mathbb{R}^2$.

8.2 (b) The function

$$G(x, y) = \frac{x^2 - y^2}{x^2 + y^2} \qquad (x, y) \neq (0, 0)$$

is not continuous at (0, 0) no matter what value we assign to G(0, 0). This is because it has been shown (in Example 8.1 (b)) that G(x, y) tends to no limit as (x, y) tends to (0, 0).

If G is continuous at every point of domain D in $\mathbb{R}^n$ then we say that G is continuous in D. In this case we will say G belongs to $\mathbb{C}(D)$, the class of functions that are continuous in D.

Theorems analogous to the Bounded Range Theorem, Extreme Value Theorem, and the Intermediate Value Theorem for continuous functions of one variable are true for continuous functions of several variables. An efficient development of these theorems requires a knowledge of the topology of $\mathbb{R}^n$ that is beyond the scope of this text.

PARTIAL DIFFERENTIATION

Partial Derivatives

For z = f(x, y) defined in a domain D in $\mathbb{R}^2$ we define the *partial derivative of f with respect* to x at the point (a, b) in D by

$$\lim_{x \to a} \frac{f(x, b) - f(a, b)}{x - a} = \frac{\partial f}{\partial x}(a, b) = \partial_x f(a, b) = f_x(a, b) = \frac{\partial z}{\partial x} = z_x$$

and define the *partial derivative of f* with *respect to* y at the point (a, b) by

$$\lim_{y \to b} \frac{f(a, y) - f(a, b)}{y - b} = \frac{\partial f}{\partial y}(a, b) = \partial_y f(a, b) = f_y(a, b) = \frac{\partial z}{\partial y} = z_y.$$

The derivatives are said to exist at the point if the limits exist. There are five notations shown for each derivative. Other notations are sometimes used but will not appear in this text. We refer to f_x as the partial derivative of f with respect to x holding y constant and refer to f_y as the partial derivative of f with respect to y holding x constant. These derivatives can be computed by the usual rules of differentiation, treating the independent variables that are being held constant as constant parameters.

The partial derivatives of functions of more than two variables are similarly defined. For example,

$$\lim_{x \to a} \frac{F(x, b, c) - F(a, b, c)}{x - a} = \frac{\partial F}{\partial x}(a, b, c) = \partial_x F(a, b, c) = f_x(a, b,c)$$

is the partial derivative of F(x, y, z) with respect to x holding y and z constant, evaluated at the point (a, b, c) in D.

THE CLASS OF $\mathbb{C}^1$ FUNCTIONS

If F is defined and continuous on domain D in $\mathbb{R}^n$ and if all of the partial derivatives

$$\frac{\partial F}{\partial x_1}, \frac{\partial F}{\partial x_2}, \ldots, \frac{\partial F}{\partial x_n}$$

exist and are also continuous throughout D, then we say that F belongs to the class of $\mathbb{C}^1$ functions on D. For convenience we will sometimes write this as $F \in \mathbb{C}^1(D)$. Then we have the following analogue of Corollary 4.8.

Theorem 8.2 A Mean Value Theorem

Theorem 8.2 Suppose $f \in \mathbb{C}^1(D)$ for D in $\mathbb{R}^2$ and that for a point (a, b) in D there exists a $\rho > 0$ such that the disc $(x - a)^2 + (y - b)^2 < \rho^2$ is contained in D. Then for each h, k such that $h^2 + k^2 < \rho^2$ there exist a pair of numbers λ_1, λ_2 with $0 < \lambda_1, \lambda_2 < 1$ and

$$f(a + h, b + k) = f(a, b) + f_x(a + \lambda_1 h, b)h + f_y(a + h, b + \lambda_2 k)k.$$

Alternatively, there exists a single number ϑ, $0 < \vartheta < 1$, such that

$$f(a + h, b + k) = f(a, b) + f_x(a + \vartheta h, b + \vartheta k)h + f_y(a + \vartheta h, b + \vartheta k)k.$$

There are theorems similar to Theorem 8.2 for functions of more than two variables.

Properties of Partial Derivatives

Partial derivatives have properties analogous to those described in Theorem 4.2 for derivatives of functions of one variable. In addition we have the following analogue of the chain rule.

Theorem 8.3 A Chain Rule for Functions of Several Variables

Theorem 8.3 Suppose z = f(x, y) belongs to $\mathbb{C}^1(D)$ for D in $\mathbb{R}^2$.

(a) If x = x(t) and y = y(t) belong to $\mathbb{C}^1(I)$ for I in $\mathbb{R}^1$ where t in I implies (x(t), y(t)) is in D, then

$$\frac{dz}{dt} = f_x(x, y)\, x'(t) + f_y(x, y) y'(t).$$

(b) If x = x(s, t) and y = y(s, t) belong to $\mathbb{C}^1(D')$ where (s, t) in D' implies (x(s, t), y(s, t)) in D, then

$$\frac{\partial z}{\partial s} = f_x \frac{\partial x}{\partial s} + f_y \frac{\partial y}{\partial s} \quad \text{and} \quad \frac{\partial z}{\partial t} = f_x \frac{\partial x}{\partial t} + f_y \frac{\partial y}{\partial t}$$

More generally, if $u = F(x_1, x_2, \ldots, x_n)$ is in $\mathbb{C}^1(D)$ for D in $\mathbb{R}^n$ and if $x_k = f(y_1, y_2, \ldots, y_n)$ is in $\mathbb{C}^1(D')$ for $k = 1, 2, \ldots, n$ and D′ in $\mathbb{R}^m$ then

$$\frac{\partial u}{\partial y_p} = F_{x_1}\frac{\partial x_1}{\partial y_p} + \cdots + F_{x_n}\frac{\partial x_n}{\partial y_p} \text{ for } p = 1, 2, \ldots, m.$$

The statement of this theorem does not include all possible cases of composed functions of several variables and the corresponding formulas for their derivatives. It is only meant to indicate the general form that chain rule expressions might take.

Differentiability

Not every function of several variables whose partial derivatives with respect to each independent variable exist necessarily belongs to $\mathbb{C}^1$. We say that $z = f(x, y)$ is *differentiable at the point* (a, b) in D in $\mathbb{C}^2$ if:

(a) $f_x(a, b)$ and $f_y(a, b)$ exist and,

(b) $f(a + h, b + k) = f(a, b) + f_x(a, b)h + f_y(a, b)k + \Delta_1(h, k)h + \Delta_2(h, k)k$ where $\Delta_1(h, k)$ and $\Delta_2(h, k)$ tend to zero as (h, k) tends to (0, 0)

If f is differentiable at each point in D then we say f is differentiable in D and we write $f \in \partial(D)$. Differentiability for functions of more than two variables is defined in a similar fashion. The set of functions which is differentiable in D is strictly larger than the set $\mathbb{C}^1(D)$ and is smaller than the set $\mathbb{C}(D)$.

GEOMETRIC INTERPRETATION OF DIFFERENTIABILITY

A function of one variable $y = f(x)$ is differentiable at a point P if and only if the graph of f(x) has a unique tangent line at P. The graph of a function of two variables $z = F(x, y)$ is a surface in $\mathbb{R}^3$ and F is differentiable at (x_0, y_0) if there is a unique plane tangent to the surface at (x_0, y_0, z_0).

Theorem 8.4

Theorem 8.4 Let f be defined in domain D in R. Then $f \in \mathbb{C}^1(D)$ implies that $f \in \partial(D)$ but not conversely. Also, $f \in \partial(D)$ implies that $f \in \mathbb{C}(D)$ but the converse is false.

Derivatives of Higher Order

The derivatives defined above are derivatives of order one; i.e., first order derivatives. Just as we can define derivatives of order higher than one for functions of one variable, we can define higher order derivatives for functions of more than one variable. For the function $z = f(x, y)$ we can define four partial derivatives of order two:

$$\frac{\partial}{\partial x}\left(\frac{\partial f}{\partial x}\right) = \frac{\partial^2 f}{\partial x^2} = \partial_{xx} f = f_{xx},$$

$$\frac{\partial}{\partial y}\left(\frac{\partial f}{\partial x}\right) = \frac{\partial^2 f}{\partial y \partial x} = \partial_{yx} f = f_{yx},$$

$$\frac{\partial}{\partial x}\left(\frac{\partial f}{\partial y}\right) = \frac{\partial^2 f}{\partial x \partial y} = \partial_{xy} f = f_{xy},$$

$$\frac{\partial}{\partial y}\left(\frac{\partial f}{\partial y}\right) = \frac{\partial^2 f}{\partial y^2} = \partial_{yy} f = f_{yy}.$$

Higher order derivatives for functions of more than two variables are similarly defined.

MIXED PARTIAL DERIVATIVES

Higher order derivatives such as f_{xy} and f_{yx} that involve differentiation with respect to more than just a single independent variable are called *mixed partial derivatives.*

Theorem 8.5 Equality of Mixed Partial Derivatives

Theorem 8.5 Suppose $z = f(x, y)$ is defined in domain D in $\mathbb{R}^2$ and that

(a) the first order partial derivatives f_x and f_y exist at each point of D

(b) the mixed partial f_{xy} exists at each point of D and f_{xy} is continuous at (a, b) in D

Then the other mixed partial derivative f_{yx} exists at (a, b) and $f_{xy}(a, b) = f_{yx}(a, b)$.

THE CLASS OF $\mathbb{C}^m$ FUNCTIONS

If a function of n variables $f = f(\underline{X})$ is continuous on domain D in $\mathbb{R}^n$ together with all its partial derivatives of order less than or equal to m, then we say f belongs to $\mathbb{C}^m(D)$.

Example 8.3 Notation for Derivatives of Higher Order

For f in $C^4(D)$ in the case n = 2, we write

$$\frac{\partial^4 f}{\partial x \partial^3 y} \text{ to indicate } \frac{\partial}{\partial x}\left(\frac{\partial^3 f}{\partial y^3}\right) \text{ and } \frac{\partial^4 f}{\partial^3 y \partial x} \text{ to indicate } \frac{\partial^3}{\partial y^3}\left(\frac{\partial f}{\partial x}\right)$$

Since the partial derivatives are all continuous throughout D, an extension of the previous theorem implies that it does not matter whether we differentiate first with respect to x and then three times with respect to y or differentiate in the opposite order. Thus when the derivatives are all continuous in D the notations seen above can be used interchangeably.

Taylor's Theorem

We have the following version of Taylor's theorem for functions of several variables. For simplicity we will state the theorem for functions of two variables.

Theorem 8.6 Taylor's Theorem in Two Variables

Theorem 8.6 Suppose $f = f(x, y)$ belongs to $\mathbb{C}^{m+1}(D)$ and that $\underline{P} = (a, b)$ is in D with $N_r[\underline{P}]$ contained in D for some $r > 0$. Then for all (x, y) in $N_{\in}[\underline{P}]$

$$f(x, y) = f(a, b) + f_x(a, b)h + f_y(a, b)k + \frac{1}{2}(h^2 f_{xx}(a, b) + 2hkf_{xy}(a, b) + k^2 f_{yy}(a, b)) + \cdots$$

$$= \sum_{j=0}^{m} \frac{1}{j!}(h\partial_x + k\partial_y)^j f(a, b) + R_m$$

where

$$h = x - a,\ k = y - b, \text{ and for some } \lambda,\ 0 < \lambda < 1,$$

$$R_m = \frac{1}{(m+1)!}(h\partial_x + k\partial_y)^{m+1} f(a + \lambda h, b + \lambda k)$$

For each positive integer j the expression $(h\partial_x + k\partial_y)^j$ is intended to be expanded formally by the binomial theorem.

Differentiation of Implicit Functions

If the equation $F(x, y, z) = 0$ can be solved explicitly for z in terms of x and y, $z = f(x, y)$, then we say z is an *explicit function* of x and y. The *Implicit Function Theorem*, Theorem 8.7(a), states conditions on F sufficient to imply the existence of a local solution for z in terms of x and y. When such a solution exists we say the function $f(x, y)$ is defined *implicitly* by the equation $F(x, y, z) = 0$. When a solution exists it is not always practical to solve for the function $f(x, y)$ but we can compute the derivatives of f with respect to x and y nevertheless.

We will illustrate the concept of implicit differentiation with examples without stating a theorem. In each case we are assuming that an implicit function is defined. Theorems 8.7(a) and 8.7(b) state conditions sufficient to ensure the existence of the assumed implicit functions.

Example 8.4 Implicit Differentiation

8.4 (a) One Function of Two Variables Suppose F is a $\mathbb{C}^1$ function of two variables and that $F(x, y) = 0$. Suppose further that f is a $\mathbb{C}^1$ function of one variable with the property that for all x in some interval J, we have

$$F(x, f(x)) = 0 \quad \text{for } x \text{ in } J.$$

Then the function f(x) is defined implicitly by the function F. According to the chain rule, for each x in J where $F_y(x, f(x))$ is not zero, the derivative of f can be computed as follows

$$F_x + F_y f'(x) = 0 \quad \text{and} \quad f'(x) = -\frac{F_x(x, f(x))}{F_y(x, f(x))}$$

8.4 (b) One Function of Three Variables Suppose F is a $\mathbb{C}^1$ function of three variables and that $F(x, y, z) = 0$. Suppose further that f is a $\mathbb{C}^1$ function of two variables with the property that for (x, y) in some domain D in $\mathbb{R}^2$

$$F(x, y, f(x, y)) = 0 \quad \text{for all } (x, y) \text{ in } D.$$

Then for each (x, y) in D where $F_z(x, y, f(x, y))$ is not zero, the implicitly defined function f can be differentiated with respect to x and y

$$F_x + F_z f_x = 0 \quad \text{and} \quad f_x = -\frac{F_x(x, y, f(x, y))}{F_z(x, y, f(x, y))} = \frac{\partial z}{\partial x}$$

$$F_y + F_z f_y = 0 \quad \text{and} \quad f_y = -\frac{F_y(x, y, f(x, y))}{F_z(x, y, f(x, y))} = \frac{\partial z}{\partial y}$$

We interpret this to mean that z is implicitly defined as a function of x and y; f_x is the partial derivative of z with respect to x, holding y constant and f_y is the partial derivative of z with respect to y holding x constant.

Suppose still further that there is another $\mathbb{C}^1$ function of two variables g = g(y, z) such that

$$F(g(y, z), y, z) = 0 \quad \text{for all y, z in some domain U in } \mathbb{R}^2.$$

Then for each (x, y) in U where $F_x(g(y, z), y, z)$ is not zero, the implicitly defined function g can be differentiated with respect to y and z

$$\partial_y F(g(y, z), y, z) = F_x g_y + F_y = 0 \quad \text{and} \quad g_y = -\frac{F_y(g(y, z), y, z)}{F_x(g(y, z), y, z)} = \frac{\partial x}{\partial y}$$

$$\partial_z F(g(y, z), y, z) = F_x g_z + F_z = 0 \quad \text{and} \quad g_z = -\frac{F_z(g(y, z), y, z)}{F_x(g(y, z), y, z)} = \frac{\partial x}{\partial z}$$

Here $g_z = \partial x/\partial z$ indicates the partial derivative of x with respect to z holding y constant. Note that

$$\frac{\partial x}{\partial z}\frac{\partial z}{\partial x} = \left(-\frac{F_z}{F_x}\right)\left(-\frac{F_x}{F_z}\right) = 1 .$$

8.4 (c) Two Functions of Five Variables Suppose F and G are $\mathbb{C}^1$ functions of five variables and that F(x, y, z, u, v) = 0 and G(x, y, z, u, v) = 0. Suppose also that f and g are $\mathbb{C}^1$ functions of three variables with the property that for all (x, y, z) in some domain D in $\mathbb{R}^3$

$$F(x, y, z, f(x, y, z), g(x, y, z)) = 0 \text{ and } G(x, y, z, f(x, y, z), g(x, y, z)) = 0.$$

According to the chain rule,

$$F_x + F_u f_x + F_v g_x = 0 \quad \text{and} \quad G_x + G_u f_x + G_v g_x = 0.$$

We can solve these two equations for f_x and g_x by means of Cramer's rule. We find

$$f_x = \left| \frac{\begin{matrix} -F_x & F_v \\ -G_x & G_v \end{matrix}}{\begin{matrix} F_u & F_v \\ G_u & G_v \end{matrix}} \right| \quad \text{and} \quad g_x = \left| \frac{\begin{matrix} F_u & -F_x \\ G_u & -G_x \end{matrix}}{\begin{matrix} F_u & F_v \\ G_u & G_v \end{matrix}} \right| .$$

If we denote the determinant in the denominators of these two expressions by J, then at each point (x, y, z) in D where J(x, y, z, f(x, y, z), g(x, y, z)) is not zero these expressions equal the derivatives f_x and g_x of the implicitly defined functions f and g. We could compute the derivatives f_y, g_y, and f_z, g_z in a similar way. We interpret f_x and g_x as the partial derivatives of u and v, respectively, with respect to x holding y and z constant.

If p and q are $\mathbb{C}^1$ functions of three variables with the property that for all (y, u, v) in some domain U in $\mathbb{R}^3$

$$F(p(y, u, v), y, q(y, u, v), u, v) = 0 \text{ and } G(p(y, u, v), y, q(y, u, v), u, v) = 0$$

then according to the chain rule

$$F_x p_u + F_z q_u + F_u = 0 \quad \text{and} \quad G_x p_u + G_z q_u + G_u = 0.$$

At each point in U where the determinant

$$J(p(y, u, v), y, q(y, u, v), u, v) = \begin{vmatrix} F_x & F_z \\ G_x & G_z \end{vmatrix}$$

is different from zero we have

$$p_u = \frac{\begin{vmatrix} -F_u & F_z \\ -G_u & G_z \end{vmatrix}}{\begin{vmatrix} F_x & F_z \\ G_x & G_z \end{vmatrix}} = \frac{\partial x}{\partial u} \quad \text{and} \quad q_u = \frac{\begin{vmatrix} F_u & -F_x \\ G_u & -G_x \end{vmatrix}}{\begin{vmatrix} F_u & F_v \\ G_u & G_v \end{vmatrix}} = \frac{\partial z}{\partial u}.$$

Note that (unlike 8.4(b)) in this case

$$\frac{\partial u}{\partial x}\frac{\partial x}{\partial u} \neq 1.$$

JACOBIAN DETERMINANTS

Determinants like those appearing in 8.4(c) occur frequently in connection with functions of several variables and are called Jacobian determinants. We adopt the following notation:

$$\begin{vmatrix} F_x & F_z \\ G_x & G_z \end{vmatrix} = \frac{\partial(F, G)}{\partial(x, z)} \qquad \begin{vmatrix} F_u & F_v \\ G_u & G_v \end{vmatrix} = \frac{\partial(F, G)}{\partial(u, v)}$$

Then the results of 8.4(c) can be expressed in this notation

$$f_x = -\frac{\partial(F, G)}{\partial(x, v)} \bigg/ \frac{\partial(F, G)}{\partial(u, v)} \quad \text{and} \quad g_x = -\frac{\partial(F, G)}{\partial(u, x)} \bigg/ \frac{\partial(F, G)}{\partial(u, v)}$$

$$p_u = -\frac{\partial(F, G)}{\partial(u, z)} \bigg/ \frac{\partial(F, G)}{\partial(x, z)} \quad \text{and} \quad q_u = -\frac{\partial(F, G)}{\partial(x, u)} \bigg/ \frac{\partial(F, G)}{\partial(x, z)}$$

We can compute Jacobian determinants for m functions of n real variables as long as $m \le n$.

Example 8.5 Jacobian Determinants

If F, G and H denote C functions of six variables x, y, z, u, v and w, then we can choose any three of the six variables and compute the corresponding Jacobian. For example, choosing x, y and u,

$$\frac{\partial(F, G, H)}{\partial(x, y, u)} = \begin{vmatrix} F_x & F_y & F_u \\ G_x & G_y & G_u \\ H_x & H_y & H_u \end{vmatrix}$$

If E, F, G and H are $\mathbb{C}^1$ functions of the same six variables then we can choose any *four* of the variables, say x, y, u and w, and compute

$$\frac{\partial(E, F, G, H)}{\partial(x, y, u, w)} = \begin{vmatrix} E_x & E_y & E_u & E_w \\ F_x & F_y & F_u & F_w \\ G_x & G_y & G_u & G_w \\ H_x & H_y & H_u & H_w \end{vmatrix}$$

Implicit Function Theorems

In Example 8.4(b) we assumed that the equation $F(x, y, z) = 0$ implicitly defined z as a function of x and y and in Example 8.4(c) we made the assumption that the simultaneous equations $F(x, y, z, u, v) = 0$ and $G(x.y, z, u, v) = 0$ contained the implicit definition of u and v as functions of x, y, and z. However, not every such equation or system of equations leads to well defined implicit functions. Conditions sufficient to ensure the existence of functions implicitly defined by the equations are the subject of implicit function theorems. We state two such theorems. Using these theorems as examples, it should be clear how to formulate analogous theorems for situations involving different numbers of equations or numbers of variables.

Theorem 8.7 (a) One Equation in n + 1 Variables

Theorem 8.7 (a) Let the function $F(X, y) = F(x_1, x_2, \dots, x_n, y)$ be defined on an open set U in $\mathbb{R}^{n+1}$ and let $(\underline{P}, q) = (p_1, \dots, p_n, q)$ denote a point of U for which

$$F(P, q) = 0 \quad \text{and} \quad F_y(P, q) \neq 0.$$

Then there exists a box B in $\mathbb{R}^n$ given by $|x_j - p_j| < \alpha$, for $j = 1, \dots, n$ and an interval I in $\mathbb{R}^n$, $|y - q| < \beta$, such that for each $\underline{X}$ in B there exists a unique y in I such that $F(\underline{X}, y) = 0$. The function $y = \varphi(X)$ thus defined is in $\mathbb{C}^1$ with

$$\frac{\partial \varphi}{\partial x_j} = -\frac{F_{x_j}(\underline{X}, \varphi(\underline{X}))}{F_y(\underline{X}, \varphi(\underline{X}))} \quad j = 1, \dots, n.$$

Theorem 8.7 (b) Two Equations in Five Variables

Theorem 8.7 (b) Let the functions $F(x, y, z, u, v)$ and $G(x, y, z, u, v)$ be defined on an open set U in $\mathbb{R}^5$. Let the point (a, b, c, p, q) in U be such that

$$F(a, b, c, p, q) = 0, \quad G(a, b, c, p, q) = 0, \text{ and}$$

$$\frac{\partial(F, G)}{\partial(u, v)} \neq 0 \text{ at } (a, b, c, p, q).$$

Then there exists a box B_3 in $\mathbb{R}^3$, $\{|x - a| < \alpha, |y - b| < \alpha, |z - c| < \alpha\}$, and a box B_2 in $\mathbb{R}^2$, $\{|u - p| < \beta, |v - q| < \beta\}$, such that for each (x, y, z) in B_3 there exists a unique (u, v) in B_2 such that

$$F(x, y, z, u, v) = 0 \quad \text{and} \quad G(x, y, z, u, v) = 0.$$

The functions $u = \varphi(x, y, z)$ and $v = \psi(x, y, z)$ so defined are in $\mathbb{C}^1$ with

$$\varphi_x = -\frac{\partial(F, G)}{\partial(x, v)} \bigg/ \frac{\partial(F, G)}{\partial(u, v)} \quad \text{and} \quad \Psi_x = -\frac{\partial(F, G)}{\partial(u, x)} \bigg/ \frac{\partial(F, G)}{\partial(u, v)}$$

and similar formulas for φ_y, φ_z, ψ_y and ψ_z.

Example 8.6 An Inverse Mapping Theorem

Consider an implicit function theorem for two functions of four variables in which the equations $F(x, y, u, v) = 0$ and $G(x, y, u, v) = 0$ have the special form

$$x - f(u, v) = 0 \quad \text{and} \quad y - g(u, v) = 0$$

for f and g defined and $\mathbb{C}^1$ on an open set U in $\mathbb{R}^2$. If (u_0, v_0) in U is such that

$$x_0 = f(u_0, v_0),\ y_0 = g(u_0, v_0), \text{ and } J = \frac{\partial(f, g)}{\partial(u, v)} \neq 0 \text{ at } (u_0, v_0)$$

then by a variant of Theorem 8.7 (b), it follows that there exist two boxes

$$B = \{|x - x_0| < \alpha, |y - y_0| < \alpha\} \text{ and } B' = \{|u - u_0| < \beta, |v - v_0| < \beta\}$$

in $\mathbb{R}^2$ such that for each (x, y) in B there exists a unique (u, v) in B′ such that

$$x - f(u, v) = 0 \quad \text{and} \quad y - g(u, v) = 0.$$

The functions $u = \varphi(x, y)$ and $v = \psi(x, y)$ so defined are $\mathbb{C}^1$ with

$$\varphi_x = g_v/J, \qquad \psi_x = -g_u/J$$

$$\varphi_y = -f_v/J, \qquad \psi_y = f_u/J.$$

This result asserts that the equations $x = f(u, v)$, $y = g(u, v)$ define a one to one correspondence between points (u, v) in B′ and points (x, y) in B.

TRANSFORMATIONS, MAPPINGS

Equations of the form $x = \varphi(u, v)$, $y = \psi(u, v)$ are said to define a transformation or mapping from $\mathbb{R}^2$ into $\mathbb{R}^2$.

We have just shown that the mapping in Example 8.6 is one to one (invertible) in a neighborhood of any point (u_0, v_0) where the Jacobian of the transformation, $J = \partial(f, g)/\partial(u, v)$, does not vanish. A result of this form is called an *inverse function theorem*.

We can state an inverse function theorem more generally for the case of n equations in 2n variables. We have stated the result here for the case n = 2. Coordinate transformations (i.e., changes of variables) are examples of transformations where it is particularly important to know if the transformation is invertible. A change of variables is not invertible in a neighborhood of any point where the Jacobian of the transformation vanishes.

APPLICATIONS OF PARTIAL DIFFERENTIATION

Leibniz Rule

Suppose that $\varphi = \varphi(x)$ and $\psi = \psi(x)$ are $\mathbb{C}^1$ on the interval $I = [a, b]$ and that $\varphi(x) \le \psi(x)$ for $a \le x \le b$. Suppose also that $f = f(x, y)$ is defined and continuous on a domain D containing the set of points $\{a \le x \le b, \varphi(x) \le y \le \psi(x)$ and that $f_x(x, y)$ exists and is continuous on D. Then it follows that the function

$$F(x) = \int_{\varphi(x)}^{\psi(x)} f(x, y)\, dy$$

is a $\mathbb{C}^1$ function of x on I and

$$F'(x) = f(x, \psi(x))\, \psi'(x) - f(x, \varphi(x))\varphi'(x) + \int_{\varphi(x)}^{\psi(x)} f_x(x, y)\, dy\,.$$

This result is known as Leibniz Rule for differentiating integrals with respect to a parameter.

Tangents and Normals

Let a surface S in $\mathbb{R}^3$ be defined as the set of points such that $F(x, y, z) = 0$ where $F(x, y, z)$ is a $\mathbb{C}^1$ function of three variables. Suppose $\underline{P} = (a, b, c)$ is a point on S (i.e., $F(a, b, c) = 0$) and that Γ denotes a smooth curve in $\mathbb{R}^3$, lying in S and passing through $\underline{P}$. If Γ has a parametric representation of the form

$$\Gamma = \{x = x(t), y = y(t), z = z(t), t_0 \le t \le t_1\} \quad x(t), y(t), z(t) \text{ in } \mathbb{C}^1[t_0, t_1]$$

then Γ lies in S if

$$F(x(t), y(t), z(t)) = 0 \quad \text{for } t_0 \le t \le t_1,$$

and Γ passes through $\underline{P}$ at time t in $[t_0, t_1]$ if $x(\tau) = a$, $y(\tau) = b$, $z(\tau) = c$.

It follows from the chain rule that for each t in $[t_0, t_1]$

$$\frac{\partial F}{\partial t} = F_x(x(t), y(t), z(t))x'(t) + F_y(x(t), y(t), z(t))y'(t) + F_z(x(t), y(t), z(t))z'(t) = 0$$

The left side of this equation has the form of a dot product of two 3–vectors and in this notation the equation takes the form

$$(\partial_x F, \partial_y F, \partial_z F) \cdot (x'(t), y'(t), z'(t)) = 0$$

It follows that the two vectors $(\partial_x F, \partial_y F, \partial_z F)$ and $(x'(t), y'(t), z'(t))$ are orthogonal.

It is known from single variable calculus that the vector $(x'(t), y'(t), z'(t))$ is *tangent* to the curve Γ and we conclude that the direction of the vector $(\partial_x F, \partial_y F, \partial_z F)$ is perpendicular to Γ. We have not assumed any special properties for Γ except that Γ lies in S. Then at each point $\underline{P}$ on the surface S the vector $(\partial_x F, \partial_y F, \partial_z F)$ is normal to any curve Γ lying in S and passing through $\underline{P}$. That is to say, $(\partial_x F, \partial_y F, \partial_z F)$ is normal to the surface S.

THE GRADIENT OF A FUNCTION

We introduce the notation

$$\textbf{grad } F(x, y, z) = (\partial_x F, \partial_y F, \partial_z F)$$

and refer to this as the *gradient of the function* $F(x, y, z)$. We have just shown that at each point on the smooth surface $S = \{F(x, y, z) = 0\}$ the vector **grad** F is normal to the surface. It follows from elementary analytic geometry that the equation of the plane that is tangent to the surface S at point $\underline{P} = (a, b, c)$ is

$$\partial_x F(a, b, c)(x - a) + \partial_y F(a, b, c)(y - b) + \partial_z F(a, b, c)(z - c) = 0.$$

The Directional Derivative

If p, q, r are real numbers such that $p^2 + q^2 + r^2 = 1$ then the directed line segment from the origin to the point (p, q, r) makes angles α, β and γ with the x, y, and z axes, respectively, where $\text{Cos } \alpha = p$, $\text{Cos } \beta = q$ and $\text{Cos } \gamma = r$. Then

$$\underline{u} = \text{Cos } \alpha \, \underline{i} + \text{Cos } \beta \, \underline{j} + \text{Cos } \gamma \, \underline{k}$$

is a *unit vector* (vector of length one) with direction numbers p, q, and r. If $f = f(x, y, z)$ belongs to $\mathbb{C}^1(D)$ for some domain D in $\mathbb{R}^3$, and if $P = (a, b, c)$ is in D, then we define

$$\nabla_u f(a, b, c) = \underset{h \to 0}{\text{Lim}} \frac{f(a + h \text{ Cos } \alpha, b + h \text{ Cos } \beta, c + h \text{ Cos } \gamma) - f(a, b, c)}{h}$$

as the *directional derivative* of $f(x, y, z)$ in the direction $\underline{u} = (\text{Cos } \alpha, \text{Cos } \beta, \text{Cos } \gamma)$ at the point $\underline{P}$.

Example 8.7 Directional Derivatives

If the unit vector u is directed along one of the coordinate axes then the directional derivative in the direction $\underline{u}$ reduces to the previously defined partial derivative with respect to the corresponding variable. That is,

if $\underline{u} = (1, 0, 0)$ then $\nabla_u f = f_x$

if $\underline{u} = (0, 1, 0)$ then $\nabla_u f = f_y$

if $\underline{u} = (0, 0, 1)$ then $\nabla_u f = f_z$

Theorem 8.8

Theorem 8.8 Suppose f belongs to $\mathbb{C}^1(D)$ for some D in $\mathbb{R}^n$ and $\underline{u}$ is unit vector in $\mathbb{R}^n$. Then at each point $\underline{P}$ in D

$$\nabla_u f(\underline{P}) = \mathbf{grad}\ f(\underline{P}) \cdot \underline{u}.$$

Corollary 8.9 Under the conditions of Theorem 8.8, at each $\underline{P}$ in D f increases most rapidly in the direction of the vector **grad** f and decreases most rapidly in the opposite direction, −**grad** f.

Maxima and Minima

Let $f = f(x_1, x_2, \ldots, x_n)$ be differentiable in a neighborhood of the point $\underline{P}$ in $\mathbb{R}^n$. We say that f has a *relative maximum* at $\underline{P}$ if $f(\underline{P}) \geq f(\underline{Q})$ for all $\underline{Q}$ in a neighborhood of $\underline{P}$, and f has a *relative minimum* at $\underline{P}$ if $f(\underline{P}) \leq f(\underline{Q})$ for all $\underline{Q}$ in a neighborhood of $\underline{P}$. We say $\underline{P}$ is a relative extreme point f or f if P is either a relative maximum or a relative minimum.

Theorem 8.10

Theorem 8.10 Suppose f is differentiable in a neighborhood of $\underline{P}$ and that $\underline{P}$ is a relative extreme point for f. Then **grad** $f(P) = (0, 0, \ldots, 0) = \underline{0}$.

A point where the gradient of f vanishes is called a *critical point* for f. Every relative extreme point for f is a critical point but a critical point need not be a relative extreme point. Determining whether a critical point for f is a relative maximum or relative minimum (or neither of these) requires knowledge of the higher order derivatives of f at the point. For functions of two variables we have the following theorem.

Theorem 8.11

Theorem 8.11 Suppose $f = f(x, y)$ is in $\mathbb{C}^2$ in a neighborhood of the point $\underline{P} = (a, b)$ and that $f_x(\underline{P}) = f_y(\underline{P}) = 0$. If, in addition,

$$f_{xx}f_{yy} - f_{xy}{}^2 > 0 \text{ at } \underline{P}$$

then f has a relative extreme point at P. If f_{xx} and f_{yy} are both negative at $\underline{P}$ then $\underline{P}$ is a relative maximum and if f_{xx}, f_{yy} are both positive at $\underline{P}$ then $\underline{P}$ is a relative minimum for f. If

$$f_{xx}f_{yy} - f_{xy}{}^2 < 0 \text{ at } \underline{P}$$

then $\underline{P}$ is a saddle point for f (a saddle point is neither a maximum nor a minimum). If $f_{xx}f_{yy} - f_{xy}{}^2$ vanishes at $\underline{P}$ then any of these alternatives is possible.

This result is the two dimensional analogue of the single variable result that is proved in Problem 4.28. See Problem 8.26 for an n-dimensional version of this theorem.

SOLVED PROBLEMS

Continuous and Differentiable Functions

PROBLEM 8.1

Give an example of a function $f = f(x, y)$ that is continuous at (0, 0) but neither $f_x(0, 0)$ nor $f_y(0, 0)$ exists.

SOLUTION 8.1

Consider the function $f(x, y) = (1 - |x|)(1 - |y|)$. The function $\varphi(x) = 1 - |x|$ is a continuous function of one variable and since $f(x, y)$ is a product of two such functions, $f(x, y)$ can be shown to be continuous. But $f(x, 0) = 1 - |x| = \varphi(x)$ is a function of one variable which is not differentiable at $x = 0$. Similarly, $f(0, y) = 1 - |y| = f(y)$ is not differentiable at $y = 0$. Then neither $f_x(0, 0)$ nor $f_y(0, 0)$ exists.

PROBLEM 8.2

Give an example of a function $f = f(x, y)$ such that $f_x(0, 0)$ and $f_y(0, 0)$ both exist but f is not continuous at (0, 0).

SOLUTION 8.2

Consider the function

$$f(x, y) = \begin{cases} \dfrac{xy}{x^2 + y^2} & \text{for } (x, y) \neq (0, 0) \\ 0 & \text{for } (x, y) = (0, 0) \end{cases}$$

Then for $(x, y) \neq (0, 0)$ we can use the rules of differentiation to compute the derivatives

$$f_x(x, y) = \frac{y(y^2 - x^2)}{(x^2 + y^2)^2} \qquad f_y(x, y) = \frac{-x(y^2 - x^2)}{(x^2 + y^2)^2}$$

At the origin we must use the definition of the derivative rather than the rules. Then we have

$$f_x(0, 0) = \lim_{x \to 0} \frac{f(x, 0) - f(0, 0)}{x - 0} = 0 \text{ and } f_y(0, 0) = \lim_{y \to 0} \frac{f(0, y) - f(0, 0)}{y - 0} = 0$$

Thus $f(x, y)$ is seen have first partial derivatives at each point. But as x tends to zero, $f(x, x)$ tends to the value 1/2 while $f(x, -x)$ tends to the value $-1/2$. In addition, $f(x, 0) = 0$ for all x and $f(0, y) = 0$ for all y. Thus $f(x, y)$ tends to no limit at (0, 0) and therefore cannot be continuous at (0, 0). For functions of one variable existence of the derivative implies continuity. This example shows that the result does not carry over to functions of several variables.

For the function $f(x, y)$ in this example, $f(x, 0)$ is a continuous function of x at $x = 0$ and $f(0, y)$ is a continuous function of y at $y = 0$ but $f(x, y)$ is not a continuous function of (x, y) at $(0, 0)$. We describe this situation by saying f is separately continuous in x and y at $(0, 0)$ but $f(x, y)$ is not jointly continuous at $(0, 0)$.

PROBLEM 8.3

Give an example of a function $f = f(x, y)$ such that $f_x(0, 0)$ and $f_y(0, 0)$ both exist but f is not differentiable at $(0, 0)$.

SOLUTION 8.3

Consider the function $f(x, y) = \sqrt{|xy|}$. Then $f_x(0, 0)$ and $f_y(0, 0)$ each exist and equal zero. But Taylor's theorem implies,

$$f(h, k) = f(0, 0) + f_x(0, 0)h + f_y(0, 0)k + \Delta_1(h, k)h + \Delta_2(h, k)k$$

Thus, $\sqrt{|hk|} = \Delta_1(h, k)h + \Delta_2(h, k)k$ which implies that $\Delta_1(h, h) + \Delta_2(h, h) = 1$. Hence $\Delta_1(h, h) + \Delta_2(h, h)$ does not tend to zero as h tends to zero as required for f to be differentiable at $(0, 0)$. Thus $f(x.y)$ is not differentiable at $(0, 0)$.

PROBLEM 8.4

Show that if f is differentiable at a point then f is continuous there.

SOLUTION 8.4

If $f = f(x, y)$ is differentiable at the point (a, b) then by definition

$$f(a + h, b + k) = f(a, b) + f_x(a, b)h + f_y(a, b)k + \Delta_1(h, k)h + \Delta_2(h, k)k$$

where $\Delta_1(h, k)$ and $\Delta_2(h, k)$ tend to zero as (h, k) tends to $(0, 0)$. But then $|f(a + h, b + k) - f(a, b)|$ tends to zero as (h, k) tends to zero which implies that $f = f(x, y)$ is continuous at (a, b). A similar argument can be used to prove that differentiability implies continuity for a function of n variables. As Problem 8.2 shows, a function f may fail to be differentiable at a point where each of its partial derivatives exists. The next problem shows that if each of the partial derivatives exists and is *continuous* at $\underline{P}$ then f is differentiable at $\underline{P}$.

PROBLEM 8.5

Show that if $f = f(x, y)$ is $\mathbb{C}^1$ in an open set D in $\mathbb{R}^2$ then f is differentiable at each point of D.

SOLUTION 8.5

For any point $\underline{P} = (a, b)$ in D we can use Theorem 8.2 to write,

$$f(a + h, b + k) = f(a, b) + f_x(a + \lambda_1 h, b)h + f_y(a, b + \lambda_2 k)k$$

for λ_1, λ_2 such that $0 < \lambda_1, \lambda_2 < 1$. Then

$$f(a + h, b + k) = f(a, b) + f_x(a, b)h + f_y(a, b)k + \Delta_1(h, k)h + \Delta_2(h, k)k$$

where

$$\Delta_1(h, k) = f_x(a + \lambda_1 h, b) - f_x(a, b) \text{ and } \Delta_2(h, k) = f_y(a, b + \lambda_2 k) - f(a, b).$$

Since f is in $\mathbb{C}^1$, f_x and f_y are continuous at (a, b) and it follows that $\Delta_1(h, k)$ and $\Delta_2(h, k)$ tend to zero as (h, k) tends to (0, 0). Then f is differentiable at (a, b). This proof extends to functions of n variables.

Mixed Partial Derivatives

PROBLEM 8.6

Give an example of a function f = f(x, y) where $f_{xy}(0, 0)$ and $f_{yx}(0, 0)$ both exist but they are not equal.

SOLUTION 8.6

Consider the function

$$f(x, y) = \begin{cases} \dfrac{xy(x^2 - y^2)}{x^2 + y^2} & \text{for } (x, y) \neq (0, 0) \\ 0 & \text{for } (x, y) = (0, 0) \end{cases}$$

For (x, y) ≠ (0, 0) we use the rules of differentiation to compute

$$f_x(x, y) = y\frac{x^4 + 4x^2y^2 - y^4}{(x^2 + y^2)^2} \qquad f_y(x, y) = x\frac{x^4 - 4x^2y^2 - y^4}{(x^2 + y^2)^2}$$

The derivatives at the origin must be computed using the definition

$$f_x(0, 0) = \lim_{x \to 0} \frac{f(x, 0) - f(0, 0)}{x - 0} = 0$$

$$f_y(0, 0) = \lim_{y \to 0} \frac{f(0, y) - f(0, 0)}{y - 0} = 0$$

Then

$$f_{xy}(0, 0) = \lim_{x \to 0} \frac{f_y(x, 0) - f_y(0, 0)}{x - 0} = \lim_{x \to 0} \frac{x(x^4/x^4) - 0}{x - 0} = 1$$

and

$$f_{yx}(0, 0) = \lim_{y \to 0} \frac{f_x(0, y) - f_y(0, 0)}{y - 0} = \lim_{y \to 0} \frac{y(-y^4/y^4) - 0}{y - 0} = -1$$

Hence

$$f_{xy}(0, 0) \neq f_{yx}(0, 0).$$

This is a result of the fact that neither f_{xy} nor f_{yx} is continuous at (0, 0).

Chain Rule

PROBLEM 8.7 THE TOTAL DERIVATIVE

Consider the function u = F(x, y, z, t) and compute the derivative of u with respect to t if x, y and z depend explicitly on t as specified by the functions x = x(t), y = y(t) and z = z(t).

SOLUTION 8.7

According to the chain rule, if u(t) = F(x(t), y(t), z(t), t) then

$$\frac{\partial u}{\partial t} = \frac{\partial F}{\partial x}\frac{\partial x}{\partial t} + \frac{\partial F}{\partial y}\frac{\partial y}{\partial t} + \frac{\partial F}{\partial z}\frac{\partial z}{\partial t} + \frac{\partial F}{\partial t}$$

$$= (\textbf{grad}\ F) \cdot \underline{V} + \frac{\partial F}{\partial t} \quad \text{where } \underline{V} = (x'(t), y'(t), z'(t)).$$

Note that if $\underline{V} = \underline{0}$ then du/dt = ∂F/∂t but when $\underline{V}$ is not zero then in general du/dt and ∂F/∂t are not equal. In fluid dynamics, du/dt is referred to as the total derivative of u(t). It reflects the change in the scalar quantity u with respect to t, following the particle whose location at time t is given by (x(t), y(t), z(t)).

PROBLEM 8.8 LAPLACE'S EQUATION IN POLAR COORDINATES

Suppose u = u(x, y) is a $\mathbb{C}^2$ function of x and y in a region D in $\mathbb{R}^2$ and that at each point in D we have

$$\nabla^2 u(x, y) = \partial_{xx}u + \partial_{yy}u = 0.$$

We say that the *Laplacian* of u(x, y) is zero and that u(x, y) satisfies *Laplace's equation* in D. For the change of variables x = r Cos ϑ, y = r Sin ϑ, show that u(r, ϑ) = u(x(r, ϑ), y(r, ϑ)) satisfies

$$\partial_{rr}u(r, \vartheta) + \frac{1}{r}\partial_r u(r, \vartheta) + \frac{1}{r^2}\partial_{\theta\theta}u(r, \vartheta) = 0 \tag{1}$$

at each point of D. Equation (1) expresses Laplace's equation in polar coordinates.

SOLUTION 8.8

According to the chain rule,

$$\frac{\partial u}{\partial x} = \frac{\partial u}{\partial r}\frac{\partial r}{\partial x} + \frac{\partial u}{\partial \vartheta}\frac{\partial \vartheta}{\partial x}, \quad \frac{\partial u}{\partial y} = \frac{\partial u}{\partial r}\frac{\partial r}{\partial y} + \frac{\partial u}{\partial \vartheta}\frac{\partial \vartheta}{\partial y}$$

In addition we have

$$F(x, y, r, \vartheta) = x - r\,\text{Cos}\,\vartheta = 0 \quad G(x, y, r, \vartheta) = y - r\,\text{Sin}\,\vartheta = 0$$

and

$$F_x + F_r r_x + F_\theta \vartheta_x = 1 - \text{Cos}\,\vartheta\, r_x + r\,\text{Sin}\,\vartheta\,\vartheta_x = 0$$

$$G_x + G_r r_x + G_\theta \vartheta_x = 0 - \text{Sin}\,\vartheta\, r_x - r\,\text{Cos}\,\vartheta\,\vartheta_x = 0.$$

This leads to r_x = Cos ϑ and ϑ_x = (–Sin ϑ)/r. Similarly,

$$F_y + F_r r_y + F_\vartheta \vartheta_y = 1 - \text{Cos}\,\vartheta\, r_y + r\,\text{Sin}\,\vartheta\,\vartheta_y = 0$$

$$G_y + G_r r_y + G_\vartheta \vartheta_y = 0 - \text{Sin}\,\vartheta\, r_y - r\,\text{Cos}\,\vartheta\,\vartheta_y = 0.$$

leads to $r_y = \text{Sin}\,\vartheta$ and $\vartheta_y = (\text{Cos}\,\vartheta)/r$. Then

$$\frac{\partial u}{\partial x} = \frac{\partial u}{\partial r}\text{Cos}\,\vartheta - \frac{1}{r}\text{Sin}\,\vartheta\frac{\partial u}{\partial \vartheta} \qquad \frac{\partial u}{\partial y} = \frac{\partial u}{\partial r}\text{Cos}\,\vartheta - \frac{1}{r}\text{Sin}\,\vartheta\frac{\partial u}{\partial \vartheta}$$

It follows from these equations that derivatives with respect to x and y can be expressed in terms of derivatives with respect to r and ϑ by

$$\frac{\partial}{\partial x} = \text{Cos}\,\vartheta\frac{\partial}{\partial r} - \frac{1}{r}\text{Sin}\,\vartheta\frac{\partial}{\partial \vartheta} \qquad \frac{\partial}{\partial y} = \text{Sin}\,\vartheta\frac{\partial}{\partial r} - \frac{1}{r}\text{Cos}\,\vartheta\frac{\partial}{\partial \vartheta}$$

Then applying these to u_x and u_y,

$$\partial_{xx}u = \frac{\partial}{\partial x}(u_x) = \text{Cos}\,\vartheta\frac{\partial}{\partial r}(u_x) - \frac{1}{r}\text{Sin}\,\vartheta\frac{\partial}{\partial \vartheta}(u_x)$$

$$= \text{Cos}\,\vartheta\frac{\partial}{\partial r}\left(\frac{\partial u}{\partial r}\text{Cos}\,\vartheta - \frac{1}{r}\text{Sin}\,\vartheta\frac{\partial u}{\partial \vartheta}\right) - \frac{1}{r}\text{Sin}\vartheta\frac{\partial}{\partial \vartheta}\left(\frac{\partial u}{\partial r}\text{Cos}\,\vartheta - \frac{1}{r}\text{Sin}\,\vartheta\frac{\partial u}{\partial \vartheta}\right)$$

$$= \text{Cos}^2\vartheta\,\partial_{rr}u - \frac{2\,\text{Sin}\,\vartheta\,\text{Cos}\,\vartheta}{r}\partial_{r\vartheta}u + \left(\frac{\text{Sin}\,\vartheta}{r}\right)^2\partial_{\vartheta\vartheta}u$$

$$+ \frac{\text{Sin}^2\vartheta}{r}\partial_r u + \frac{2\,\text{Sin}\,\vartheta\,\text{Cos}\,\vartheta}{r^2}\partial_\vartheta u$$

$$\partial_{yy}u = \text{Sin}^2\vartheta\,\partial_{rr}u + \frac{2\,\text{Sin}\,\vartheta\,\text{Cos}\,\vartheta}{r}\partial_{r\vartheta}u + \left(\frac{\text{Cos}\,\vartheta}{r}\right)^2\partial_{\vartheta\vartheta}u$$

$$+ \frac{\text{Cos}^2\vartheta}{r}\partial_r u + \frac{2\,\text{Sin}\,\vartheta\,\text{Cos}\,\vartheta}{r^2}\partial_\vartheta u$$

Adding the expressions for u_{xx} and u_{yy} leads to (1).

PROBLEM 8.9 LAPLACE'S EQUATION IN SPHERICAL COORDINATES

Suppose $u = u(x, y, z)$ is a $\mathbb{C}^2$ function of x, y and z in a region D in $\mathbb{R}^3$ and that at each point in D we have $\partial_{xx}u + \partial_{yy}u + \partial_{zz}u = 0$. If we introduce the change of variable

$$x = \rho\,\text{Sin}\,\varphi\,\text{Cos}\,\vartheta,\ y = \rho\,\text{Sin}\,\varphi\,\text{Sin}\,\vartheta,\ z = \rho\,\text{Cos}\,\varphi \tag{1}$$

then show that $u(\rho, \vartheta, \varphi) = u(x(\rho, \vartheta, \varphi), y(\rho, \vartheta, \varphi), z(\rho, \vartheta, \varphi))$ satisfies

$$\partial_{\rho\rho}u + \rho^{-2}\partial_{\varphi\varphi}u + (\rho\,\text{Sin}\,\varphi)^{-2}\partial_{\vartheta\vartheta}u + 2\rho^{-1}\partial_\rho u + \text{Cot}\,\varphi\,\rho^{-2}\partial_{\varphi\varphi}u = 0 \tag{2}$$

i.e.,

$$(\rho\,\text{Sin}\,\varphi)^{-2}((\text{Sin}\,\varphi)^2\partial_\rho(\rho^2\partial_\rho u) + \text{Sin}\varphi\,\partial_\varphi(\text{Sin}\,\varphi\partial_\varphi u) + \partial_{\vartheta\vartheta}u) = 0$$

SOLUTION 8.9

The chain rule implies that

$$u_x = u_\rho \rho_x + u_\vartheta \vartheta_x + u_\varphi \varphi_x$$

$$u_y = u_\rho \rho_y + u_\vartheta \vartheta_y + u_\varphi \varphi_y \tag{3}$$

$$u_z = u_\rho \rho_z + u_\vartheta \vartheta_z + u_\varphi \varphi_z.$$

Now the coordinate change is described by

$$x = x(\rho, \vartheta, \varphi),\ y = y(\rho, \vartheta, \varphi),\ z = z(\rho, \vartheta, \varphi)$$

with

$$J = \frac{\partial(x, y, z)}{\partial(\rho, \vartheta, \varphi)} = -\rho^2 \operatorname{Sin} \varphi.$$

At each point where $J \neq 0$, the coordinate transformation can be inverted to obtain functions

$$\rho = \rho(x, y, z),\ \vartheta = \vartheta(x, y, z),\ \varphi = \varphi(x, y, z).$$

Differentiating (1) with respect to x yields

$$1 = x_\rho \rho_x + x_\vartheta \vartheta_x + x_\varphi \varphi_x$$
$$0 = y_\rho \rho_x + y_\vartheta \vartheta_x + y_\varphi \varphi_x$$
$$0 = z_\rho \rho_x + z_\vartheta \vartheta_x + z_\varphi \varphi_x$$

Using Cramer's rule we obtain

$$r = \begin{vmatrix} 1 & x_\vartheta & x_\varphi \\ 0 & y_\vartheta & y_\varphi \\ 0 & z_\vartheta & z_\varphi \end{vmatrix} /J, \quad \vartheta = \begin{vmatrix} x_\rho & 1 & x_\varphi \\ y_\rho & 0 & y_\varphi \\ z_\rho & 0 & z_\varphi \end{vmatrix} /J, \quad \varphi = \begin{vmatrix} x_\rho & x_\vartheta & 1 \\ y_\rho & y_\vartheta & 0 \\ z_\rho & z_\vartheta & 0 \end{vmatrix} /J$$

By differentiating (1) with respect to y and z and solving in this way we can obtain the partial derivatives of r, ϑ and φ with respect to y and z. We use these in (3) to find u_x, u_y and u_z in terms of u_ρ, u_ϑ and u_φ. As we did in the previous problem, we then apply the operators suggested by (3)

$$\partial_x = \rho_x \partial_\rho + \vartheta_x \partial_\vartheta + \varphi_x \partial_\varphi$$

$$\partial_y = \rho_y \partial_\rho + \vartheta_y \partial_\vartheta + \varphi_y \partial_\varphi$$

$$\partial_z = \rho_z \partial_\rho + \vartheta_z \partial_\vartheta + \varphi_z \partial_\varphi$$

to the expressions for u_x, u_y and u_z in terms of u_ρ, u_ϑ and u_φ. After a considerable amount of manipulation we are led to (2).

The Mean Value Theorem and Taylor's Theorem

PROBLEM 8.10

Let D be a region in $\mathbb{R}^2$ with the property that any point of D can be joined to any other point of D by a polygonal path lying within D. Such a region is said to be connected. If $f = f(x, y)$ is $\mathbb{C}^1$ in D and $f_x = f_y = 0$ at each point of D, then show that f is constant in D.

SOLUTION 8.10

Let $\underline{P}_1 = (p_1, q_1)$ and $\underline{P}_n = (p_n, q_n)$ denote two points of D and let $\underline{P}_n, \ldots, \underline{P}_{n-1}$ denote the vertices of a polygonal path lying within D and joining $\underline{P}_1$ to $\underline{P}_n$. Then $\underline{P}_2 = (p_1 + h, q_1 + k)$ for some h and k and by Theorem 8.2

$$\begin{aligned} f(p_2, q_2) &= f(p_1 + h, q_1 + k) = f(p_1, q_1) + f_x(p_1 + \vartheta h, q_1 + \vartheta k)h + \\ &\quad + f_y(p_1 + \vartheta h, q_1 + \vartheta k)k \\ &= f(p_1, q_1) \end{aligned}$$

Here we used the fact that $f_x = f_y = 0$ at each point of D to conclude that $f(\underline{P}_1) = f(\underline{P}_2)$. In the same way we can show $f(\underline{P}_1) = f(\underline{P}_2) = \cdots = f(\underline{P}_n)$ and in general for any two points $\underline{P}$ and $\underline{Q}$ in D we have $f(\underline{P}) = f(\underline{Q})$; f is constant on D.

PROBLEM 8.11

Suppose $u = u(x, y)$ is $\mathbb{C}^1$ in a region D in $\mathbb{R}^2$. Then use Taylor's theorem to show that for any point (x, y) in D and for $h > 0$ sufficiently small that $(x \pm h, y)$ is in D we have

$$\left| \frac{\partial u}{\partial x} - \frac{u(x + h, y) - u(x - h, y)}{2h} \right| \le Ch^2 \quad \text{for some } C > 0. \tag{1}$$

SOLUTION 8.11

According to Theorem 8.6, for u in $\mathbb{C}^3$

$$u(x + h, y) = u(x, y) + u_x(x, y)h + u_{xx}(x, y)\frac{h^2}{2} + u_{xxx}(x + \lambda h, y)\frac{h^3}{6} \tag{2}$$

and

$$u(x - h, y) = u(x, y) - u_x(x, y)h + u_{xx}(x, y)\frac{h^2}{2} - u_{xxx}(x - \mu h, y)\frac{h^3}{6} \tag{3}$$

for $0 < \lambda, \mu < 1$. Subtracting (3) from (2) and solving for u_x leads to

$$u_x(x, y) - \frac{u(x + h, y) - u(x - h, y)}{2h} = -(u_{xxx}(x + \lambda h, y) + u_{xxx}(x - \mu h, y))\frac{h^2}{12}$$

For u in $\mathbb{C}^3$ it follows that there exists a positive constant C such that

$$|\, u_{xxx}(x + \lambda h, y) + u\,(x - \mu h, y) \,| \le 12\,C$$

PROBLEM 8.12

Suppose $u = u(x, y)$ is $\mathbb{C}^4$ in a region D in $\mathbb{R}^2$. Then use Taylor's theorem to show that for any point (x, y) in D and for $h > 0$ sufficiently small that $(x \pm h, y \pm h)$ is in D we have

$$\left| \frac{\partial^2 u}{\partial x^2} - \frac{u(x+h, y) - 2u(x, y) + u(x-h, y)}{h^2} \right| < C_1 h^2 \quad \text{for some } C_1 > 0 \ (1)$$

and

$$\left| \frac{\partial^2 u}{\partial y^2} - \frac{u(x, y+h) - 2u(x, y) + u(x, y-h)}{h^2} \right| < C_2 h^2 \quad \text{for some } C_2 > 0 \ (2)$$

hence for $C_3 = C_1 + C_2$

$$| u_{xx} + u_{yy} - h^{-2}(u(x, y+h) + u(x+h, y) - 4u(x, y) + \\ + u(x, y-h) + u(x-h, y)) | < C_3 h^2 \qquad (3)$$

SOLUTION 8.12

According to Taylor's theorem, for u in $\mathbb{C}^4$

$$u(x+h, y) = u(x, y) + u_x(x, y)h + \\ + u_{xx}(x, y)\frac{h^2}{2} + u_{xxx}(x, y)\frac{h^3}{6} + u_{xxxx}(x + \lambda h, y)\frac{h^4}{24}$$

$$u(x-h, y) = u(x, y) - u_x(x, y)h + \\ + u_{xx}(x, y)\frac{h^2}{2} - u_{xxx}(x, y)\frac{h^3}{6} + u_{xxxx}(x - \mu h, y)\frac{h^4}{24}$$

Adding these two equations and solving for $u_{xx}(x, y)$ leads to

$$u_{xx}(x, y) - \frac{u(x+h, y) - 2u(x, y) + u(x-h, y)}{h^2} = \\ = -(u_{xxxx}(x + \lambda h, y) + u_{xxxx}(x - \mu h, y))\frac{h^2}{12}$$

Then for u in $\mathbb{C}^4$ (1) follows with

$$| u_{xxxx}(x + \lambda h, y) + u_{xxxx}(x - \mu h, y) | < 12C_1$$

A similar argument leads to

$$u_{yy}(x, y) - \frac{u(x, y+h) - 2u(x, y) + u(x, y-h)}{h^2} = \\ = -(u_{yyyy}(x, y + \nu h) + u_{yyyy}(x, y - \eta h))\frac{h^2}{12}$$

and this yields (2). Then (1) and (2) together imply (3). The result (3) says that the *local truncation error* for the centered difference approximation for the Laplacian on a square grid is of order h^2.

PROBLEM 8.13 TAYLOR'S THEOREM IN N VARIABLES

State a version of Taylor's theorem, Theorem 8.6, for $f = f(x_1, \ldots, x_n)$.

SOLUTION 8.13

We first must introduce some additional notation for a function $f = f(x_1, \ldots, x_n)$ that is $\mathbb{C}^m$ in a domain D in $\mathbb{R}^n$. For

$$\underline{x} = (x_1, \ldots, x_n) \quad \text{and} \quad \underline{h} = (h_1, \ldots, h_n)$$

Let

$$df(\underline{x};\, \underline{h}) = \sum_{i=1}^{n} \frac{\partial f}{\partial x_i}(\underline{x})h_i$$

$$d^2f(\underline{x};\, \underline{h}) = \sum_{i=1}^{n}\sum_{j=1}^{n} \frac{\partial f}{\partial x_i \partial x_j}(\underline{x})h_i h_j$$

$$d^3f(\underline{x};\, \underline{h}) = \sum_{i=1}^{n}\sum_{j=1}^{n}\sum_{k=1}^{n} \frac{\partial f}{\partial x_i \partial x_j \partial x_k}(\underline{x})h_i h_j h_k$$

with similar definitions up to $d^m f(\underline{x};\underline{h})$. Then we can state the following version of Taylor's theorem for functions of n variables.

Theorem 8.12 Suppose $f = f(x_1, \ldots, x_n)$ belongs to $\mathbb{C}^m(D)$ in some domain D in $\mathbb{R}^n$. Then for every $\underline{x} = (x_1, \ldots, x_n)$ and $\underline{h} = (h_1, \ldots, h_n)$ such that $\underline{x}$ and $\underline{x} + \underline{h}$ are in D and the line segment joining $\underline{x}$ to $\underline{x} + \underline{h}$ lies within D, there exists some λ, $0 < \lambda < 1$, such that

$$f(\underline{x} + \underline{h}) = f(\underline{x}) + df(\underline{x};\underline{h}) + \frac{1}{2!}d^2f(\underline{x};\underline{h}) + \frac{1}{3!}d^3f(x;\, h) + \cdots + \frac{1}{m!}d^m f(x + \lambda\underline{h}{:}\underline{h})$$

Implicit Differentiation

PROBLEM 8.14

Let $F = F(p, V, T)$ denote a $\mathbb{C}^1$ function of three variables. Suppose the equation

$$F(p, V, T) = 0 \tag{1}$$

defines implicit functions $p = p(V, T)$, $V = V(p, T)$ and $T = T(p, V)$. Show that

$$\frac{\partial p}{\partial V}\frac{\partial V}{\partial T}\frac{\partial T}{\partial p} = -1\,. \tag{2}$$

SOLUTION 8.14

It follows from (1) that $F(p(V, T), V, T) = 0$ for all V, T and so by the chain rule

$$F_p \frac{\partial p}{\partial V} + F_V = 0.$$

Then at each point where F_p does not vanish,

$$\frac{\partial p}{\partial V} = -\frac{F_V}{F_p}.$$

Similarly,

$$\frac{\partial V}{\partial T} = -\frac{F_T}{F_V} \quad \text{and} \quad \frac{\partial T}{\partial p} = -\frac{F_P}{F_T}.$$

Then

$$\frac{\partial p}{\partial V}\frac{\partial V}{\partial T}\frac{\partial T}{\partial p} = \left(-\frac{F_V}{F_p}\right)\left(-\frac{F_T}{F_V}\right)\left(-\frac{F_P}{F_T}\right) = -1.$$

This example illustrates that the symbols ∂p, ∂V, ∂T cannot be cancelled as if they were numbers in fractions.

PROBLEM 8.15

Find the set of points where the equations

$$F(x, y, z, u) = x^2 + y^2 + z^2 + 4u^2 - 24 = 0 \quad (1)$$

$$G(x, y, z, u) = u^3 - 4xyz - 100 = 0 \quad (2)$$

determine implicit functions $x = x(u, z)$ and $y = y(u, z)$ and compute the partial derivatives x_u, x_z, y_u and y_z of these functions.

SOLUTION 8.15

Since we have two equations and four variables, it is appropriate to treat two of the variables as dependent variables and treat the other two as independent variables. The problem calls for x and y to be dependent and for u and z to be independent variables. In order to compute the partial derivatives of x and y with respect to u, we hold z constant and differentiate (1) and (2) with respect to u. By the chain rule, we find

$$F_x x_u + F_y y_u + F_u = 0$$

$$G_x x_u + G_y y_u + G_u = 0.$$

We can solve these equations for x_u and y_u at each point where the Jacobian determinant

$$J = \frac{\partial(F, G)}{\partial(x, y)} = \begin{vmatrix} F_x & F_y \\ G_x & G_y \end{vmatrix} = \begin{vmatrix} 2x & 2y \\ 4yz & 4xz \end{vmatrix} = 8z(x^2 - y^2)$$

is different from zero. At such points we have

$$x_u = \frac{\partial(F, G)}{\partial(u, y)} /J = 1/J \begin{vmatrix} -F_u & F_y \\ -G_u & G_y \end{vmatrix} = 1/J \begin{vmatrix} -8u & 2y \\ -3u^2 & xz \end{vmatrix}$$

$$= -\frac{u(4xz - 3yu)}{4z(x^2 - y^2)}$$

Similarly,

$$y_u = -\frac{\partial(F, G)}{\partial(x, u)} /J = -\frac{u(x - 2yz)}{4z(x^2 - y^2)}.$$

To compute x_z and y_z we differentiate (1) and (2) with respect to z, holding x constant. This produces

$$F_x x_z + F_y y_z + F_z = 0$$

$$G_x x_z + G_y y_z + G_z = 0$$

and

$$x_z = -\frac{\partial(F, G)}{\partial(z, y)} /J, \qquad y_z = -\frac{\partial(F, G)}{\partial(x, z)} /J$$

$$= -\frac{x(z^2 + y^2)}{4z(x^2 - y^2)} \qquad = -\frac{y(x^2 - z^2)}{z(x^2 - y^2)}$$

PROBLEM 8.16

Suppose that F is a $\mathbb{C}^2$ function of three variables and that $F(x, y, z) = 0$ in $\mathbb{R}^3$. Suppose also that $f(x, y)$ is a $\mathbb{C}^2$ function of two variables such that

$$F(x, y, f(x, y)) = 0 \quad \text{for all } (x, y) \text{in some D in } \mathbb{R}^3.$$

Then compute the following second order derivatives of the implicitly defined function $z = f(x, y)$: f_{xx}, f_{xy}, f_{yy}, f_{yx}.

SOLUTION 8.16

According to Example 8.2(b), we have

$$f_x = -F_x/F_z \quad \text{and} \quad f_y = -F_y/F_z.$$

Then since $F_x = F_x(x, y, f(x, y))$ etc., we can use the chain rule to compute

$$f_{xx} = -\left(\frac{(F_{xx} + F_{xz}z_x)F_z - F_x(F_{zx} + F_{zz}z_x)}{(F_z)^2}\right)$$

$$f_{xy} = -\left(\frac{(F_{xy} + F_{xz}z_y)F_z - F_x(F_{zy} + F_{zz}z_y)}{(F_z)^2}\right)$$

$$f_{yy} = -\left(\frac{(F_{yy} + F_{yz}z_y)F_z - F_y(F_{zy} + F_{zz}z_y)}{(F_z)^2}\right)$$

$$f_{yx} = -\left(\frac{(F_{yx} + F_{yz}z_x)F_z - F_y(F_{zx} + F_{zz}z_x)}{(F_z)^2}\right)$$

Note that if we substitute the expression for z_y into f_{xy} and the expression for z_x into f_{yx}, we obtain

$$f_{xy} = -(F_{xy}F_z - F_yF_{xz} - F_xF_{zy} + F_xF_yF_{zz}/F_z)/(F_z)^2$$

$$f_{yx} = -(F_{yx}F_z - F_xF_{yz} - F_yF_{zx} + F_xF_yF_{zz}/F_z)/(F_z)^2.$$

For F in $\mathbb{C}^2$ we have $F_{xy} = F_{yx}$, $F_{xz} = F_{zx}$ and $F_{zy} = F_{yz}$ and it follows that $f_{xy} = f_{yx}$.

PROBLEM 8.17

At all points where the function

$$x^2 + y^4 + z^6 = 1$$

determines z implicitly as a function of x and y, compute z_{xx} and z_{xy}.

SOLUTION 8.17

We compute

$$2x + 6z^5z_x = 0$$

or

$$z_x = -\frac{x}{3z^5} \quad \text{for } z \neq 0.$$

Then

$$z_{xx} = -\left(\frac{1 \cdot 3z^5 - x(15z^4z_x)}{9z^{10}}\right) = -\left(\frac{3z^6 + 5x^2}{9z^{11}}\right) \quad z \neq 0$$

and

$$z_{xy} = \frac{5x}{3z^6}z_y = -\frac{10xy^3}{9z^{11}} \quad z \neq 0$$

Jacobians and Transformations

PROBLEM 8.18

Consider the transformation from $\mathbb{R}^2$ into $\mathbb{R}^2$ given by

$$T: \begin{array}{l} x = x(u, v) \\ y = y(u, v) \end{array} \qquad (1)$$

with Jacobian

$$J = \frac{\partial(x, y)}{\partial(u, v)} \neq 0 \quad \text{for all } (u, v) \in \mathbb{R}^2.$$

Show that the Jacobian j of the inverse transformation

$$T^{-1}: \quad \begin{aligned} u &= u(x, y) \\ v &= v(x, y) \end{aligned} \tag{2}$$

satsifies $j = 1/J$; i.e., the Jacobian of T^{-1} is J^{-1}.

SOLUTION 8.18

The Jacobian J is given by

$$J = \frac{\partial(x, y)}{\partial(u, v)} = x_u y_v - x_v y_u \tag{3}$$

and the Jacobian j of the inverse transformation is

$$j = \frac{\partial(u, v)}{\partial(x, y)} = u_x v_y - u_y v_x \tag{4}$$

But by differentiating (1) with respect to x we find

$$1 = x_u u_x + x_v v_x$$

$$0 = y_u u_x + y_v v_x$$

and solving for u_x and v_x by Cramer's rule produces

$$u_x = \frac{\begin{vmatrix} 1 & x_v \\ 0 & y_v \end{vmatrix}}{J} = \frac{y_v}{J} \qquad v_x = \frac{\begin{vmatrix} x_u & 1 \\ y_u & 0 \end{vmatrix}}{J} = \frac{-y_u}{J} \tag{5}$$

Similarly, if we differentiate (1) with respect to y and solve for u_y and v_y we find

$$u_y = -\frac{x_v}{J} \quad \text{and} \quad v_y = \frac{x_u}{J} \tag{6}$$

Now using (5) and (6) in (4) leads to

$$j = \left(\frac{y_v}{J}\frac{x_u}{J} - \frac{x_v}{J}\frac{y_u}{J}\right) = \frac{x_u y_v - x_v y_u}{J^2} = \frac{1}{J}$$

At each point where the Jacobian J is different from zero, the transformation (1) is invertible. We have shown that the inverse transformation (2) has Jacobian j equal to the reciprocal of J. This result extends to transformations from $\mathbb{R}^n$ into $\mathbb{R}^n$.

PROBLEM 8.19

Compute the Jacobians for the transformations

(a) $x = r \cos \vartheta \quad 0 \le r, 0 \le \vartheta < 2\pi$
$y = r \sin \vartheta$

(b) $x = \rho \sin \varphi \cos \vartheta$
$y = \rho \sin \varphi \sin \vartheta \quad 0 < r, 0 < i < 2p, 0 < f < p$
$z = \rho \cos \varphi$

Are these transformations one to one in a neighborhood of every point?

SOLUTION 8.19

8.18 (a) For transformation (a)

$$J = \frac{\partial(x, y)}{\partial(r, \vartheta)} = \begin{vmatrix} \cos \vartheta & -r \sin \vartheta \\ \sin \vartheta & r \cos \vartheta \end{vmatrix} = r$$

This transformation is one to one near every point except the point $r = 0$, the origin, where $J = 0$. The inverse for transformation (a) is the transformation

$$r = \sqrt{x + y} \quad \text{and} \quad \vartheta = \text{Arctan}\,\frac{y}{x}.$$

As (x, y) approaches $(0, 0)$ along the line $y = kx$, then (r, ϑ) tends to $(0, \text{Arctan } k)$. Then $x = 0$, $y = 0$ corresponds to $r = 0$, $\vartheta = \text{Arctan } k$ for all k and the transformation is not one to one near this point.

8.18(b) For transformation (b)

$$J = \frac{\partial(x, y, z)}{\partial(\rho, \vartheta, \varphi)} = \begin{vmatrix} \sin \vartheta \cos \varphi & \rho \cos \vartheta \cos \varphi & -\rho \sin \vartheta \sin \varphi \\ \sin \vartheta \sin \varphi & \rho \cos \vartheta \sin \varphi & \rho \sin \vartheta \cos \varphi \\ \cos \vartheta & -\rho \sin & 0 \end{vmatrix}$$
$$= r^2 \sin \varphi$$

This transformation is one to one near every point except the origin and points on the z axis. The inverse for transformation (b) is the transformation

$$\rho = \sqrt{x^2 + y^2 + z^2}, \quad \vartheta = \text{Arctan}\,\frac{y}{x}, \quad \varphi = \text{Arctan}\,\frac{\sqrt{x^2 + y^2}}{z}$$

Then as (x, y, z) approaches $(0, 0, 0)$ along the line $x = t$, $y = \alpha t$, $z = \beta t$, $0 \le t$, $(\rho, \vartheta, \varphi)$ tends to the point $(0, \text{Arctan } \sqrt{1 + \alpha^2}/\beta, \text{Arctan } \alpha/\beta)$. Similarly, as (x, y, z) moves along the line $x = \alpha t$, $y = \beta t$, $z = z_0$ toward a point $(0, 0, z_0)$ on the z axis, $(\rho, \vartheta, \varphi)$ tends to the point $(z_0, \text{Arctan } \beta/\alpha, \text{Arctan } \sqrt{\alpha^2 + \beta^2}/z_0)$. Thus transformation (b) is not one to one near the origin nor near any point on the z axis.

Applications of Partial Derivatives

PROBLEM 8.20 LEIBNIZ' RULE

Let g(x) be a $\mathbb{C}^1$ function of one variable and for fixed constant a, define

$$u(x, t) = \int_{x-at}^{x+at} g(s)\, ds$$

Show that u(x, t) satisfies

$$\frac{\partial^2 u}{\partial t^2} = a^2 \frac{\partial^2 u}{\partial x^2} \tag{1}$$

SOLUTION 8.20

Let $\varphi(x, t) = x + at$ and $\psi(x, t) = x - at$. Then

$$u(x, t) = \int_{\psi(x,t)}^{\varphi(x,t)} g(s)\, ds$$

and according to Leibniz' rule,

$$\begin{aligned} \partial_x u(x, t) &= g(\varphi(x, t))\varphi_x(x, t) - g(\psi(x, t))\psi_x(x, t) \\ &= g(\varphi(x, t)) \cdot 1 - g(\psi(x, t)) \cdot 1 \end{aligned}$$

$$\begin{aligned} \partial_t u(x, t) &= g(\varphi(x, t))\varphi_t(x, t) - g(\psi(x, t))\psi_t(x, t) \\ &= g(\varphi(x, t)) \cdot a - g(\psi(x, t)) \cdot (-a). \end{aligned}$$

Differentiating once again, using the chain rule

$$\begin{aligned} \partial_{xx} u(x, t) &= g'(\varphi(x, t))\varphi_x(x, t) - g'(\psi(x, t))\psi_x(x, t) \\ &= g'(\varphi(x, t)) \cdot 1 - g'(\psi(x, t)) \cdot 1 \end{aligned}$$

$$\begin{aligned} \partial_{tt} u(x, t) &= ag'(\varphi(x, t))\varphi_t(x, t) + ag(\psi(x, t))\psi_t(x, t) \\ &= ag(\varphi(x, t)) \cdot a + g(\psi(x, t) \cdot (-a) = a^2 \partial_{xx} u(x, t) \end{aligned}$$

we see at once that u(x, t) satisfies (1).

PROBLEM 8.21 LEIBNIZ' RULE

Compute F′(x) and G′(x) if

$$F(x) = \int_{\varphi(x)}^{\psi(x)} \frac{e^{-x(1+t^2)}}{1 + t^2}\, dt \quad \text{and} \quad G(x) = \int_0^x e^{-\gamma(x)z^2}\, dz\,.$$

SOLUTION 8.21

Applying Leibniz' rule to F(x) leads to

$$F'(x) = \frac{e^{-x(1+\psi(x)^2)}}{1 + \psi(x)^2}\, \psi'(x) - \frac{e^{-x(1+\varphi(x)^2)}}{1 + \varphi(x)^2}\, \varphi'(x) - \int_{\varphi(x)}^{\psi(x)} e^{-x(1+t^2)}\, dt$$

For G(x) we find

$$G'(x) = e^{-x^2\gamma(x)} - \int_0^x z^2\gamma'(z)e^{-\gamma(x)z^2}dz.$$

PROBLEM 8.22 TANGENTS AND NORMALS

Show that the spherical surface $F(x, y, z) = x^2 + y^2 + z^2 - R^2 = 0$ contains both of the circles

$$C_1: \begin{cases} x = R \operatorname{Cos} t \\ y = R \operatorname{Sin} t \\ z = 0 \end{cases} \quad 0 \le t \le 2\pi \qquad C_2: \begin{cases} x = R \operatorname{Cos} t \\ y = 0 \\ z = R \operatorname{Sin} t \end{cases} \quad 0 \le t \le 2\pi$$

For $k = 1, 2$ if $\underline{T}_k(t)$ denotes the vector that is tangent to C_k at the point $(x(t), y(t), z(t))$, then show that the vector grad $F(t)$ is normal to $\underline{T}_k$ at each point of C_k. In particular the point $(R, 0, 0)$ is on both C_1 and C_2. Conclude that **grad** F is normal to the surface of the sphere at this point. Conclude more generally that **grad** F is normal to the surface of the sphere at each of its points.

SOLUTION 8.22

Since

$$F(R \operatorname{Cos} t, R\operatorname{Sin} t, 0) = R^2\operatorname{Cos}^2 t + R^2\operatorname{Sin}^2 t + 0^2 - R^2 = 0$$

is satisfied identically for all t in $(0, 2\pi)$ it follows that the circle C_1 lies in the spherical surface $F(x, y, z) = 0$. In fact, C_1 is the equator of the sphere. Likewise, the circle C_2 lies in the sphere and contains the north and south poles of the sphere. C_1 and C_2 are called great circles on the sphere since they have the same center as the sphere. The vectors

$$\underline{T}_1(t) = -R \operatorname{Sin} t\, \underline{i} + R \operatorname{Cos} t\, \underline{j} + 0\underline{k}$$

and

$$\underline{T}_2(t) = -R \operatorname{Sin} t\, \underline{i} + 0\underline{j} + R \operatorname{Cos} t\, \underline{k}$$

are tangent to the circles C_1 and C_2, respectively, at each of their points. Note that $\underline{T}_1 \cdot \underline{T}_2 = 0$ at $t = 0$ which is a reflection of the fact that C_1 intersects C_2 at right angles at the point $(R, 0, 0)$ which corresponds to the value $t = 0$ on both circles.

Applying the chain rule to $F(t) = F(x(t), y(t), z(t))$ yields

$$\begin{aligned} F'(t) &= F_x x'(t) + F_y y'(t) + F_z z'(t) \\ &= 2x(t)x'(t) + 2y(t)y'(t) + 2z(t)z'(t) \\ &= (2x(t), 2y(t), 2z(t)) \cdot (x'(t), y'(t), z'(t)). \end{aligned}$$

For both C_1 and C_2 this reduces to

$$F'(t) = 2R^2(-\operatorname{Cos} t \operatorname{Sin} t + \operatorname{Sin} t \operatorname{Cos} t) = 0 \quad \text{for } 0 \le t \le 2\pi.$$

This implies that the vector **grad** $F(t) = (F_x, F_y, F_z)(t) = (2x(t), 2y(t), 2z(t))$ is orthogonal to $\underline{T}_1(t)$ at each point of C_1 and is orthogonal to $\underline{T}_2(t)$ at each point of C_2. In particular, at $t = 0$ **grad** $F(0)$ is orthogonal to both $\underline{T}_1(0)$ and to $\underline{T}_2(0)$. Since $\underline{T}_1(0)$ and $\underline{T}_2(0)$ are each tangent to the surface of the sphere and are orthogonal to one another, it follows that **grad** $F(0)$ must be perpendicular to the plane of $\underline{T}_1(0)$ and $\underline{T}_2(0)$; i.e., **grad** $F(0)$ is perpendicular to the plane that is tangent to the sphere at $(R, 0, 0)$.

More generally we can show in the same way that **grad** $F(t)$ is orthogonal to the tangent vector to every great circle in the sphere. Through each point on a great circle there passes another great circle intersecting the first at right angles. Then **grad** $F(t)$ is orthogonal to both great circles at this point of intersection and it follows that grad F is normal to the surface of the sphere at that point.

We have just shown a special case of the following more general fact: at each point of the surface $F(x, y, z) = 0$, the vector **grad** F has the direction of the normal to the surface.

PROBLEM 8.23 THE DIRECTIONAL DERIVATIVE

Suppose $f = f(x, y, z)$ is in $\mathbb{C}^1$ (D) for some domain D in $\mathbb{R}^3$ and that $P = (a, b, c)$ is in D. Then show that $\nabla_u f(P)$, the directional derivative of f at P in the direction of the unit vector $u = (\text{Cos}\,\alpha, \text{Cos}\,\beta, \text{Cos}\,\gamma)$, is equal to **grad** $F(P) \cdot u$.

SOLUTION 8.23

By definition of the directional derivative

$$\nabla_u f(a, b, c) = \lim_{h\to 0} \frac{f(a + h\,\text{Cos}\,\alpha, b + h\,\text{Cos}\,\beta, c + h\,\text{Cos}\,\gamma) - f(a, b, c)}{h}$$

But Theorem 8.2 rephrased for a function of three variables implies that for some $\lambda_1, \lambda_2, \lambda_3$, $0 < \lambda_1, \lambda_2, \lambda_3 < 1$,

$$\begin{aligned} &f(a + h\text{Cos}\alpha, b + h\text{Cos}\beta, c + h\text{Cos}\gamma) - f(a, b, c) \\ &= f_x(a + \lambda_1 h\,\text{Cos}\,\alpha, b, c)h\,\text{Cos}\,\alpha \\ &+ f_y(a + h, b + \lambda_2 h\,\text{Cos}\,\beta, c)h\,\text{Cos}\,\beta + f_y(a + h, b + h, c + \lambda_3 h\,\text{Cos}\gamma)h\,\text{Cos}\,\gamma \end{aligned}$$

Then dividing by h and letting h tend to zero leads to

$$\begin{aligned} \nabla_u f(a, b, c) &= f_x(a, b, c)\text{Cos}\alpha + f_y(a, b, c)\text{Cos}\beta + f_z(a, b, c)\text{Cos}\gamma \\ &= \textbf{grad}\, f(a, b, c).(\text{Cos}\alpha, \text{Cos}\beta, \text{Cos}\gamma) \end{aligned}$$

Note that the formula, $\nabla_u f(\underline{P}) = \textbf{grad}\, f(\underline{P}) \cdot \underline{u}$ implies that if f has first partials at $\underline{P}$ with respect to each independent variable then f has a directional derivative at $\underline{P}$ in *every* direction $\underline{u}$. But we have already seen in Problem 8.2 an example of a function f that has first partials at $\underline{P}$ with respect to each independent variable and yet f is not continuous at $\underline{P}$. Thus for a function f of several variables, the existence of a directional derivative at $\underline{P}$ for every direction $\underline{u}$ is not sufficient to imply continuity of f at $\underline{P}$.

PROBLEM 8.24

Show that the C^1 function $f = f(x, y, z)$ increases most rapidly in the direction of grad f and decreases most rapidly in the opposite direction, −grad f.

SOLUTION 8.24

By the definition of the directional derivative, $\nabla_u f$ equals the rate of change of f in the direction of the unit vector $\underline{u}$. By the result of the previous problem, this equals the dot product, **grad** $f \cdot \underline{u}$. But by a well known property of the dot product,

$$\mathbf{grad}\, f \cdot \underline{u} = \|\mathbf{grad}\, f\| \, \|\underline{u}\| \operatorname{Cos} \vartheta,$$

where ϑ is the angle between the vectors **grad** f and $\underline{u}$. But Cos ϑ assumes its maximum value 1 for $\vartheta = 0$ and assumes its minimum value −1 for $\vartheta = \pi$. Thus $\nabla_u f$ assumes its maximum value when the unit vector $\underline{u}$ has the direction **grad** f and assumes its minimum value when $\underline{u}$ has the direction of the vector −**grad** f.

PROBLEM 8.25

Suppose that $f = f(x, y)$ is differentiable in a neighborhood of the point (a, b). If (a, b) is a relative extreme point for f then show that $f_x(a, b) = f_y(a, b) = 0$.

SOLUTION 8.25

Note first that if f is differentiable in a neighborhood of (a, b) and (a, b) is a relative extreme point for f then we are dealing with an interior extreme point and not a *boundary* extreme point. The derivatives of f need not vanish at an extreme point that occurs on the boundary of the domain of f.

If f(x, y) is differentiable in a neighborhood of (a, b) then $\varphi(x) = f(x, b)$ is a differentiable function of the single variable x in a neighborhood of $x = a$. Moreover, by its definition it is clear that $\varphi(x)$ has a relative extreme point at $x = a$. Then by Theorem 4.6 it follows that $\varphi'(x) = f_x(x, b)$ must vanish at $x = a$. Similarly, $\psi(y) = f(a, y)$ is a differentiable function of y in a neighborhood of b. Since $\psi(y)$ has an interior extreme point at $y = b$, it follows that $\psi'(b) = f_y(a, b) = 0$. This is Theorem 8.10 in the special case $n = 2$.

PROBLEM 8.26 SECOND DERIVATIVE TEST FOR EXTREMA

State and prove an n-dimensional version of Theorem 8.11

SOLUTION 8.26

We shall use the notation of Problem 8.13. In particular we use the notation

$$d^2f(\underline{x}; \underline{h}) = \sum_{i=1}^{n} \sum_{j=1}^{n} \frac{\partial^2 f}{\partial x_i \partial x_j}(\underline{x})\, h_i h_j$$

for the second derivative of $f = f(x_1, \ldots, x_n)$. Then we can state the following theorem.

Theorem 8.13 Suppose that $f = f(x_1, \ldots, x_n)$ is C^2 in a neighborhood of a point $\underline{c} = (c_1, \ldots, c_n)$ in $\mathbb{R}^n$ where we have

$$f_{x_1}(\underline{c}) = \cdots = f_{x_n}(\underline{c}) = 0.$$

Suppose that for all $\underline{h}$ in $\mathbb{R}^n$ such that $h \neq 0$:

(a) $d^2f(\underline{c};\underline{h}) > 0$ then f has a relative minimum at $\underline{x} = \underline{c}$

(b) $d^2f(\underline{c};\underline{h}) < 0$ then f has a relative maximum at $\underline{x} = \underline{c}$

(c) if $d^2f(\underline{c};\underline{h})$ assumes both positive and negative values for $\underline{h}$ in $\mathbb{R}^n$ then $\underline{x} = \underline{c}$ is neither a maximum nor a minimum but is called a saddle point for f.

Note that $d^2f(\underline{c};\underline{h})$ can be expressed in matrix notation as $d^2f(\underline{c};\underline{h}) = \underline{h}^T A\, \underline{h}$ where

$$A_{ij} = (a_{ij}) \text{ with } a_{ij} = \frac{\partial^2 f}{\partial x_i \partial x_j}(\underline{c})$$

denotes the n by n matrix of second derivatives of f evaluated at $\underline{c}$. Note that since f is in C^2, $a_{ij} = a_{ji}$; i.e., A is symmetric. Any matrix A with the property:

$\underline{h}^T A\, \underline{h} > 0$ for all $\underline{h} \neq \underline{0}$ in $\mathbb{R}^n$ is said to be *positive definite*

$\underline{h}^T A\, \underline{h} < 0$ for all $\underline{h} \neq \underline{0}$ in $\mathbb{R}^n$ is said to be *negative definite*

Then the critical point $\underline{x} = \underline{c}$ is a:

relative maximum for f if A is negative definite at $\underline{x} = \underline{c}$

relative minimum for f if A is positive definite at $\underline{x} = \underline{c}$

Proof of Theorem 8.13(a) By assumption $d^2f(\underline{x};\, \underline{h}) > 0$ for all h in $\mathbb{R}^n$ such that $\|\underline{h}\| = 1$. Since f is in C^2 and the set $\{\underline{h} \in \mathbb{R}^n : \|h\| = 1\}$ is compact, it follows that there exists a positive constant μ such that

$$d^2f(\underline{c};\, \underline{h}) \geq \mu \quad \text{for } \|\underline{h}\| = 1,$$

and there exists a second positive constant δ such that

$$d^2f(\underline{z};\, \underline{h}) \geq \mu/2 \quad \text{for } \|\underline{h}\| = 1, \quad \text{and } \|\underline{z} - \underline{c}\| < \delta.$$

Theorem 8.12 implies that for each ϑ, $0 \leq \vartheta \leq 1$, there exists a λ, $0 < \lambda < 1$, such that

$$f(\underline{c} + \vartheta\underline{h}) = f(\underline{c}) + df(\underline{c};\, \vartheta\underline{h}) + \frac{1}{2!}\, d^2f(\underline{c} + \lambda\vartheta\underline{h};\, \vartheta\underline{h})$$

Since $\underline{c}$ is a critical point for f, $df(\underline{c};\, \vartheta\underline{h}) = 0$. Then for $\|\underline{h}\| = 1$ and $0 < \vartheta < d$ we have

$$f(\underline{c} + \vartheta\underline{h}) - f(\underline{c}) = \frac{\vartheta^2}{2!}\, d^2f(\underline{c} + \lambda\vartheta\underline{h};\, \underline{h}) > \frac{\mu\vartheta^2}{4}$$

and it follows that $\underline{c}$ is a relative minimum for f. The proof of (b) is similar.

To prove part (c), note that by assumption there exist unit vectors $\underline{h}$ and $\underline{k}$ in $\mathbb{R}^n$ such that

$$d^2f(\underline{c}; \underline{h}) > 0 \quad \text{and} \quad d^2f(\underline{c}; \underline{k}) < 0.$$

Then Theorem 8.12 can be used again to show that for ϑ sufficiently small and positive,

$$f(\underline{c} + \vartheta\underline{h}) > f(\underline{c}) \quad \text{and} \quad f(\underline{c} + \vartheta\underline{k}) < f(\underline{c}).$$

Then $\underline{c}$ is neither a relative maximum nor a relative minimum for f. We say that $\underline{c}$ is a *saddle point.*

In the special case n = 2, we have

$$d^2f(\underline{c}; \underline{h}) = \underline{h}^T \begin{bmatrix} f_{xx}(c) & f_{xy}(c) \\ f_{xy}(\underline{c}) & f_{yy}(\underline{c}) \end{bmatrix} \underline{h}$$

and this matrix can be shown to be:

negative definite if $f_{xx}f_{yy} - f_{xy}{}^2 > 0$ and $f_{xx} < 0, f_{yy} < 0$ at $\underline{c}$

positive definite if $f_{xx}f_{yy} - f_{xy}{}^2 > 0$ and $f_{xx} > 0, f_{yy} > 0$ at $\underline{c}$

indefinite if $f_{xx}f_{yy} - f_{xy}{}^2 < 0$ at $\underline{c}$

Then Theorem 8.11 is seen to be a special case of Theorem 8.13.

PROBLEM 8.27 QUADRATIC FORMS

A *quadratic form* on $\mathbb{R}^n$ is a function of n variables of the form

$$Q(x) = Q(x_1, \ldots, x_n) = \frac{1}{2}\,\underline{x} \cdot A\,\underline{x} - \underline{b} \cdot \underline{x} + C \tag{1}$$

where A, $\underline{b}$, and C denote respectively, an n by n matrix, a vector in $\mathbb{R}^n$ and a scalar constant. Show that if $A = A^T$ then

$$\textbf{grad}\ Q(x_1, \ldots, x_n) = A\underline{x} - \underline{b} \tag{1}$$

$$d^2Q(\underline{x}; \underline{h}) = \underline{h} \cdot A\,\underline{h} \quad \text{for all } \underline{h} \text{ in } \mathbb{R}^n. \tag{2}$$

SOLUTION 8.27

Recall that if $\underline{C}_j$ denotes the jth column of the matrix A, then for $x = (x_1, \ldots, x_n)$ in $\mathbb{R}^n$

$$A\underline{x} = \underline{x}_1\underline{C}_1 + \cdots + \underline{x}_n\underline{C}_n$$

Recall also that $\underline{x}$ in $\mathbb{R}^n$ can be expressed

$$\underline{x} = x_1\underline{e}_1 + \cdots + x_n\underline{e}_n$$

where $\underline{e}_k$ denotes the kth element of the standard basis for $\mathbb{R}^n$; i.e., $\underline{e}_k$ has a 1 as its kth component while all other components are zeroes.

Then for j = 1, …, n

$$\partial_{x_j}Q(\underline{x}) = \frac{1}{2}\,\partial_{x_j}\underline{x} \cdot A\underline{x} + \frac{1}{2}\underline{x} \cdot \partial_{x_j}Ax - \underline{b} \cdot \partial_{x_j}x + \partial_{x_j}C \tag{3}$$

But C is a constant and for each j,

$$\partial_{x_j}\underline{x} = \partial_{x_j}(x_1\underline{e}_1 + \cdots + x_n\underline{e}_n) = \underline{e}_j \tag{4}$$

$$\partial_{x_j}Ax = \partial_{x_j}(x_1\underline{C}_1 + \cdots + x_n\underline{C}_n) = \underline{C}_j \tag{5}$$

Since $A = A^T$ we have $\underline{C}_j = \underline{R}_j^T$ where $\underline{R}_j$ denote the jth row of A. Then using (4) and (5) in (3) leads to

$$\partial_{x_j}Q(\underline{x}) = \frac{1}{2}(\underline{e}_j \cdot A\underline{x} + \underline{x} \cdot \underline{R}_j) - \underline{e}_j \cdot \underline{b} + 0.$$

But now recall that the product $A\underline{x}$ can also be expressed as

$$A\underline{x} = (\underline{x} \cdot \underline{R}_1, \ldots, \underline{x} \cdot \underline{R}_n)^T = (\underline{x} \cdot \underline{R}_1)\underline{e}_1 + \cdots + (\underline{x} \cdot \underline{R}_n)\underline{e}_n$$

hence $\underline{x} \cdot \underline{R}_j = \underline{e}_j \cdot A\underline{x}$ and

$$\partial_{x_j}Q(\underline{x}) = \frac{1}{2}(\underline{e}_j \cdot A\underline{x} + \underline{x} \cdot \underline{R}_j) - \underline{e}_j \cdot \underline{b}$$

$$= \frac{1}{2}(\underline{e}_j \cdot A\underline{x} + \underline{e}_j \cdot A\underline{x}) - \underline{e}_j \cdot \underline{b} = \underline{e}_j \cdot (A\underline{x} - \underline{b}) \tag{6}$$

Since $\partial_{x_j}Q(\underline{x}) = \underline{e}_j \cdot \textbf{grad}\ Q(\underline{x})$, the result (1) follows. Next, differentiating (6) with respect to x_k and using (5), we see that

$$\partial_{x_k}(\partial_{x_j}Q(\underline{x})) = \underline{e}_j \cdot (\partial_{x_k}Ax) = \underline{e}_j \cdot \underline{C}_k = a_{jk} = a_{kj} = \partial_{x_j}(\partial_{x_k}Q(\underline{x}))$$

Then

$$d^2Q(\underline{x}; \underline{h}) = h^T(Q_{x_jx_k})\underline{h} = \underline{h} \cdot A\,\underline{h}\,.$$

PROBLEM 8.28 LEAST SQUARES SOLUTION OF AN OVERDETERMINED SYSTEM

Let B denote an m by n matrix with m > n = rank B. Then the system $B\underline{u} = \underline{f}$ is a system of m equations in n unknowns. Such a system of linear equations is said to be *overdetermined.* In general, a solution for an overdetermined sytem does not exist. Show that the following are equivalent:

1. $\underline{u}$ in $\mathbb{R}^n$ satisfies $B^TB\underline{u} = B^T\underline{f}$
2. $\| B\underline{u} - \underline{f} \|^2 < \| B\underline{v} - \underline{f} \|^2$ for all $\underline{v} \neq \underline{u}$

SOLUTION 8.28

Note that

$$\| B\underline{u} - \underline{f} \| = (B\underline{u} - \underline{f}) \cdot (B\underline{u} - \underline{f}) = B\underline{u} \cdot B\underline{u} - 2B\underline{u} \cdot \underline{f} + \underline{f} \cdot \underline{f}$$
$$= \underline{u} \cdot B^T B\underline{u} - 2\underline{u} \cdot B^T\underline{f} + \| \underline{f} \|^2$$

Then by the result of the previous problem, for

$$Q(\underline{u}) = \frac{1}{2} \| B\underline{u} - \underline{f} \|^2$$

$$\textbf{grad } Q(\underline{u}) = B^T B\underline{u} - B^T\underline{f}$$

$$d^2Q(\underline{u}; \underline{h}) = \underline{h}B^T B\underline{h}.$$

Note that the n by n matrix B^TB satisfies

$$(B^TB)^T = B^TB \quad \text{and} \quad \underline{h} \cdot B^TB\underline{h} = B\underline{h} \cdot B\underline{h} = \| B\underline{h} \|^2 > 0.$$

Since we have assumed rank B = n, $B\underline{h} = \underline{0}$ if and only if $\underline{h} = \underline{0}$. Thus B^TB is symmetric and positive definite.

Now if $\underline{u} = (u_1, \ldots, u_n)$ is a minimum for the $\mathbb{C}^2$ function $Q(\underline{u}) = Q(u_1, \ldots, u_n)$ then

$$\textbf{grad } Q(\underline{u}) = \underline{0} \quad \text{and } \underline{u} \text{ satisfies } B^TB\underline{u} = B^T\underline{f}.$$

On the other hand, if $B^TB\underline{u} = B^T\underline{f}$ then $\underline{u}$ is a critical point for $Q(\underline{u})$. But B^TB is positive definite so $d^2Q(\underline{u}; \underline{h}) > 0$ for all $\underline{h}$ in $\mathbb{R}^n$. Then it follows from Theorem 8.13 that $\underline{u}$ is a minimum for $Q(\underline{u})$.

The equations in $B^TB\underline{u} = B^T\underline{f}$ are called the *normal equations* associated with the overdetermined system $B\underline{u} = \underline{f}$ and the solution $\underline{u}$ is called the *least squares solution* for $B\underline{u} = \underline{f}$.

Optimization with Constraints, Lagrange Multipliers

PROBLEM 8.29

Minimize the function $f(x, y, z) = x^2 + y^2 + z^2$ subject to the constraint $g(x, y, z) = x^2 - z^2 - 1 = 0$. This is equivalent to the problem of finding the point on the surface $S = \{g(x, y, z) = 0\}$ that is closest to the origin.

SOLUTION 8.29

We will solve the problem by two methods.

Solution by elimination Here we use the constraint to solve for x in terms of the other variables

$$x^2 = 1 + z^2.$$

Then we can eliminate x^2 from the expression to be minimized

$$f(x(y, z), y, z) = (1 + z^2) + y + z = 1 + y^2 + 2z^2.$$

At a point where f is minimal, we have

$$f_y = 2y = 0 \quad \text{and} \quad f_z = 4z = 0.$$

Then $y = z = 0$, and substituting these values into the constraint equation leads to

$$g(x, 0, 0) = x^2 - 1 = 0$$

which implies $x = +1, -1$. Thus f is minimal at the points (1, 0, 0) and (–1, 0, 0).

Solution by Lagrange Multipliers In this method we first form the function

$$F(x, y, z, \lambda) = f(x, y, z) - \lambda g(x, y, z)$$

and then seek to minimize F as a function of the *four* variables (x, y, z, λ). In the next problem we will provide an explanation of why the method works. In this example we simply apply it.

At a minimum of F we have **grad** $F = \underline{0}$. That is,

$$F_x = f_x - \lambda g_x = 2x - 2\lambda x = 0 \quad \text{or} \quad x(\lambda - 1) = 0$$

$$F_y = f_y - \lambda g_y = 2y = 0 \quad \text{or} \quad y = 0$$

$$F_z = f_z - \lambda g_x = 2z + 2\lambda z = 0 \quad \text{or} \quad z(\lambda - 1) = 0$$

$$F_\lambda = 0 - g(x, y, z) = 0 \quad \text{or} \quad x^2 - z^2 - 1 = 0$$

From the first of these equations we have

$$x^2(\lambda - 1)^2 = 0$$

and then we can use the constraint to eliminate x^2. This produces

$$(z^2 + 1)(\lambda - 1)^2 = 0 \quad \text{or} \quad \lambda = 1.$$

It follows then from $F_z = 0$ that $z = 0$ and hence the constraint implies $x^2 = 1$. This is the same solution we obtained by elimination. However the Lagrange multiplier method may be applied in problems where elimination is not viable. In the next problem we explain the rationale behind the method.

PROBLEM 8.30

Suppose that f and g are $\mathbb{C}^1$ in a domain D in $\mathbb{R}^n$ and that at the point P in D the function f has a relative extreme point subject to the constraint $g(x, y, z) = 0$. If **grad** $g(x) \neq 0$ in a neighborhood of P then prove that there exists a real number λ such that

$$\textbf{grad}\,(f - \lambda g) = \underline{0} \quad \text{at } \underline{P}.$$

i.e.

$$\textbf{grad}\,F(\underline{x}; \lambda) = \underline{0} \quad \text{at } \underline{P}$$

where $F(\underline{x}; \lambda) = f(\underline{x}) - \lambda g(\underline{x})$ is a $\mathbb{C}^1$ function of $(x_1, \ldots, x_n; \lambda)$.

SOLUTION 8.30

For convenience we will give the proof for n = 3. The proof for general n is very similar. Note that the method of Lagrange multipliers provides a means for finding admissible critical points for f (critical points at which the constraints are satisfied). Whether the critical points are extreme points must be checked separately.

Suppose here that f(x, y, z) has a critical point at $\underline{P} = (a, b, c)$ and that the constraint function g(x, y, z) is such that **grad** $g \neq \underline{0}$ in a neighborhood of $\underline{P}$. For the sake of discussion suppose

$$g_z(a, b, c) \neq 0.$$

Then g defines z implicitly as a function of x and y near $\underline{P}$. That is,

$$z = \varphi(x, y) \quad \text{and} \quad g(x, y, \varphi(x, y)) = 0 \quad \text{near } \underline{P}.$$

Since $\underline{P}$ is a critical point for the function f(x, y, f(x, y)) we have at $\underline{P}$

$$f_x + f_z\varphi_x = 0 \tag{1}$$

$$f_y + f_z\varphi_y = 0. \tag{2}$$

In addition, in a neighborhood of $\underline{P}$,

$$g_x + g_z\varphi_x = 0 \tag{3}$$

$$g_y + g_z\varphi_y = 0. \tag{4}$$

From (3) and (4) we have

$$\varphi_x = -\frac{g_x}{g_z} \quad \text{and} \quad \varphi_y = -\frac{g_y}{g_z}$$

and substituting these into (1) and (2) leads to

$$f_x - \frac{g_x}{g_z} f_z = 0 \quad \text{and} \quad f_y - \frac{g_y}{g_z} f_z = 0.$$

Then

$$f_x - g_x \frac{f_z}{g_z} = f_x - \lambda g_x = 0$$

$$f_y - g_y \frac{f_z}{g_z} = f_y - \lambda g_y = 0$$

$$f_z - g_z \frac{f_z}{g_z} = f_z - \lambda g_z = 0$$

for $\lambda = f_z / g_z$. The last of these three equations is an obvious identity.

We have shown that if $\underline{P}$ is a critical point for f subject to the constraint g(x, y, z) = 0, we must have

$$\mathbf{grad}\,(f - \lambda g) = \underline{0}.$$

This condition together with the constraint equation implies $F = f - \lambda g$ satisfies

$$\mathbf{grad}\,F(x, y, z, \lambda) = 0 \quad \text{at } \underline{P}.$$

The Lagrange multiplier method generalizes to problems with more than one constraint. For example, to minimize $f = f(x, y, z)$ subject to the two constraints $g(x, y, z) = 0$ and $h(x, y, z) = 0$, we form the function

$$F(x, y, z, \lambda, \mu) = f(x, y, z) - \lambda g(x, y, z) - \mu h(x, y, z).$$

Then at a relative extreme point for f, we have **grad** $F = \underline{0}$.

PROBLEM 8.31

Minimize the function

$$f(x, y, z) = 2x^2 + 3x + 2xy + 2y + y^2 + 5z + 6xz + 8yz + 4z^2$$

subject to the constraint:

$$g(x, y, z) = y^2 - x - 1.$$

SOLUTION 8.31

We form the function $F(x, y, z, \lambda) = f(x, y, z) - \lambda g(x, y, z)$ and note that **grad** F is zero if and only if

$$\begin{aligned}
&4x + 3 + 2y + 6z - \lambda(-1) = 0\\
&2x + 2 + 2y + 8z - \lambda(2y) = 0\\
&5 + 6x + 8y + 8z - \lambda(0) = 0\\
&-(y^2 - x - 1) = 0.
\end{aligned}$$

We solve the constraint equation for x in terms of y and substitute into the remaining three equations. This leaves

$$\begin{aligned}
&4y^2 + 2y - 1 + 6z + \lambda = 0\\
&2y^2 + 2y + 8z - 2\lambda = 0\\
&6y^2 + 8y - 1 + 8z = 0.
\end{aligned}$$

The third equation can be solved for z in terms of y. Substituting z into the second equation allows that equation to be solved for λ in terms of y. Finally $\lambda(y)$ and $z(y)$ are substituted into the first equation leaving a quadratic equation in y

$$10y^2 + 28y - 4 = 0.$$

Then

$$\begin{aligned}
&y = 2.93, .136\\
&x = y^2 - 1 = 7.58, -.98\\
&z = (1 - 8y - 6y^2)/8 = -9.24, -.1608
\end{aligned}$$

are the only two points where the gradient of $F(x, y, z, \lambda)$ vanishes. There is no standard method for determining whether the critical points in this example are indeed minima for f.

PROBLEM 8.32

Minimize the function $f(x, y, z) = x^2 + y^2 + z^2$ subject to the two constraints:

$$g = ax + by + cz - d = 0$$
$$h = px + qy + rz - t = 0$$

SOLUTION 8.32

Here a, b, c, d, p, q, r and t denote given, nonzero constants. The set of points where $g(x, y, z) = h(x, y, z) = 0$ is the intersection of the two planes $g = 0$ and $h = 0$. Thus the points form a straight line in $\mathbb{R}^3$ and since d and t are not zero, the line does not pass through the origin. The function f equals the distance from the origin to a point on the line and it follows that there exists a unique point on the line where f is minimized. We can now locate that point by the Lagrange multiplier method.

Form the function

$$F(x, y, z, \lambda, \mu) = x^2 + y^2 + z^2 - \lambda(ax + by + cz - d) - \mu(px + qy + rz - t)$$

Then **grad** $F = \underline{0}$ if and only if

$$2x - \lambda a - \mu p = 0 \quad \text{or } x = \frac{\lambda a + \mu p}{2} \tag{1}$$

$$2y - \lambda b - \mu q = 0 \quad \text{or } x = \frac{\lambda b + \mu q}{2} \tag{2}$$

$$2z - \lambda b - \mu r = 0 \quad \text{or } x = \frac{\lambda c + \mu r}{2} \tag{3}$$

$$ax + by + cz - d = 0$$

$$px + qy + rz - t = 0.$$

We substitute the expressions for x, y and z in terms of λ and μ into the two constraint equations. After simplifying we have

$$\lambda(a^2 + b^2 + c^2) + \mu(pa + qb + rc) = 2d$$

$$\lambda(pa + qb + rc) + \mu(p^2 + q^2 + r^2) = 2t.$$

This set of two equations for the two unknowns λ and μ can be solved so long as the determinant does not vanish. This is the case if the two planes $g = 0$ and $h = 0$ intersect in a line. Then the values of λ and μ are substituted into equations (1), (2), and (3) to determine the unique point on the line which is closest to the origin.

For a function of one variable the limit of f(x) as x tends to c exists if the limits as x approaches c from the left and right both exist and if these limits are equal. For a function of several variables the limit of F($\underline{x}$) as $\underline{x}$ tends to $\underline{c}$ exists if the limiting value as $\underline{x}$ approaches $\underline{c}$ along every path exists and if these values are all equal. If there exist two paths to $\underline{c}$ which produce unequal limits, F($\underline{x}$) tends to no limit at $\underline{c}$. The situation with respect to continuity of a function and existence of derivatives is similarly complicated for functions of several variables. A function of several variables may have partial derivatives with respect to each independent variable at a point $\underline{c}$ and still fail to be continuous at the point. However, if the partial derivatives exist and are also continuous in a neighborhood of $\underline{c}$ then the function is said to belong to the class $\mathbb{C}^1$ in a neighborhood of $\underline{c}$. In this case the function is continuous at $\underline{c}$ and it has the following properties as well:

- *the function satisfies a mean value theorem*
- *various versions of the chain rule apply*

A class of functions somewhat larger than the class $\mathbb{C}^1$ is the class of differentiable functions. Functions that are $\mathbb{C}^1$ are differentiable and functions that are differentiable are continuous but the converse statements are both false. A differentiable function of two variables has for its graph a surface in $\mathbb{R}^3$ having a unique tangent plane at each point on the surface.

It is possible to define derivatives of higher order for multivariable functions including mixed partial derivatives like f_{xy} and f_{yx}. Mixed partials like this need not equal each other at a point where they are not continuous. However at a point where the mixed partials are continuous, the differentiation will produce the same result regardless of the order in which the derivatives are applied. Functions that are continuous together with all their partial derivatives of order up to and including the order m are said to belong to the class $\mathbb{C}^m$. Here m denotes a positive integer. A multivariable version of Taylor's theorem is true for $\mathbb{C}^m$ functions of several variables.

Systems of several (not necessarily linear) equations in several variables may define some of the variables as implicit functions of others. Conditions under which this is indeed the case are described by implicit function theorems. These theorems state that an implicit function is well defined near a point where certain determinants known as Jacobian determinants are different from zero. The derivatives of these implicit functions can be computed by differentiating the equations in the system directly without explicitly solving for the dependent variables in terms of the independent variables. In fact the choice as to which variables are dependent and which are independent is arbitrary. For example in a set of two equations in five variables (see Example 8.4) we can theoretically solve the equations to obtain two of the variables as functions of the other three. Which two variables are chosen to be dependent and which are then independent is subject only to the nonvanishing of the appropriate Jacobians.

In a similar vein, we can view N equations in 2N variables as defining a mapping or transformation from $\mathbb{R}^N$ *to* $\mathbb{R}^N$*. Then the mapping is one to one or invertible in a neighborhood of any point where the associated Jacobian determinant is not zero. Thus the nonvanishing of the Jacobian implies the existance of the inverse for the transformation, even though we may not be able to explicitly solve for the inverse. Coordinate transformations are one example of this situation.*

There are numerous applications of partial differentiation. Some of them are:

- *Leibniz' Rule for differentiating an integral dependent on a parameter*
- *geometrical applications*
  - *tangents and normals*
  - *directional derivatives*
  - *gradient of a function*
- *optimization of functions of several variables*
  - *location of critical points*
  - *classification of critical points*
  - *optimization with constraints—Lagrange multipliers*

9

Multiple Integrals

In Chapter 5 we considered the Riemann integral of a function $f = f(x)$ over an interval of the real line. In this chapter we are going to extend this concept to functions of several variables. When defining the Riemann integral over domains in $\mathbb{R}^n$ we find that we may consider domains of integration that are much more complicated than simple rectangular boxes (the n-dimensional analogue of an interval). Thus we begin the chapter by discussing point sets and describing those sets that are allowable as domains of integration. In order to avoid unnecessary complication we initially limit our discussion to sets in $\mathbb{R}^2$.

After describing those sets in $\mathbb{R}^2$ over which the integral can be defined, we define the notion of the Riemann integral of a function of two variables, the so called double integral. We define the double integral in two ways, first in terms of a limit and then in terms of upper and lower sums. These definitions are shown to be equivalent. For convenience we first define the integral over rectangular sets and then show how to generalize to more general sets.

There is no analogue of the fundamental theorem of calculus for multiple integrals but in its place we have Fubini's theorem which states that a multiple integral is equal to a succession of single integrals. This sequence of single integrals is called an iterated integral. The parts of the iterated integral can be evaluated by means of the fundamental theorem.

Following the development of the double integral we show how each of the results extends in a natural way to triple integrals and generalization to integrals over regions in $\mathbb{R}^2$ for $n > 3$ is completely similar. Finally we show how to carry out changes of variables in multiple integrals.

DOUBLE INTEGRALS

In Chapter 5 we considered integration for functions of one variable. We found that conditions for the existence of the integral of a function f over an interval I concern only the function f and not the interval I. In extending the concept of the integral to a function $f = f(x, y)$ of two variables over a set G in $\mathbb{R}^2$ we shall find the existence of the integral to depend both on the function f and on the set G.

Point Sets in $\mathbb{R}^2$

OPEN SETS, CLOSED SETS

Let A denote a set of points in $\mathbb{R}^2$. A point p in A is an interior point if there is a neighborhood of p that is contained in A (i.e., the neighborhood contains no points that are not in A). If A consists entirely of interior points then A is said to be an open set. A point q in $\mathbb{R}^2$ is said to be an accumulation point or cluster point for A if every neighborhood of q contains points of A different from q. Note that q may or may not belong to A. The set A is said to be closed if A contains all its accumulation points.

BOUNDARY OF A SET

A point b in $\mathbb{R}^2$ is said to be a *boundary* point for A if every neighborhood of b contains points of A and points that are not in A. Then a boundary point of A is an accumulation point of A that is not an interior point of A. The *boundary* of A is the set of all the boundary points of A.

CONNECTED SETS, DOMAINS AND REGIONS

The set of points A in $\mathbb{R}^2$ is said to be *connected* if any two points of A can be joined by a polygonal path consisting of finitely many segments with each segment containing only points of A. A set in $\mathbb{R}^2$ which is both open and connected is called a *domain*.

The domain A is said to be *bounded* if there is some $R > 0$ for which A lies inside the disc $D_R = \{(x, y): x^2 + y^2 \le R^2\}$ or equivalently if there is some $M > 0$ for which A lies inside the square $S_M = \{(x, y): |x| \le M, |y| \le M\}$. By a *region* we shall mean a bounded domain together with all its boundary points. Thus a region is always a closed, bounded and hence compact set.

Example 9.1 Point Sets in $\mathbb{R}^2$

9.1 (a) Consider the sets

$$D_r(a, b) = \{(x, y): (x - a)^2 + (y - b)^2 < r^2,\}$$

and

$$R = \{(x, y): p < x < q; c < y < d\} \quad \text{with } p < q \text{ and } c < d.$$

Each of these sets is bounded. For example, if we have $r^2 > a^2 + b^2$, then it follows that $D_r(a.b)$ is contained in D_R for $R = 2r$. The boundary of the set $D_r(a, b)$ is the circumference of the disc; i.e., the circle $(x - a)^2 + (y - b)^2 = r^2$. Every neighborhood of each of these boundary points contains points inside the disc as well as points that are outside the disc. Since every neighborhood of a boundary point contains points inside the disc, the boundary points are all accumulation points. Since the inequalities in the definition of the set $D_r(a, b)$ are strict inequalities, none of the points on the circumference of the disc belongs to the disc and so the disc is not closed. Each point of the disc can be surrounded by a neighborhood lying entirely inside the disc and so the disc is open. We refer to $D_r(a, b)$ as the open disc of radius r centered at (a, b). The set consisting of $D_r(a, b)$ together with all its boundary points is a closed set. Since this set is also bounded, it is compact.

The boundary of the set R is the perimeter of the rectangle. Since all the points of the rectangle are interior points, the rectangle is open and since R does not include the accumulation points on the boundary, R is not closed. Note that the plane $\mathbb{R}^2$ is an example of a set that is both open and closed.

Finally note that any two points in $D_r(a, b)$ can be joined by a straight line lying inside the disc. Thus the disc is a connected set and since it is open, it is an example of a domain. The closed disc is a region.

Similarly the set R is connected and open, hence it is a domain. If $D_r(a, b)$ and R have no points in common then their union is an example of a set that is not connected. The closed rectangle formed by adding the perimeter of the rectangle to the set R is another example of a region.

9.1 (b) Consider the sets

$$E = \{(x, y): (x - a)^2 + (y - b)^2 > r^2\} \quad \text{and} \quad Q = \{(x, y): x > c, y > d\}.$$

Neither of these sets is bounded. E is the set of points exterior to the disc $D_r(a, b)$ hence E is contained in no disc of finite radius. Q is the quadrant lying above the line x = c and to the right of the line y = d. Thus Q is contained in no square of finite side length. Since these sets are not bounded neither is compact. The boundary of E is the same circle that forms the boundary of $D_r(a, b)$ and the boundary of Q consists of the two lines x = c and y = d. Since these boundaries do not belong to the respective sets, the sets are not closed. Each of the sets is connected since any point in either set can be joined to any other point of the set by a polygonal path that lies within the set. Each of the sets is open so each is a domain. Since neither is bounded, the closed sets obtained by adding the boundary points are not regions.

We now describe those sets which are suitable as domains of integration for double integrals.

Sets with Area

For $a_k < b_k$, $k = 1, 2$, consider the rectangle $R = \{(x, y): a_1 \le x \le b_1, a_2 \le y \le b_2\}$. We define the area of R to be $A(R) = (b_1 - a_1)(b_2 - a_2)$. More generally, let f(x) and g(x) denote two functions that are continuous on the closed interval [a, b] with $f(x) \le g(x)$ for x in [a, b]. Then the set

$$\Omega_x = \{(x, y): a \le x \le b, f(x) \le y \le g(x)\}$$

has area equal to

$$A(\Omega_x) = \int_a^b (g(x) - f(x))dx.$$

Similarly, $\Omega_y = \{(x, y): c \le y \le d, p(y) \le x \le q(y)\}$

has area equal to

$$A(\Omega_y) = \int_c^d (q(y) - p(y))dy$$

where p(y), q(y) denote continuous functions of y on the closed interval [c.d] with $p(y) \le q(y)$ for y in [c, d]. Then all of the sets R, Ω_x and Ω_y are examples of *sets with area*. These are not the most general sets having area that we could consider nor does every set in $\mathbb{R}^2$ have area. We have the following sufficient condition for a region to be a set with area.

Theorem 9.1

Theorem 9.1 Suppose the boundary of the region Ω consists of the union of finitely many segments, each of the form

$$\Gamma_x = \{(x, y): a \le x \le b, y = f(x)\} \quad \text{for some } a < b \text{ and } f \in \mathbb{C}[a, b]$$

or

$$\Gamma_y = \{(x, y): c \le y \le d, x = g(y)\} \quad \text{for some } c < d \text{ and } g \in \mathbb{C}[c, d]$$

or

$$\Gamma_s = \{(x, y): x = \varphi(s), y = \psi(s), a \le s \le b\} \text{ for } a < b \text{ and } \varphi, \psi \in \mathbb{C}^1[a, b]$$

Then Ω is a set with area.

There is a more general way to characterize sets having area.

SETS OF CONTENT ZERO

Let U denote a set in $\mathbb{R}^2$ with the property that for every e > 0 there is a finite number of rectangles $R_1, \ldots, R_n$ such that U is contained in the union of the rectangles and

$$A(R_1) + \cdots + A(R_n) < \epsilon.$$

Then we say that U is a set with *(Jordan) content zero*. Note that the union of finitely many sets of content zero is again a set of content zero. A set has positive content if it contains some rectangle of positive area.

Note that a set has zero content if it can be covered by finitely many rectangles with arbitrarily small total area. The set U is a set of measure zero if it can be covered by a countable collection of rectangles with arbitrarily small total area. The notion of measure is more general than that of content in the sense that there are sets of measure zero for which the content is not even defined. However, compact sets have measure zero if and only if they have content zero. Since we will be dealing only with compact domains of integration, the simpler notion of content will be sufficient for our purposes.

Theorem 9.2

Theorem 9.2 The region Ω in $\mathbb{R}^2$ is a set with area if the boundary of Ω has content zero.

Partitions

Consider a closed rectangle in $\mathbb{R}^2$, $R = \{a \le x \le b, c \le y \le d\}$. We can define partitions

$$\prod_x \text{ for the interval } [a, b]: a = x_0 < x_1 < \cdots < x_n = b$$

and

$$\prod_y \text{ for the interval } [c, d]: c = y_0 < y_1 < \cdots < y_m = d.$$

Then R is divided into $p = mn$ subrectangles $R_{ij} = \{x_{i-1} \le x \le x_i; y_{j-1} \le y \le y_j\}$. Then subrectangle R_{ij} has area $A_{ij} = (x_i - x_{i-1})(y_j - y_{j-1})$.

We refer to the collection $\Delta = \{R_{11}, \ldots, R_{mn}\}$ of subrectangles as a *partition* for the rectangle R. A partition Δ' for R is said to be a *refinement* for the partition Δ if the partitions $\prod'_x$ and $\prod'_y$ are refinements respectively of $\prod_x$ and $\prod_y$. We write $\Delta = \prod_x \times \prod_y$ and $\Delta' = \prod'_x \times \prod'_y$.

We can also define partitions for sets that are not rectangles. In this case the partition will contain at least some subsets that are not rectangles.

DIAMETER OF A SET, MESH SIZE

Let Ω denote a region in $\mathbb{R}^2$. Then the diameter of Ω is defined to mean the length of the longest line segment joining two points of Ω. If Δ denotes a partition of Ω then we define the mesh size $\|\Delta\|$ of Δ to be the maximum of the diameters of the sets S_k forming the partition.

The Limit Definition of the Double Integral

We shall first define the double integral over a rectangle and then extend the definition to more general sets.

RIEMANN SUM

Let $f = f(x, y)$ be defined and bounded on the closed rectangle R and let $\Delta = \{R_{11}, \ldots, R_{mn}\}$ denote a partition for R. Then for evaluation points $\underline{z}_{ij} = (\xi_i, \nu_j)$ arbitrarily chosen in R_{ij} we can define the *Riemann sum* for f based on the partition Δ and evaluation points $\underline{z} = (\xi, \nu)$

$$RS[f; \Delta; \underline{z}] = \sum_{i=1}^{n}\sum_{j=1}^{n} f(\xi_i, \nu_j)A_{ij}$$

We say that f is *Riemann integrable on R* if the following limit exists

$$\lim_{\|\Delta\| \to 0} RS[f; \Delta; \underline{z}].$$

This limit exists and equals A if for every $\in > 0$ there exists a $\delta > 0$ such that

$$|RS[f; \Delta'; \underline{z}'] - A| < \in$$

for all partitions $\Delta' = \{R'_{ij}\}$ with $\|\Delta'\| < \delta$ and z'_{ij} in R'_{ij}. In this case we refer to the limiting value as the Riemann double integral of f on R and write

$$\lim_{\|\Delta\|\to 0} RS[f; \Delta; z] = \iint_R f(x, y)$$

Theorem 9.3

Theorem 9.3 Suppose $f = f(x, y)$ is defined on the closed rectangle R and that the limit of $RS[f; \Delta; \underline{z}]$ as $\|\Delta\|$ tends to zero exists. Then the limiting value is unique. Moreover, f must then be bounded on R.

Upper and Lower Sums

Let $f = f(x, y)$ be defined and bounded on the closed rectangle R and let m, M denote respectively the greatest lower bound and least upper bound for the set of values of f on R. Let $\Delta = \{R_{11}, \ldots, R_{nm}\}$ denote a partition for R and let m_{ij}, M_{ij} denote respectively the greatest lower bound and least upper bound for the set of values for f(x, y) on the subrectangle R_{ij}. Then we define:

LOWER (DARBOUX) SUM

The lower sum for f based on the partition Δ is

$$s[f; \Delta] = \sum_{i=1}^{n}\sum_{j=1}^{m} m_{ij}A_{ij} \quad \text{where } A_{ij} = \text{area of } R_{ij}$$

UPPER (DARBOUX) SUM

The upper sum for f based on the partition Δ is

$$S[f; \Delta] = \sum_{i=1}^{n}\sum_{j=1}^{m} M_{ij}A_{ij}$$

Then we have the analogue of Theorem 5.1

Theorem 9.4

Theorem 9.4 Let $f = f(x, y)$ be defined and bounded on closed rectangle R.

(a) for every partition Δ for R and any choice of evaluation points $\underline{z}$

$$mA(R) \le s[f; \Delta] \le RS[f; \Delta; \underline{z}] \le S[f; \Delta] \le MA(R)$$

(b) if Δ′ is a refinement for Δ then

$$s[f; \Delta] \le s[f; \Delta'] \quad \text{and} \quad S[f; \Delta'] \le S[f; \Delta]$$

(c) for all partitions Δ_1 and Δ_2 of R, $s[f; \Delta_1] < S[f; \Delta_2]$

(d) if we define $s[f] = \text{Lub } s[f; \Delta]$ and $S[f] = \text{Glb } S[f; \Delta]$ where the Lub and Glb are taken over all partitions of R, then

$$mA(R) \le s[f] \le S[f] \le MA(R)$$

Definition of the Double Integral in Terms of Sums

If $f = f(x, y)$ is defined and bounded on closed rectangle R, then we say that f is *integrable over* R if

$$s[f] = S[f].$$

As in the case of the integral in $\mathbb{R}^1$, the limit definition is equivalent to the definition in terms of upper and lower sums.

Theorem 9.5 Equivalence of the Definitions of the Double Integral

Theorem 9.5 Let $f = f(x, y)$ be defined and bounded on closed rectangle R. Then the following are equivalent:

(a) $s[f] = S[f] = A$

(b) the limit of $RS[f; \Delta; \underline{z}]$ as $\|\Delta\|$ tends to zero exists and equals A

(c) for each $\in > 0$ there exists a partition Δ for R such that

$$\sum_i \sum_j (M_{ij} - m_{ij})A_{ij} < \in$$

(d) for each $\in > 0$ there exists a partition Δ for R such that

$$\sum_i \sum_j \lambda_{ij}A_{ij} < \in \quad \text{where } \lambda_{ij} = \text{Lub}\,\{f(\underline{x}) - f(\underline{z}): \underline{x}, \underline{z} \in R_{ij}\}$$

Example 9.2 A Nonintegrable Function

Let Ω denote the unit square $\{0 \le x \le 1, 0 \le y \le 1\}$ and suppose $f = f(x, y)$ equals 0 if x and y are both rational numbers and equals 1 otherwise. Then f is defined over Ω and for every partition of Ω, the lower sum for f on Ω is zero while the upper sum equals one. Then $s[f] = 0 < S[f] = 1$ and f is not integrable on Ω.

We can now extend the definition of the double integral to sets more general than rectangles.

The Double Integral Over General Regions

EXTENSION BY ZERO

Let $f = f(x, y)$ be defined and bounded on a region Ω in $\mathbb{R}^2$. Then we define the function

$$F(x, y) = \begin{cases} f(x, y) & \text{if } (x, y) \text{ lies in } \Omega \\ 0 & \text{if } (x, y) \text{ is not in } \Omega \end{cases}$$

and refer to F as the *extension by zero* for f.

THE DOUBLE INTEGRAL OVER W

Let R denote any closed rectangle containing Ω. Then we say that f is integrable over Ω if and only if F is integrable over R. The value of the integral of f over Ω equals the value of the integral of F over R.

Theorem 9.6

Theorem 9.6 The definition of the integral of f over W is independent of the choice of the rectangle R.

AREA OF A SET

Let Ω be a bounded set in $\mathbb{R}^2$ and let

$$f(x, y) = \begin{cases} 1 & \text{if } (x, y) \text{ is in } \Omega \\ 0 & \text{if } (x, y) \text{ is not in } \Omega \end{cases}$$

Then f is bounded on any bounded rectangle R containing Ω and by Theorem 9.4 $0 \leq s[f] \leq S[f] \leq A(R)$. We refer to s[f] and S[f] respectively as the inner area and outer area of Ω. When these two values are equal we call their common value the area of Ω.

Theorem 9.7 Area of a Set

Theorem 9.7 For every bounded set Ω in $\mathbb{R}^2$ the inner area is less than or equal to the outer area and equality occurs if and only if the boundary of Ω is a set with content zero. In this case we refer to the common value of the inner and outer areas as the area of Ω.

Note that f is continuous at each point of any rectangle containing Ω except at the boundary points of Ω. Thus f is integrable if this set of points has content zero. More generally we have

Theorem 9.8 Existence of the Double Integral

Let $f = f(x, y)$ be defined and bounded on a region Ω in $\mathbb{R}^2$. Suppose that the boundary of Ω and the set of discontinuities for f each have content zero in $\mathbb{R}^2$. Then f is integrable over Ω.

Properties of the Double Integral

Let Ω denote a set with area in $\mathbb{R}^2$, and suppose f and g are functions defined and continuous on Ω. Then according to Theorem 9.8 the double integrals of f and g over Ω exist. We have the following properties for the double integral:

1. $\iint_\Omega 1 = A(\Omega)$

2. $\iint_\Omega \alpha f + \beta g = \alpha \iint_\Omega f + \beta \iint_\Omega g$ for all constants α and β

3. If $\Omega = \Omega_1 + \Omega_2$ then $\iint_\Omega f = \iint_{\Omega_1} f + \iint_{\Omega_2} f$

Here Ω, Ω_1 and Ω_2 all have area and $\Omega = \Omega_1 + \Omega_2$ means that every point of Ω is a point of Ω_1 or Ω_2 or their boundary points. In addition Ω_1 and Ω_2 have at most some of their boundary points in common.

4. if $f(x, y) \leq g(x, y)$ on Ω then $\iint_\Omega f \leq \iint_\Omega g$

5. $\left|\iint_\Omega f\right| \leq \iint_\Omega |f|$

6. $\left|\iint_\Omega f\right| \leq \max_\Omega |f| \, A(\Omega)$

7. If m, M denote respectively the minimum and maximum values of f on Ω, then there exists a value μ, $m \leq \mu \leq M$, such that

$$mA(\Omega) \le \iint_\Omega f = \mu A(\Omega) \le MA(\Omega)$$

Property 7 is a version of the *mean value theorem* for double integrals.

Evaluation of Double Integrals

For double integrals there is no analogue of the fundamental theorem of calculus for evaluating the integrals. Instead under certain conditions we can replace a double integral by two single integrals performed in succession.

ITERATED INTEGRALS

An *iterated integral* is an integral of the form

$$\int_a^b dx \int_{\varphi(x)}^{\psi(x)} f(x, y)dy = \int_a^b F(x)\, dx$$

where

$$F(x) = \int_{\varphi(x)}^{\psi(x)} f(x, y)dy$$

or it can be of the form

$$\int_c^d dy \int_{p(y)}^{q(y)} f(x, y)dx = \int_c^d G(y)dy$$

where

$$G(y) = \int_{p(y)}^{q(y)} f(x, y)dx.$$

In computing F(x) we integrate f(x, y) as a function of y holding x constant and to compute G(y) we integrate f as a function of x holding y constant. In either case we are doing a single variable integration and so we have the fundamental theorem of calculus at our disposal.

Theorem 9.9

Theorem 9.9 Let $\Omega_x = \{(x, y): a \le x \le b, \psi(x) \le y \le \varphi(x)\}$ where ψ and φ are continuous on [a, b] with $\varphi < \psi$ and suppose $f = f(x, y)$ is continuous on Ω_x. Then f is integrable on Ω_x and

$$\iint_{\Omega_x} f = \int_a^b dx \int_{\varphi(x)}^{\psi(x)} f(x, y)dy \tag{9.1}$$

Similarly, if $\Omega_y = \{(x, y): c \le y \le d, p(y) \le y \le q(y)\}$ where p and q are continuous on [c, d] with $p < q$ and if f is continuous on Ω_y then f is integrable on Ω_y and

$$\iint_{\Omega_y} f = \int_c^d dx \int_{p(y)}^{q(y)} f(x, y)dx \tag{9.2}$$

Example 9.3 Iterated Integrals

9.3 (a) Consider the set of points in the xy-plane contained between the two curves $y = x + 2$ and $y = 4 - x^2$. The curves can be seen to intersect at $x = -2$ and $x = 1$ and it follows that this set of points is the region

$$\Omega_x = \{(x, y): -2 \le x \le 1, x + 2 \le y \le 4 - x^2\}.$$

This is a set with area and the double integral of the continuous function $f(x, y) = 4x^2y$ over Ω_x exists. According to Theorem 9.9

$$\iint_{\Omega_x} f = \int_{-2}^{1} dx \int_{x+2}^{4-x^2} 4x^2y\, dy$$

$$= \int_{-2}^{1} (2x^2y^2)\Big|_{x+2}^{4-x^2} dx$$

$$= \int_{-2}^{1} (2x^2(4-x^2) - 2x^2(2+x)^2)dx$$

$$= \int_{-2}^{1} (2x^6 - 14x^4 + 8x^3 + 4x^2)dx$$

$$= \frac{1}{7}x^7 - \frac{14}{5}x^5 + 2x^4 + \frac{4}{3}x^3 \Big|_{-2}^{1} = -\frac{621}{7} - 42$$

9.3(b) Consider the set of points in the xy-plane bounded by the lines $x = 0$, $y = 0$ and $x = 2 - 2y$. This is a seen to be a region of the form

$$\Omega_y = \{(x, y): 0 \le y \le 1, 0 \le x \le 2 - 2y\}.$$

Then double integral of the continuous function $f(x, y) = 4 - x^2 - 4y^2$ over Ω_y exists and according to Theorem 9.9 we have

$$\iint_{\Omega} 4 - x^2 - 4y^2 = \int_0^1 dy \int_0^{2-2y} (4 - x^2 - 4y^2)dx$$

$$= \int_0^1 (8(1-y) - \frac{8}{3}(1-y)^3 - 8y^2 + 8y^3)dy = \frac{8}{3}$$

Fubini's Theorem

We should point out that Theorem 9.9 is not the most general theorem regarding iterated integrals that we could state. If the function f is integrable over the region Ω, then the double integral is equal to an appropriate iterated integral as indicated in (9.1) and (9.2) but it is not necessary for the function f to be continuous in order to be integrable. The most general theorem of this sort is known as Fubini's theorem and must be expressed in the setting of the Lebesgue theory of integration. A version of this theorem suitable for the Riemann integral can be stated as follows. For convenience we take the domain of integration to be a rectangle.

Theorem 9.10

Theorem 9.10 If $f = f(x, y)$ is integrable over the closed rectangle $R = \{a \le x \le b,\ c \le y \le d\}$ and if $f(x, y)$ is integrable with respect to the variable y for each fixed x in [a, b] then

$$\iint_R f = \int_a^b F(x)\, dx$$

where

$$F(x) = \int_c^d f(x, y)dy$$

is integrable on [a, b].

VOLUMES

Note that the integrals in Example 9.3 can be interpreted as volumes of solids. The solid in Example 9.3(a) has Ω_x as its base and generators perpendicular to the plane of Ω_x. The solid is bounded above by the surface $z = 4x^2y$. Similarly, the solid in Example 9.3(b) is a cylindrical solid having Ω_y for its base and the surface $z = 4 - x^2 - 4y^2$.

MULTIPLE INTEGRALS

The development of double integrals generalizes immediately to multiple integrals in $\mathbb{R}^n$. Since integrals in $\mathbb{R}^3$, triple integrals, occur frequently in applications, we will indicate how the discussion of double integrals must be modified to extend to the case n = 3. The development for larger n proceeds similarly.

Sets With Volume

We define the *volume* of the rectangular box $B = \{(x, y, z): a_1 < x \le b_1, a_2 \le y \le b_2, a_3 \le z \le b_3\}$ where $a_k < b_k$ for k = 1, 2, 3 to be $V(B) = (b_1 - a_1)(b_2 - a_2)(b_3 - a_3)$. More generally let $\varphi(x, y)$ and $\psi(x, y)$ denote functions defined and continuous on the domain Ω in the plane where Ω denotes a set with area and $\varphi(x, y) \le \psi(x, y)$. Then the set

$$\Omega_{xy} = \{(x, y, z):(x, y) \in \Omega, \varphi(x, y) < z < \psi(x, y)\}$$

has volume equal to

$$V(\Omega_{xy}) = \iint_\Omega (\psi(x, y) - \varphi(x, y))dx\,dy.$$

The set Ω_{xy} is not the most general example of a set with volume. Using the notion of the volume of a rectangular box we can define a set U in $\mathbb{R}^3$ to have content zero if, for each $\in > 0$ there is a finite set of boxes $B_1, \ldots, B_n$ such that U is contained in the union of the boxes and $V(B_1) + \cdots + V(B_n) < \in$. The notionof measure extends to higher dimensions as well and the relationship between measure and content is the same as it was in two dimensions.

Theorem 9.11

Theorem 9.11 The region Ω in $\mathbb{R}^3$ is a set with volume if the boundary of Ω has content zero.

Example 9.4 Sets With Content Zero

A set of points $C = \{(x, y): x = x(t), y = y(t), a \leq t \leq b\}$ is said to form a *smooth curve* in $\mathbb{R}^2$ if x and y belong to $C^1[a, b]$ for real numbers $a < b$. We can show that any smooth curve is a set of content zero in $\mathbb{R}^2$.

Similarly, a smooth curve $C = \{(x, y, z): x = x(t), y = y(t), z = z(t), a \leq t \leq b\}$ is a set of content zero in $\mathbb{R}^3$. The concept of a smooth surface in $\mathbb{R}^3$ can be defined in a similar fashion and likewise shown to be a set of content zero in $\mathbb{R}^3$. Note, however, that a plane is a smooth surface and hence it has content zero in $\mathbb{R}^3$. But a plane does not have content zero in $\mathbb{R}^2$. In general a set of content zero in $\mathbb{R}^n$ will have content zero in $\mathbb{R}^m$ for any $m > n$ but the converse is false. A *compact* set in $\mathbb{R}^n$ has n-dimensional measure zero if and only if its n-dimensional content equals zero.

The Limit Definition of the Triple Integral

We define partitions of rectangular boxes in a fashion similar to the definition of partitions of rectangles in $\mathbb{R}^2$. The mesh size of the partition, $\| \Delta \|$, is defined as the maximum diameter of the sub-boxes in the partition.

For $f = f(x, y, z)$ defined and bounded on the closed box B, for partition $\Delta = \{B_{111}, \ldots, B_{pqr}\}$ of B, and for evaluation points $\underline{s}_{jk} = (\xi_i, \nu_j, \upsilon_k)$ arbitrarily chosen in B_{ijk} we define the Riemann sum for f based on the partition Δ and the evaluation points $\underline{s}_{ijk}$,

$$RS[f, \Delta, \underline{s}] = \sum_i \sum_j \sum_k f(\xi_i, \nu_j, \upsilon_k) V(B_{ijk})$$

Then we say that f is *Riemann integrable* on B if the following limit exists

$$\lim_{\|\Delta\| \to 0} RS[f, \Delta, \underline{s}].$$

In this case, we write

$$\iiint_B f = \lim_{\|\Delta\| \to 0} RS[f, \Delta, \underline{s}].$$

There is an equivalent definition of the triple integral in terms of upper and lower sums. We will omit this detail.

The Triple Integral On General

Using the notion of extension by zero, we can define what is meant by the triple integral over a general region in $\mathbb{R}^3$. Then we have the analogue of Theorem 9.8 for triple integrals.

Theorem 9.12

Theorem 9.12 Let $f = f(x, y, z)$ be defined and bounded on region Ω in $\mathbb{R}^3$. If the boundary of Ω and the set of discontinuities for f on Ω each have content zero in $\mathbb{R}^3$ then f is integrable over Ω.

Iterated

As was the case for double integrals, multiple integrals are generally evaluated by evaluating equivalent iterated integrals. The following is one possible version of Theorem 9.9 stated for triple integrals.

Theorem 9.13

Theorem 9.13 Let the region Ω in $\mathbb{R}^2$ be a set with area and let $\varphi = \varphi(x, y)$, $\psi = \psi(x, y)$ be continuous on Ω with $\varphi(x, y) < \psi(x, y)$ for (x, y) in Ω. Suppose $f = f(x, y, z)$ is continuous on the set $\Omega = \{(x, y, z): (x, y) \in \Omega, \varphi(x, y) \le z \le \psi(x, y)\}$. Then

$$\iiint_{\Omega_{xy}} f = \iint_{\Omega} \int_{\varphi(x,y)}^{\psi(x,y)} f(x, y, z)dz.$$

Example 9.5

Consider the set $\Omega_{xy} = \{1 \le x \le 2, 0 \le y \le x, 1 \le z \le xy\}$. Then $f(x, y, z) = 16xy$ is continuous on Ω_{xy} and

$$\iiint_{\Omega_{xy}} f = \iint_{\Omega} \int_1^{xy} 16xy\,dz = \int_1^2 dx \int_0^x dy \int_0^{xy} 16xy\,dz$$

$$= \int_1^2 dx \int_0^x 16xyz\Big|_{z=1}^{z=xy} dy$$

$$= \int_1^2 dx \int_0^x (16x^2y^2 - 16xy)dy$$

$$= \int_1^2 \left(\frac{16}{3}x^2y^3 - 8xy^2\right)\Bigg|_{y=0}^{z=xy} dx$$

$$= \int_1^2 \left(\frac{16}{3}x^5 - 16x^3\right) dx = 26$$

Change of Variables in Multiple Integrals

Theorem 5.16 explains the change of variable procedure for single integrals. We will describe the analogous procedure for triple integrals. The change of variables for double integrals is similar.

Theorem 9.14

Theorem 9.14 Suppose that Φ is a transformation from $\mathbb{R}^2$ into $\mathbb{R}^2$ given by

$$x = f(u, v, w)$$
$$y = g(u, v, w)$$
$$z = h(u, v, w)$$

for f, g and h defined and C^1 on some open set U in $\mathbb{R}^3$. Suppose further that the Jacobian

$$J = \frac{\partial(f, g, h)}{\partial(u, v, w)}$$

of the transformation Φ does not vanish at any point of U. Let Ω denote a region contained in U and let $h = h(x, y, z)$ denote a function which is integrable on the set $\Phi[\Omega]$, the image of Ω under the transformation Φ. Then the function

$$h(x(u, v, w), y(u, v, w), z(u, v, w))\frac{\partial(f, g, h)}{\partial(u, v, w)}$$

is integrable on Ω. Moreover,

$$\iiint_{\Phi[\Omega]} h(x, y, z)dx\,dy\,dz = \iiint_{\Omega} h(x(u, v, w), y(u, v, w), z(u, v, w))$$

$$\times \frac{\partial(f, g, h)}{\partial(u, v, w)} dudvdw$$

Similar statements apply to double integrals and to multiple integrals in n variables for $n > 3$.

Example 9.6 Change of Variables in Multiple Integrals

9.6 (a) Consider the double integral of the function $f(x, y) = y$ over the set $\{(x, y):a^2 < x^2 + y^2 < b^2, y > 0\}$. This set is the upper half of the annular region between the circle of radius a and the circle of radius b. We are supposing here that $0 < a < b$. If we let Φ denote the tranformation

$$x = r \operatorname{Sin} \vartheta \qquad y = r \operatorname{Cos} \vartheta$$

then it was shown in Problem 8.19(a) that the Jacobian J is given by

$$J = \frac{\partial(x, y)}{\partial(r, i)} = r.$$

Moreover, Φ maps the rectangle $\Omega = (r, \vartheta)$: $0 < \vartheta < \pi$, $a < r < b$ onto the set $\Phi(\Omega) = \{(x, y):a^2 < x^2 + y^2 < b^2, y > 0\}$. Since f is $\mathbb{C}^1$ and J is nowhere zero on Ω, it follows from the double integral version of Theorem 9.14 that

$$\iint_{\Phi[\Omega]} y\,dx\,dy = \int_0^{\pi}\int_a^b r^2 \operatorname{Sin} \vartheta r\,dr\,d\vartheta$$

$$= \int_0^{\pi} \operatorname{Sin} \vartheta\,d\vartheta \int_a^b r^2 dr = \frac{2}{3}(b^3 - a^3)$$

The transformation Φ here amounts to a change of variables in the double integral from Cartesian to polar coordinates.

9.6 (b) Compute the volume of the spherical cap cut off from the sphere $x^2 + y^2 + z^2 = 4R^2$ by the plane $z = R$. This volume can be computed by integrating the function $f = 1$ over this three dimensional region. However, the integration is more easily executed in spherical coordinates. Thus consider the transformation Φ given by

$$x = \rho \operatorname{Sin} \varphi \operatorname{Cos} \vartheta$$
$$y = \rho \operatorname{Sin} \varphi \operatorname{Sin} \vartheta$$
$$z = \rho \operatorname{Cos} \varphi$$

We can see be examining Figure 9.1 that the spherical cap is the image under the transformation Φ of the set

$$\Omega = \{(\rho, \vartheta, \varphi): R \operatorname{Sec} \varphi < \rho < 2R, 0 < \vartheta < 2\pi, 0 < \varphi < \pi/3\}.$$

That is, the angle φ varies within the cap from zero up to angle AOC which equals $\pi/3$. The radial variable ρ varies as a function of φ such that when φ equals the angle AOB, ρ varies between the values OB and OD. The radial distance OB

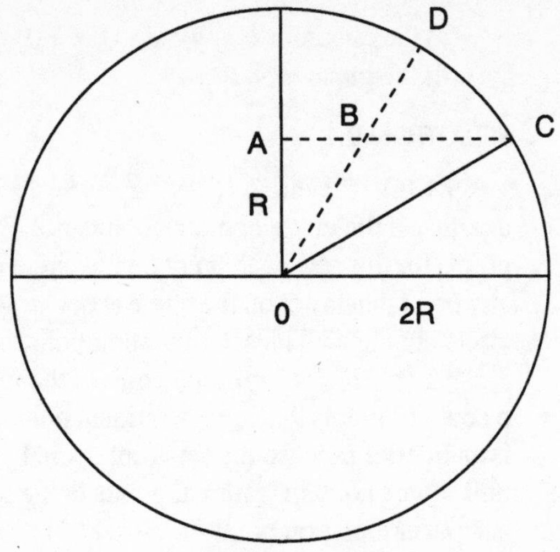

Figure 9.1
Example 9.6(b)

equals R Sec φ while OD equals 2R. The angle ϑ varies from 0 to 2π.

It was shown in Problem 8.19(b) that the Jacobian of the transformation Φ is

$$J = \frac{\partial(x, y, z)}{\partial(\rho, \vartheta, \varphi)} = \rho^2 \text{Sin}\ \varphi$$

and since J does not vanish at any point of Ω, it follows from Theorem 9.14 that

$$\iiint_{\Phi(\Omega)} 1\ dx\,dy\,dz = \int_0^{\pi/3}\int_0^{2\pi}\int_{R\text{Sec}\varphi}^{2R} 1\rho^2 \text{Sin}\ \varphi\ d\rho\ d\vartheta\ d\varphi$$

$$= \int_0^{\pi/3}\int_0^{2\pi} \frac{1}{3}\text{Sin}\ \varphi\ (8R^8 - R^3\text{Sec}^3\varphi)d\vartheta\ d\varphi$$

$$= \frac{2\pi R^3}{3}\int_0^{\pi/3} (8 - \text{Sec}^3\varphi)\ \text{Sin}\ \varphi\ d\varphi = \frac{5\pi R^3}{3}$$

SOLVED PROBLEMS

Point Sets in $\mathbb{R}^2$

PROBLEM 9.1

Which of the following points sets is open and which is closed?

(a) the circumference of the unit disc, $\{(x, y): x^2 + y^2 = 1\}$

(b) the rational points of the unit square $\{(x, y): 0 \le x \le 1, 0 \le y \le 1, x$ and $y \in \mathbb{Q}$

(c) the quadrant $\{(x, y): x > 0, y \ge 0\}$

(d) the plane $\mathbb{R}^2$

SOLUTION 9.1

9.1 (a) The circle $\{(x, y): x^2 + y^2 = 1\}$ is not open since any neighborhood of a point on the circle contains points not on the circle. Thus there are no interior points for this set. Each point on the circle is an accumulation point of the set and any point that is not on the circle is not an accumulation point for the set. Thus the circle contains all its accumulation points so it is a closed set.

9.1 (b) If p is a rational point of the unit square then every neighborhood of p contains points that are not rational points of the square. Then no point of the set is an interior point so the set is not open. Every neighborhood of each point of the unit square contains rational points in the square so every point in the unit square is an accumulation point of the set of rational points in the square. Since the set fails to contain all its limit points it is not closed.

9.1 (c) This set contains points of the form (x, 0) for $x > 0$. No such point is an interior point of the set so the set is not open. Each point of the form (0, y) for $y > 0$ is a limit point of the set but does not belong to the set. Thus the set is not closed.

9.1 (d) Each point in $\mathbb{R}^2$ is surrounded by a neighborhood that lies in $\mathbb{R}^2$ so all points of the set are interior points. Thus $\mathbb{R}^2$ is open. Since there are no points that are not in $\mathbb{R}^2$, there are no accumulation points of $\mathbb{R}^2$ that are not in $\mathbb{R}^2$. Thus $\mathbb{R}^2$ is closed.

PROBLEM 9.2

Describe the boundary of each of the sets from Problem 9.1

SOLUTION 9.2

9.2 (a) The complement of the circle consists of the open unit disc together with the exterior of the unit disc. Every neighborhood of each point on the circle contains points of its complement. Then every point of the circle is a boundary point of the set.

The circle forms the boundary for the unit disc $\{(x, y): x^2 + y^2 < 1\}$ and is the boundary for the exterior of the disc, $\{(x, y): x^2 + y^2 > 1\}$ as well.

9.2 (b) Every neighborhood of each point in the unit square contains both points of the square that are rational points and points that are not rational points. Then every point of the closed unit square is a boundary point for the set of rational points of the square.

9.2 (c) The boundary of the quadrant consists of all points (x, 0) with $x \ge 0$ together with all points (0, y) with $y \ge 0$.

9.2 (d) The set $\mathbb{R}^2$ is a set without a boundary since there are no points with neighborhoods containing points not in $\mathbb{R}^2$.

PROBLEM 9.3

Which of the following sets is connected?
(a) $\{(x-1)^2 + y^2 < 1\} \cup \{(x+1)^2 + y^2 < 1\}$
(b) $\{1 < x^2 + y^2 < 4\}$

SOLUTION 9.3

9.3 (a) This set consists of the union of two open discs of radius 1. One of the discs is centered at (1, 0), the other at (–1, 0). Note that the origin (0, 0) is the point where the discs are tangent to one another but this point belongs to neither disc. Any polygonal path joining a point of one disc to a point of the other without going outside both discs must pass through this point of tangency.But since this point is not in either disc, all such polygonal paths must go out of the set and the set is not connnected.

9.3 (b) This set consists of the annular region lying outside the disc of radius 1 centered at the origin and inside the disc of radius 4 with center at the same point. Clearly any point in this set can be joined to any other point of the set by a polygonal path lying within the set. This set is connected.

Sets of Zero Content

PROBLEM 9.4

Suppose $y = f(x)$ is continuously differentiable on the closed interval [a, b]. Then show that the curve $\Gamma = \{(x, y): y = f(x), a < x < b\}$ is a set of content zero in $\mathbb{R}^2$.

SOLUTION 9.4

Since $f \in \mathbb{C}^1[a, b]$, $f'(x)$ is continuous on the compact set [a, b] and it follows from the bounded range theorem that there exists $M > 0$ such that $|f'(x)| \le M$ for all x in [a, b].

Let $\in > 0$ be given and choose $N > \frac{M}{\in}(b-a)$. Then define

$$z_n = a + n\frac{b-a}{N} \quad \text{for } n = 0, 1, \ldots, N$$

$$H = M\frac{b-a}{N}$$

and let R_n denote the rectangle $\{z_n < x < z_{n+1}, f(z_n) - 2H < y < f(z_n) + 2H\}$. Note that for any two points s, t in (z_n, z_{n+1}) we have

$$|f(s) - f(t)| < |f'(x)||s - t| < M\frac{b-a}{N} = H < 2H$$

Then $(x, f(x))$ lies inside R_n for all x such that $z_n < x < z_{n+1}$ and it follows that the union of the rectangles $R_0, R_1, \ldots, R_{N-1}$ covers the set of points in the curve $\Gamma = \{(x, y): y = f(x), a \le x \le b\}$. But

$$A(R_0) + A(R_1) + \cdots + A(R_{N+1}) = N \cdot 4H \cdot \frac{b-a}{N}$$

$$= 4H(b-a) = M\frac{(b-a)}{N} < \in$$

Since $\in > 0$ was arbitrary, it follows that Γ is a set of content zero in $\mathbb{R}^2$. This result shows that a set whose boundary consists of finitely many arcs of the form $y = f(x)$ or $x = g(y)$ for $\mathbb{C}^1$ functions f and g is a set whose boundary has content zero. In particular, discs and rectangles are sets whose boundary has content zero.

Double Integrals

PROBLEM 9.5

Let $f = f(x, y)$ be defined and bounded on region Ω in $\mathbb{R}^2$. Let R denote a bounded rectangle containing Ω in its interior and define

$$F(x, y) = \begin{cases} f(x, y) & \text{if } (x, y) \in \Omega \\ 0 & \text{if } (x, y) \in R\backslash\Omega \end{cases}$$

Then we define the integral of f over Ω by

$$\iint_\Omega f = \iint_R F$$

Show that this definition does not depend on the choice of R.

SOLUTION 9.5

Let R_1 and R_2 denote two bounded rectangles containing Ω in their interiors and let F_1, F_2 denote the extension of f by zero on R_1, R_2 respectively. Then let R_3 denote a third bounded rectangle containing R_1 and R_2 in its interior and let F_3 denote the extension of f to R_3. Let Δ denote a partition of the rectangle R_3 and note that this induces partitions Δ_1 and Δ_2 on R_1 and R_2. Since F_1 and F_3 are equal where both are defined and F_3 is zero elsewhere, we have that

$$s[F_1; \Delta_1] = s[F_3; \Delta_3] \quad \text{and} \quad S[F_1; \Delta_1] = S[F_3; \Delta_3].$$

Similarly, $s[F_2; \Delta_2] = s[F_3; \Delta_3]$ and $S[F_2; \Delta_2] = S[F_3; \Delta_3]$. Then

$$\iint_{R_1} F_1 = \iint_{R_2} F_3 = \iint_{R_2} F_2 = \iint_\Omega f.$$

PROBLEM 9.6

Let Ω be a bounded set in $\mathbb{R}^2$ and let

$$f(x, y) = \begin{cases} 1 & \text{if } (x, y) \text{ is in } \Omega \\ 0 & \text{if } (x, y) \text{ is not in } \Omega \end{cases}$$

Show that $s[f] \le S[f]$,

SOLUTION 9.6

Since Ω is bounded, Ω is contained in some rectangle $R = \{a \le x \le b, c \le y \le d\}$. Let Δ denote a partition for R consisting of rectangles $\{R_{11}, \ldots, R_{mn}\}$. Let m_{ij} and

M_{ij} denote the minimum and maximum values respectively for f(x, y) on R_{ij}. Then

i) if R_{ij} lies inside Ω, $m_{ij} = M_{ij} = 1$
ii) if R_{ij} lies in the complement of Ω, $m_{ij} = M_{ij} = 0$
iii) if R_{ij} contains points of Ω and points of its complement, $m_{ij} = 0$, $M_{ij} = 1$

It follows that $s[f; \Delta] < S[f; \Delta]$. Clearly this holds for every partition Δ of the rectangle R. Now if Δ' denotes a refinement of the partition Δ then in just the same way as we did in Problem 5.2, we can show that

$$s[f; \Delta] < s[f; \Delta'] \quad \text{and} \quad S[f; \Delta'] < S[f; \Delta] \tag{1}$$

For $\Delta_1 = \prod_{x,1} \times \prod_{y,1}$ and $\Delta_2 = \prod_{x,2} \times \prod_{y,2}$ any two partitions of R, let Δ' denote the partition $\Delta' = \prod_x' \times \prod_y'$ where $\prod_x'$ ' is obtained by combining the points of $\prod_{x,1}$ with those of $\prod_{x,2}$ and $\prod_y'$ is obtained in a similar fashion. Then Δ' is a refinement for both Δ_1 and Δ_2. Then (1) implies

$$s[f;\Delta_1] < s[f;\Delta'] \leq S[f;\Delta'] < S[f;\Delta_2]. \tag{2}$$

Now it follows from (2) that

$$s[f] = \text{GLB } s[f;\Delta] \leq \text{LUB } S[f;\Delta] = S[f] \tag{3}$$

where the GLB and LUB are taken over the collection of all possible partitions of R. This proves that the inner area, s[f], is less than or equal to the outer area, S[f].

PROBLEM 9.7

Let Ω be a bounded set in $\mathbb{R}^2$ and let f be as defined in the previous problem. Then show that s[f] = S[f] if and only if the boundary of Ω is set of content zero.

SOLUTION 9.7

(a) If s[f] = S[f] then the boundary of Ω has content zero.

Since Ω is bounded it is contained in some rectangle R. Then s[f] = S[f] implies that for $\in > 0$ fixed, we can choose a partition Δ for R such that $S[f; \Delta] - s[f; \Delta] < \in$. Let $\{\hat{R}_{ij}\}$ denote the set of rectangles in Δ that contain points of Ω and points of its complement. Then the sum of the areas of these rectangles is equal to $S[f; \Delta]-s[f; \Delta]$ since these are the only rectangles of the partition Δ where m_{ij} differs from M_{ij}. If we can show that each point of the boundary of Ω belongs to one of the rectangles $\hat{R}_{ij}$ then we will have shown that the boundary of Ω can be covered by rectangles whose areas sum to less than $\in$. Since $\in$ is arbitrary, it will follow that the boundary of Ω has content zero.

All points on the boundary of Ω that lie on an edge of a rectangle from Δ are a subset of a set of content zero. Then these points are a set of content zero and we need consider only those points on the boundary of Ω that lie inside some rectangle. But if p is a point of the boundary of Ω and p belongs to the interior of some rectangle R_{pq} in D then R_{pq} must be one of the $\hat{R}_{ij}$ since every neighborhood of a point on the boundary of Ω contains points in Ω and points that are not in Ω. Then the boundary of Ω must have content zero.

(b) If the boundary of Ω has content zero then $s[f] = S[f]$.

Fix $\in > 0$ and let Δ denote a partition of R such that the sum of the areas of those rectangles containing some point of the boundary of Ω is less than $\in$. We can show that any rectangle from Δ that contains points of Ω and points not in Ω must also contain points of the boundary of Ω. But then the areas of these rectangles sums to less than $\in$ and since the sum of these areas is equal to $S[f; \Delta] - s[f; \Delta]$ it follows that for each $\in > 0$ there is a partition Δ such that $S[f; \Delta] - s[f; \Delta] < \in$. Then $s[f] = S[f]$.

To see that any R_p in Δ containing points of Ω and points not in Ω must also contain a point of the boundary of Ω, suppose R_p contains points $\underline{P} = (a, b)$ in Ω and $\underline{Q} = (c, d)$ in the complement of Ω. Then the straight line

$$\underline{X}(t) = (x(t), y(t)): \; x(t) = a + (b - a)t, \;\; y(t) = b + (d - b)t, \quad 0 \le t \le 1$$

joins $\underline{P}$ to $\underline{Q}$ and there exists a value τ, $\tau = \text{LUB}\,\{t \in [0, 1] : \underline{X}(t) \in \Omega\}$. But then $\underline{X}(\tau)$ must belong to the boundary of Ω since by the definition of LUB, if $\underline{X}(t)$ belongs to Ω then $\underline{X}(t)$ is an accumulation point of the complement of Ω and if it belongs to the complement then it must be an accumulation point for Ω.

This proves that $s[f] = S[f]$ if and only if the boundary of Ω is a set of content zero. In this case Ω is a set with area and we refer to the common value of the inner and outer area of Ω as the *area* of Ω.

PROBLEM 9.8

Suppose that regions Ω, Ω_1 and Ω_2 in $\mathbb{R}^2$ all have area and $\Omega = \Omega_1 + \Omega_2$; i.e., every point of Ω is a point of Ω_1 or Ω_2 or their boundary points and Ω_1 and Ω_2 have at most some of their boundary points in common. Then show that if f is continuous on Ω then

$$\iint_\Omega f = \iint_{\Omega_1} f + \iint_{\Omega_2} f$$

SOLUTION 9.8

Since Ω is bounded there is a bounded rectangle R containing Ω in its interior. Let F denote the extension by zero for f to the rectangle R. In addition, let

$$F_k(x, y) = \begin{cases} f(x, y) & \text{if } (x, y) \in \Omega \\ 0 & \text{if } (x, y) \in R \backslash \Omega_k \end{cases} \qquad \text{for } k = 1, 2.$$

Since f is continuous on Ω f is integrable over Ω and,

$$\iint_\Omega f = \iint_R F.$$

F is continuous on R except possibly on the boundary of Ω. Similarly F_1 and F_2 are continuous on R except possibly on the boundary of Ω_1 and Ω_2 respectively. But these sets have all been assumed to have area and so their boundaries have content zero. Then the sets of discontinuities for F_1 and F_2 have content zero and by Theorem 9.8

$$\iint_{\Omega_1} f = \iint_{\Omega_1} F_1 = \iint_R F_1 \quad \text{and} \iint_{\Omega_2} f = \iint_{\Omega} F_2 = \iint_R F_2$$

Now let

$$g(x, y) = F(x, y) - F_1(x, y) - F_2(x, y) \quad \text{for } (x, y) \text{ in R.}$$

Since $F(x, y) = F_1(x, y) + F_2(x, y)$ for (x, y) in $\Omega\backslash(\Omega_1 \cap \Omega_2)$, it follows that $g = 0$ on $\Omega\backslash(\Omega_1 \cap \Omega_2)$. By assumption, this is a set of zero content and since $g(x, y)$ is continuous on Ω except possibly on this set, we have

$$\iint_{\Omega} g = 0.$$

Finally, let

$$G(x, y) = \begin{cases} g(x, y) & \text{if } (x, y) \in \Omega \\ 0 & \text{if } (x, y) \in R\backslash\Omega \end{cases}$$

Then

$$\iint_{\Omega} g = \iint_R G = \iint_R F - \iint_R F_1 - \iint_R F_2 = \iint_{\Omega} f - \iint_{\Omega_1} f - \iint_{\Omega_2} f = 0$$

This proves the result.

PROBLEM 9.9

Compute the value of the double integral of the linear function $f(x, y) = Ax + By$ over the rectangle $R = \{0 \le x \le 2, 0 \le y \le 3\}$. Here A, B denote given constants.

SOLUTION 9.9

Note that R is a set with area and $f(x, y)$ is continuous on R. Then by Theorem 9.8 the double integral of f over R exists. For n in $\mathbb{N}$, let $\prod_x$ denote the partition of $[0, 2]$ into 2n intervals of equal length and let $\prod_y$ denote the partition of $[0, 3]$ into 3n equal intervals. Then $\Delta = \prod_x \times \prod_y$ is a partition of R into $6n^2$ equal squares each of side length $1/n$.

Then

$$S[f; \Delta] = \sum_{i=1}^{2n} \sum_{j=1}^{3n} M_{ij} A_{ij} = \frac{1}{n^2} \sum_{i=1}^{2n} \sum_{j=1}^{3n} \left(A\frac{i}{n} + B\frac{j}{n}\right)$$

$$= \frac{1}{n^2} \frac{A}{n} 3n \sum_{i=1}^{2n} i + \frac{1}{n^2} \frac{B}{n} 2n \sum_{j=1}^{3n} j$$

$$= \frac{3A}{n^2} \frac{2n(2n+1)}{2} + \frac{2B}{n^2} \frac{3n(3n+1)}{2} = A\left(6 + \frac{3}{n}\right) + B\left(9 + \frac{3}{n}\right)$$

and since $\| \Delta \|$ tends to zero as n tends to infinity, we find

$$\lim_{\|\Delta\| \to 0} S[f; \Delta] = 6A + 9B$$

PROBLEM 9.10

Compute the double integral from the previous problem by computing iterated integrals.

SOLUTION 9.10

We will first calculate the integral

$$\int_0^2 dx \int_0^3 (Ax + By)dy = \int_0^2 \left(Axy + \frac{1}{2}By^2\right)\Big|_0^3 dx$$

$$= \int_0^2 \left(3Ax + \frac{9}{2}B\right)dx = \frac{3}{2}Ax^2 + \frac{9}{2}Bx\Big|_0^2 = 6A + 9B.$$

As predicted by (9.1) in Theorem 9.9, the value of this iterated integral is equal to the value of the double integral.

Next, we calculate

$$\int_0^3 dy \int_0^2 (Ax + By)dx = \int_0^3 \left(\frac{1}{2}Ax^2 + Bxy\right)\Big|_0^2 dy$$

$$= \int_0^3 (2A + 2By)dy = 2Ay + By^2\Big|_0^3 = 6A + 9B.$$

This interated integral also equals the double integral as predicted by (9.2).

PROBLEM 9.11 FUBINI'S THEOREM

Show that if $f = f(x, y)$ is integrable over the closed rectangle $R = \{a \le x \le b, c \le y \le d\}$ and if $f(x, y)$ is integrable with respect to the variable y for each fixed x in [a, b] then

$$\iint_R f = \int_a^b F(x)\, dx$$

where

$$F(x) = \int_c^d f(x, y)dy$$

is integrable on [a, b].

SOLUTION 9.11

Let $\prod_x = \{x_1, \ldots, x_n\}$ and $\prod_y = \{y_1, \ldots, y_m\}$ denote partitions of [a, b] and [c, d] respectively and let $\Delta = \prod_x \times \prod_y$ denote the corresponding partition of R. Let R_{ij} denote the subrectangle $[x_{i-1}, x_i] \times [y_{j-1}, y_j]$ and let m_{ij}, M_{ij} denote the minimum and maximum value of f in R_{ij} respectively.

For $i = 1, \ldots, n$ let ξ_i be chosen in $[x_{i-1}, x_i]$. Then by the mean value theorem for integrals there exists a number μ_{ij}, $m_{ij} \le \mu_{ij} \le M_{ij}$, such that

$$\int_{y_{j-1}}^{y_j} f(\xi_i, y)dy = \mu_{ij}(y_j - y_{j-1}).$$

Then

$$\sum_{i=1}^{n}\sum_{j=1}^{m}\mu_{ij}A(R_{ij}) = \sum_{i=1}^{n}\sum_{j=1}^{m}\mu_{ij}(x_i - x_{i-1})(y_j - y_{j-1})$$

$$= \sum_{i=1}^{n}\sum_{j=1}^{m}\int_{y_{j-1}}^{y_j} f(\xi_i, y)dy(x_i - x_{i-1})$$

$$= \sum_{i=1}^{n}\int_{c}^{d} f(\xi_i, y)dy(x_i - x_{i-1})$$

$$= \sum_{i=1}^{n} F(\xi_i)(x_i - x_{i-1})$$

Now $\| \prod_x \|$ tends to zero as $\| \Delta \|$ tends to zero and it follows that

$$\iint_R f = \lim_{\|\Delta\|\to 0}\sum_{i=1}^{n}\sum_{j=1}^{m}\mu_{ij}A(R_{ij})$$

$$= \lim_{\|\Delta\|\to 0}\sum_{i=1}^{n} F(\xi_i)(x_i - x_{i-1}) = \int_a^b F(x)dx.$$

In a similar way we can show that if f is integrable with respect to x for each fixed y in [c, d], and if

$$G(y) = \int_a^b f(x, y)dx$$

then G(y) is integrable on [c, d] and

$$\iint_R f = \int_c^d G(y)dy.$$

Iterated Integrals

PROBLEM 9.12

For $0 < r < R$, let $\Omega = \{(x, y, z): x^2 + y^2 \le r^2, x^2 + z^2 \le R^2\}$. Then find the volume of Ω by evaluating an appropriate triple integral.

SOLUTION 9.12

Since $x^2 + y^2 \le r^2$ describes a cylinder of radius r with generators parallel to the z axis, and $x^2 + z^2 \le R^2$ describes a cylinder of radius R with generators parallel to the y axis, it follows that Ω must be the volume common to these two cylinders. Figure 9.2 shows the part of Ω lying in the first octant. By symmetry the volume of Ω is equal to eight times the volume of the first octant portion. The volume of the part of Ω in the first octant is given by the following iterated integral

$$I = \int_0^r\int_0^{y(x)}\int_0^{z(x,y)} dz\,dy\,dx$$

where

$$z(x, y) = \sqrt{R^2 - x^2} \quad \text{and} \quad y(x) = \sqrt{r^2 - x^2}.$$

Then

$$I = \int_0^r\int_0^{y(x)}\sqrt{R^2 - x^2}\,dy\,dx = \int_0^r\sqrt{R^2 - x^2}\sqrt{r^2 - x^2}\,dx$$

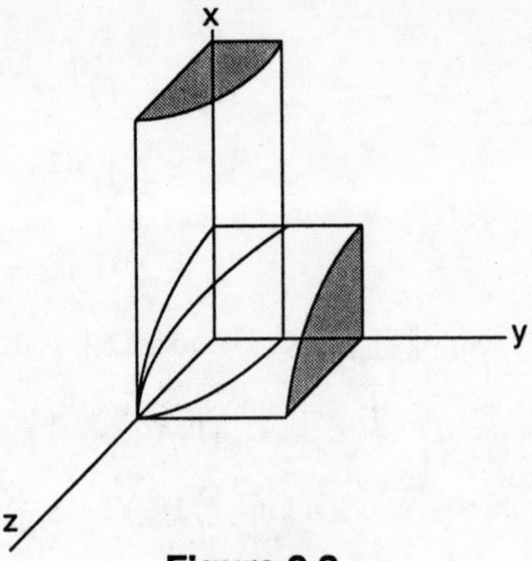

Figure 9.2
Problem 9.12

This last integral can not be evaluated in terms of elementary functions if $R > r$. In the case that $R = r$ the integral reduces to

$$I(R = r) = \int_0^R R^2 - x^2\, dx = \frac{2}{3} R^3.$$

Then

$$\iiint_\Omega 1 = 8I = \frac{16}{3} R^3 \quad \text{for } R = r.$$

PROBLEM 9.13 INTERCHANGING THE ORDER OF INTEGRATION

Integrate the continuous function $f(x, y) = x^2 \text{Sin } xy$ over the region $\Omega_y = \{y \le x \le 1, 0 \le y \le 1\}$. Integrate first with respect to y and then with respect to x.

SOLUTION 9.13

The region Ω_y is pictured in Figure 9.3 from which it becomes clear that Ω_y is identical to the region $\Omega_x = \{0 \le y \le x, 0 \le x \le 1\}$. Then

$$\iint_{\Omega_x} f = \int_0^1 \int_0^x x^2 \text{Sin } xy\, dy\, dx = \int_0^1 x^2 \left. \frac{-\text{Cos } xy}{x} \right|_0^x dx$$

$$= \int_0^1 x(1 - \text{Cos } x^2) dx = \left. \frac{x^2 - \text{Sin } x^2}{2} \right|_0^1 = \frac{1 - \text{Sin } 1}{2}$$

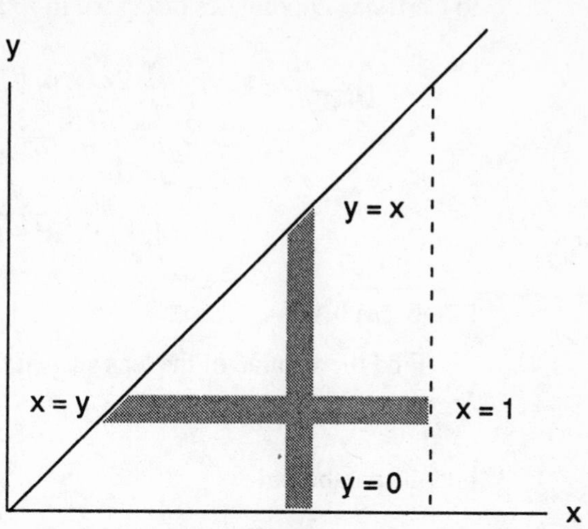

Figure 9.3
Problem 9.13

$$= \iint_{\Omega_y} f$$

$$= \int_0^1 dy \int_0^y x^2 \text{Sin } xy \, dx.$$

Note that while the two iterated integrals are equal, integrating first with respect to x leads to a more difficult integration.

PROBLEM 9.14

Find the volume removed if a cylindrical hole of radius r is drilled through the center of a sphere of radius $R > r > 0$.

SOLUTION 9.14

We note that $x^2 + y^2 + z^2 \leq R^2$ and $x^2 + y^2 \leq r^2$ are the equations of the solid sphere and cylinder respectively. Then the volume of the hole through the sphere is equal to

$$\iiint_D 1 = \iint_\Omega dx\, dy \int_{z_1(x,y)}^{z_2(x,y)} dz = \iint_\Omega z_2 - z_1 \, dxdy$$

where

$$z_2(x, y) = \sqrt{R^2 - x^2 - y^2} \quad z_1(x, y) = -\sqrt{R^2 - x^2 - y^2} = -z_2(x,y)$$

and

$$\Omega = \{x^2 + y^2 \leq r^2\}.$$

Note that the disc Ω is the image under the mapping Φ of the rectangle $U = \{0 \le \vartheta \le 2\pi, 0 \le \rho \le r\}$. By Φ we mean the coordinate transformation from polar to Cartesian coordinates discussed in Example 9.6(a). Then $\Omega = \Phi[U]$ and

$$\iint_{\Omega[U]} 2z_2(x, y) = \iint_U 2z_2(x(\rho, \vartheta), y(\rho, \vartheta)) \frac{\partial(x, y)}{\partial(\rho, \vartheta)} d\rho\, d\vartheta$$

$$= \int_0^{2\pi} \int_0^r 2\sqrt{R^2 - \rho^2}\, \rho\, d\rho\, d\vartheta$$

$$= 4p\int_0^r \rho\sqrt{R^2 - \rho^2}\, d\rho = \frac{8\pi}{3}(R^3 - (R^2 - r^2)^{3/2}).$$

PROBLEM 9.15

Find the volume of the lens shaped region common to the sphere

$$x^2 + y^2 + z^2 \le 2$$

and the paraboloid

$$z \ge x^2 + y^2.$$

SOLUTION 9.15

The boundaries of these two solids intersect along the curve

$$z = x^2 + y^2 = 2 - z^2;$$

i.e.,

$$z = 1 \quad \text{and} \quad x^2 + y^2 = 1.$$

Now let Ω denote the region $\{x^2 + y^2 \le 1\}$ in the xy–plane. Then the lens shaped region is described by

$$D = \{(x, y, z): (x, y) \in \Omega, \quad x^2 + y^2 < z < \sqrt{2 - x^2 - y^2}\}$$

and the volume equals

$$\iiint_D 1 = \iint_\Omega dx\, dy \int_{\varphi(x,y)}^{\psi(x,y)} dz$$

where

$$\varphi(x, y) = x^2 + y^2 \quad \text{and} \quad \psi(x, y) = \sqrt{2 - x^2 - y^2}.$$

Because of the symmetry of the region D, the integration is more easily carried out in cylindrical coordinates where

$$x = \rho \operatorname{Cos} \vartheta, \quad y = \rho \operatorname{Sin} \vartheta, \quad z = z \quad \text{and } J = \frac{\partial(x, y, z)}{\partial(\rho, \vartheta, z)} = \rho.$$

This change of variables reduces the integral over D to

$$\iiint_D 1 = \int_0^{2\pi} d\vartheta \int_0^1 \rho\, d\rho \int_{\varphi(\rho,\vartheta)}^{\psi(\rho,\vartheta)} dz$$

$$= \int_0^{2\pi} d\vartheta \int_0^1 (\sqrt{2-\rho^2} - \rho^2)\rho\, d\rho = \frac{2\pi}{3}\left(2^{2/3} - \frac{7}{4}\right)$$

PROBLEM 9.16

Find the volume of the solid region D common to the sphere

$$x^2 + y^2 + z^2 \le R^2$$

and the cone

$$x^2 + y^2 \le z^2, \quad z \ge 0.$$

SOLUTION 9.16

The boundary surfaces of these two solids intersect in the curve

$$z^2 = x^2 + y^2 = R^2 - z^2, \quad z \ge 0;$$

i.e.,

$$z = R/\sqrt{2} \quad \text{and} \quad x^2 + y^2 = R^2/2.$$

Then the volume of D is obtained by computing

$$\iiint_D 1 = \iint_\Omega dx\, dy \int_{z_1(x,y)}^{z_2(x,y)} dz$$

where

$$\Omega = \{x^2 + y^2 \le R^2/2\}$$

$$z_2(x, y) = \sqrt{R^2 - x^2 - y^2} \quad \text{and} \quad z_1(x, y) = \sqrt{x^2 - y^2}.$$

Note that the cone meets the xz-plane in the lines $z^2 = x^2$ from which it is clear that the generators of the cone are at an angle of $\pi/4$ to the z-axis. Then in spherical coordinates the region D is described by

$$\{(\rho, \vartheta, \varphi): 0 \le \rho \le R, 0 \le \vartheta \le 2\pi, 0 \le \varphi \le \pi/4\}$$

That is, $D = \Phi[B]$ where Φ denotes the coordinate transformation from spherical to Cartesian coordinates and B denotes the rectangular box $B = 0 \le \rho \le R$, $0 \le \vartheta \le 2\pi$, $0 \le \varphi \le \pi/4\}$.The transformation Φ is described in Example 9.6(b) where we found the Jacobian $J = \rho^2 \text{Sin}\, \varphi$. Then

$$\iiint_D 1 = \iint_0^{2\pi} d\vartheta \int_0^{\pi/4} \text{Sin}\, \varphi\, d\varphi \int_0^R \rho\, d\rho$$

$$= 2p(-\text{Cos}\, \varphi)_0^{\pi/4} \left(\frac{1}{3}\rho^3\right)_0^R = \frac{2}{3}\pi R^3 \left(1 - \text{Cos}\frac{\pi}{}\right)_4$$

PROBLEM 9.17

Calculate the double integral of $f(x, y) = x^2 - y^2$ over the region

$$\Omega = \{(x, y): |x| + |y| = 1\}.$$

SOLUTION 9.17

The region Ω is the square with sides:

$$y = 1-x, 0 \leq x \leq 1,$$
$$y = x-1, 0 \leq x \leq 1,$$
$$y = -1-x, -1 \leq x \leq 0$$
$$y = 1-x, -1 \leq x \leq 0.$$

If we introduce the change of variables

$$x = \frac{1}{2}(u + v) \qquad y = \frac{1}{2}(u - v)$$

then $u = x + y$ and $v = x - y$ and it is easy to show that Ω is the image under this transformation of the square $U = \{-1 \leq u \leq 1, -1 \leq v \leq 1\}$. The Jacobian of this transformation is

$$\frac{\partial(x, y)}{\partial(u, v)} = \begin{vmatrix} 1/2 & 1/2 \\ 1/2 & -1/2 \end{vmatrix} = -1/2$$

Thus

$$\iint_\Omega x^2 - y^2 = \iint_U uv = \int_{-1}^{1} dv \int_{-1}^{1} uv(-1/2)du$$
$$= -1/2 \int_{-1}^{1} u^2v/2 \Big|_{-1}^{1} dv = 0.$$

PROBLEM 9.18

Find the volume of the ellipsoid

$$\left(\frac{x}{a}\right)^2 + \left(\frac{y}{b}\right)^2 + \left(\frac{z}{c}\right)^2 \leq 1$$

SOLUTION 9.18

The volume of the ellipsoid is eight times the volume of the portion lying in the first octant. If we denote the first octant portion of the ellipsoid by D then the volume of D is equal to

$$\iiint_D 1 = \iint_\Omega dx\, dy \int_0^{z(x,y)} dz$$

where Ω denotes the plane region

$$\left(\frac{x}{a}\right)^2 + \left(\frac{y}{b}\right)^2 \leq 1$$

and $\quad z(x, y) = c\sqrt{1 - (x/a)^2 - (y/b)^2}.$

Now we shall introduce the change of variables Φ:

$$x = a\rho \operatorname{Cos} \vartheta, \quad y = b\rho \operatorname{Sin} \varphi, z = z.$$

Then

$$\frac{\partial(x, y, z)}{\partial(\rho, \vartheta, z)} = ab\rho$$

and

$$\iiint_D 1(x, y, z) = \iiint_U 1(\rho, \vartheta, z) = \int_0^{2\pi} d\vartheta \int_0^1 ab\rho \, d\rho \int_0^{z(\rho)} dz$$

where

$$U = \{0 \leq \rho \leq 1, 0 \leq \vartheta \leq 2\pi, 0 \leq z \leq z(\rho) = c\sqrt{1 - \rho^2}.$$

Note that D is the image of U under the transformation Φ. Thus

$$\iiint_D = abc \int_0^{2\pi} \int_0^1 \sqrt{1 - \rho^2} \rho \, d\rho \, d\vartheta = \frac{4}{3} \pi abc.$$

While the Riemann integral of a function of one variable is defined only over an interval in $\mathbb{R}^1$, the Riemann integral of a function of several variables may defined over sets more general and much more complicated than rectangular boxes (n-dimensional intervals). Those sets in $\mathbb{R}^2$ that are suitable domains of integration are referred to as sets with area. Sets with area include sets whose boundary is composed of finitely many smooth curves but more generally they are sets whose boundary is a set of zero content. A set Ω in $\mathbb{R}^2$ is said to have zero content if for each $\in > 0$ there exists a collection of finitely many rectangles whose union contains Ω and whose areas sum to less than $\in$.

If Ω in $\mathbb{R}^2$ is a set with area then $f = f(x, y)$ is integrable over Ω if f is continuous on Ω. More generally, f is integrable on Ω if the set of discontinuities for f in Ω is a set of content zero.

The integral of f over Ω is defined first in the case that Ω is a rectangle. We define partitions for rectangular sets and in turn define the integral over such sets in terms of a limit of Riemann sums as the mesh size of the partition tends to zero. Alternatively we can define upper and lower sums for f on each partition and by taking the GLB and LUB respectively over all partitions of the set we obtain the upper and lower sums for f over Ω. If these values are equal then their common value is defined to be the integral of f over Ω. The limit definition is equivalent to the definition of the integral in terms of sums.

The double integral of f over a more general set Ω in $\mathbb{R}^2$ is obtained by choosing a rectangle R enclosing Ω and extending f from Ω to R using the extension by zero. The definition is independent of the choice of the enclosing rectangle R.

Evaluating a double integral using partitions and sums is not convenient but there is no double integral analogue of the fundamental theorem of calculus which was the principal method for evaluating single variable integrals. Instead it is shown that a double integral is equal to two single integrals performed in succession. The pair of single integrals is called an iterated integral. The fundamental theorem may be applied to each of the single integrals in the iterated integral.

Integrals over sets in $\mathbb{R}^3$ are called triple integrals. Their development exactly parallels the development of double integrals. In fact the development of integrals over sets in $\mathbb{R}^n$ for any $n > 2$ may be accomplished in a similar manner. Since many applications involve only double or triple integrals, these two cases are emphasized.

Changes of variables in multiple integrals are carried out in terms of coordinate transformations and their Jacobians. Of particular interest are the transformations from Cartesian to polar coordinates in $\mathbb{R}^2$ and from Cartesian to cylindrical or spherical coordinates in $\mathbb{R}^3$.

10

Vector Differential Calculus

In this chapter we combine the techniques of vector analysis with topics from the differential calculus of functions of one and several variables. This combination of concepts leads to notational efficiency and clarity of discussion.

We begin by recalling basic notation and definitions from vector analysis. Then vector methods are introduced to the treatment of curves in $\mathbb{R}^3$ which are viewed as vector valued functions of one variable. Analysis of motion on such space curves is considerably clarified by the use of vector concepts.

The notation for differential calculus of functions of several variables is made simpler by the introduction of the vector differential operator known as the del operator. This operator is the basis for the gradient, divergence, Laplacian and curl operators. Various identities involving these operators are introduced and we conclude by stating the so-called maximum-minimum principles for the Laplacian. These principles are of considerable value in the study of certain problems in partial differential equations.

VECTOR ALGEBRA

Vectors and Scalars

COMPONENTS AND DIRECTION NUMBERS

By a *vector* $\underline{V}$ we mean an ordered triple of real numbers (v_1, v_2, v_3). We can think of the vectors as directed line segments from the origin to a point in $\mathbb{R}^3$. Alternatively we may think of the vector $\underline{V}$ as a directed line from an arbitrary point $\underline{P} = (p_1, p_2, p_3)$ to the point $\underline{Q} = (q_1, q_2, q_3)$ where $q_i - p_i = v_i$ for $i = 1, 2, 3$. We refer to the numbers v_i as the *components* or *direction numbers* for the vector $\underline{V}$. Thus the direction and length of a vector are uniquely determined by its components but the initial point of the vector may be abitrarily chosen.

UNIT VECTOR, DIRECTION COSINES

A vector of length 1 is called a *unit vector*. In the case of a unit vector the three components are equal to the cosine of the angle between the unit vector and the respective coordinate axes. Hence we refer to the components of a unit vector as *direction cosines*.

SCALARS

We will use the term *scalar* to refer to the real numbers. Scalars are single real numbers as opposed to vectors which are triples of real numbers. We will use the notation $\underline{V}$, $\underline{W}$, $\underline{X}$ to indicate vectors and we will generally denote scalars by lower case letters.

Vector Addition and Scalar Multiplication

We can define the following operations for vectors $\underline{V} = (v_1, \dots, v_n)$, $\underline{W} = (w_1, \dots, w_n)$ and any scalar α:

Vector Addition $\underline{V} + \underline{W} = (v_1 + w_1, \dots, v_n + w_n)$

Scalar Multiplication $\alpha\underline{V} = (\alpha v_1, \dots, \alpha v_n)$

The Inner Product

In addition to the operations defined above, we can define the *inner product* of two vectors. The inner product is also often called the *dot product*. The inner product of vectors $\underline{V}$ and $\underline{W}$ will be denoted by $(\underline{V}, \underline{W})$ or $\underline{V} \cdot \underline{W}$ and is defined as follows:

$$(\underline{V}, \underline{W}) = v_1w_1 + v_2w_2 + v_3w_3. \tag{10.1}$$

Note that

$$(\underline{V}, \underline{W}) = (\underline{W}, \underline{V})$$

and

$$(\alpha\underline{X} + \beta\underline{V}, \underline{W}) = \alpha(\underline{X}, \underline{W}) + \beta(\underline{V}, \underline{W}).$$

LENGTH, ZERO VECTOR

The *length* of the vector $\underline{V}$ will be denoted by $\|\underline{V}\|$ and can then be expressed as

$$\|\underline{V}\| = (\underline{V}, \underline{V})^{1/2} = \sqrt{v_1^2 + v_2^2 + v_3^2}$$

The *zero vector*, all of whose components are zero, is the only vector having zero length. For all nonzero vectors the length is positive.

ORTHOGONALITY

We can show that

$$(\underline{V}, \underline{W}) = \|\underline{V}\| \, \|\underline{W}\| \operatorname{Cos} \vartheta \tag{10.2}$$

where ϑ denotes the angle between the vectors $\underline{V}$ and $\underline{W}$. It follows from (10.2) that two nonzero vectors $\underline{V}$ and $\underline{W}$ are *orthogonal* if their inner product equals zero. The zero vector is orthogonal to every vector.

The Vector Product

The inner product is a scalar valued product of two vectors. The *vector product* or *cross product* is a product of two vectors with a vector result. The vector product is defined as follows:

$$\underline{V} \times \underline{W} = \begin{vmatrix} \underline{i} & \underline{j} & \underline{k} \\ v_1 & v_2 & v_3 \\ w_1 & w_2 & w_3 \end{vmatrix} \qquad (10.3)$$

Since interchanging two rows of a determinant changes the sign of the determinant, it follows that

$$\underline{V} \times \underline{W} = -\underline{W} \times \underline{V}\,.$$

Then the vector product is *not commutative*. We can show also that for all vectors $\underline{V}$, $\underline{W}$ and $\underline{X}$

$$\underline{X} \times (\underline{V} \times \underline{W}) = (\underline{W} \cdot \underline{X})\underline{V} - (\underline{V} \cdot \underline{X})\underline{W} \qquad (10.4)\,(a)$$

and

$$(\underline{X} \times \underline{V}) \times \underline{W} = (\underline{X} \cdot \underline{W})\underline{V} - (\underline{V} \cdot \underline{W})\underline{X} \qquad (10.4)\,(b)$$

hence $\underline{X} \times (\underline{V} \times \underline{W}) \neq (\underline{X} \times \underline{V}) \times \underline{W}$ i. e., the vector product is not associative.

We can show that

$$\underline{V} \times \underline{W} = \|\underline{V}\|\,\|\underline{W}\| \operatorname{Sin} \vartheta\, \underline{N} \qquad (10.5)$$

where ϑ denotes the angle between $\underline{V}$ and $\underline{W}$ and $\underline{N}$ denotes a unit vector that is orthogonal to both $\underline{V}$ and $\underline{W}$. Thus nonzero vectors whose vector product equals zero are parallel and the vector product of $\underline{V}$ and $\underline{W}$ is normal to the plane spanned by $\underline{V}$ and $\underline{W}$.

VECTOR DIFFERENTIAL CALCULUS

Differentiation of Vector Functions

A vector function $\underline{U}(t)$ has derivative equal to

$$\underline{U}'(t) = \lim_{h \to 0} \frac{\underline{U}(t+h) - \underline{U}(t)}{h}$$

if the limit exists. In this case we say $\underline{U}(t)$ is a *differentiable vector function* and if $\underline{U}(t) = u_1(t)\underline{i} + u_2(t)\underline{j} + u_3(t)\underline{k}$ then $\underline{U}'(t) = u_1'(t)i + u_2'(t)\underline{j} + u_3'(t)\underline{k}$.

RULES FOR DIFFERENTIATING VECTOR FUNCTIONS

The rules for differentiating combinations of vector functions are just simple extensions of the rules for differentiating scalar functions with the exception of the rule for differentiating the vector product. Since this product is not commutative, care must be taken to preserve the order of the factors. For differentiable vector functions $\underline{U}(t)$, $\underline{V}(t)$ and $\underline{W}(t)$ and differentiable scalar function f(t):

1. $\dfrac{d}{dt}(\underline{U}(t) + \underline{V}(t)) = \underline{U}'(t) + \underline{V}'(t)$

2. $\dfrac{d}{dt}(f(t)\,\underline{U}(t)) = f'(t)\underline{U}(t) + f(t)\,\underline{U}'(t)$

3. $\dfrac{d}{dt}(\underline{U}(t) \cdot \underline{V}(t)) = \underline{U}'(t) \cdot \underline{V}(t) + \underline{U}(t) \cdot \underline{V}'(t)$

4. $\dfrac{d}{dt}(\underline{U}(t) \times \underline{V}(t)) = \underline{U}'(t) \times \underline{V}(t) + \underline{U}(t) \times \underline{V}'(t)$

Space Curves

A curve Γ in $\mathbb{R}^3$ can be represented parametrically in the form

$$\Gamma : \begin{cases} x = x(t) \\ y = y(t) \quad a \le t \le b \\ z = z(t) \end{cases}$$

Then the vector $\underline{P}(t) = (x(t), y(t), z(t))$ is referred to as the *position vector* for the curve Γ. As t varies over the interval [a, b] $\underline{P}(t)$ traces out the curve Γ.

Example 10.1 Space Curves

10.1 (a) Consider Γ given by

$$\Gamma : \begin{cases} x = (x_1 - x_0)t + x_0 \\ y = (y_1 - y_0)t + y_0 \quad 0 \le t \le 1 \\ z = (z_1 - z_0)t + z_0 \end{cases}$$

Then $\underline{P}(t) = (x(t), y(t), z(t))$ moves along Γ from $\underline{P}(0) = (x_0, y_0, z_0)$ at t = 0, to $\underline{P}(1) = (x_1, y_1, z_1)$ at t = 1. Note that for each t, $0 \le t \le 1$,

$$\underline{P}(t) - \underline{P}(0) = ((x_1 - x_0)t, (y_1 - y_0)t, (z_1 - z_0)t) = t(\underline{P}(1) - \underline{P}(0)).$$

Clearly for each t, $\underline{P}(t) - \underline{P}(0)$ is a scalar multiple of $\underline{P}(1) - \underline{P}(0)$ and it follows that these vectors have the same direction for all t in [0, 1]. Thus the path Γ from $\underline{P}(0)$ to $\underline{P}(1)$ is a straight line.

10.1 (b) Consider the curve

$$\Gamma : \begin{cases} x = R\,\text{Cos}\,\Omega t + x_0 \\ y = R\,\text{Sin}\,\Omega t + y_0 \quad 0 \le t \le 2\pi/\Omega \\ z = kt \end{cases}$$

Note that

$$(x(t) - x_0)^2 + (y(t) - y_0)^2 = R^2(\text{Cos}^2\Omega t + \text{Sin}^2\Omega t) = R^2.$$

Then the projection of Γ into the xy-plane is a circle of radius R and center at (x_0, y_0). We refer to the space curve Γ as a *circular helix*.

VELOCITY VECTOR

The *velocity vector* for a curve Γ is defined to be

$$\underline{P}'(t) = \lim_{\Delta t \to 0} \frac{1}{\Delta t}(\underline{P}(t + \Delta t) - \underline{P}(t))$$

if the limit exists. When the limit exists, we have

$$\underline{P}'(t) = (x'(t), y'(t), z'(t)) \tag{10.6}$$

and we can show that the direction of the velocity vector is tangent to the curve Γ.

ARCLENGTH AND SPEED

$\underline{P}(t + \Delta t)$ and $\underline{P}(t)$ are two points on Γ and $\underline{P}(t + \Delta t) - \underline{P}(t)$ is the vector from $\underline{P}(t)$ to $\underline{P}(t + \Delta t)$; the chord joining the two points. If we let Δs denote the length of this chord then

$$\Delta s = \| \underline{P}(t + \Delta t) - \underline{P}(t) \| = \sqrt{\Delta x^2 + \Delta y^2 + \Delta z^2}$$

where

$$\Delta x = x(t + \Delta t) - x(t),\ \Delta y = y(t + \Delta t) - y(t),\ \Delta z = z(t + \Delta t) - z(t).$$

Then Δs approximates ds, the element of arclength and

$$\frac{ds}{dt} = \lim_{\Delta t \to 0} \frac{\Delta s}{\Delta t} = \| \underline{P}'(t) \| = \sqrt{x'(t) + y'(t) + z'(t)} \tag{10.7}$$

Then ds/dt represents the rate of change of position on the curve with respect to t and has the interpretation of being the speed at which the point $\underline{P}(t)$ moves along the curve Γ. Thus the *direction* of the velocity vector is *tangent* to the curve and the *magnitude* of the velocity vector is equal to the *speed* of the point $\underline{P}(t)$ as it moves along Γ. The positive sense of the velocity vector is referred to as the *direction of increasing arc length.*

UNIT TANGENT VECTOR

If we define the unit tangent vector to the curve Γ by

$$T(t) = \frac{\underline{P}'(t)}{\| \underline{P}'(t) \|}$$

then

$$\underline{P}'(t) = \frac{ds}{dt} T(t). \tag{10.8}$$

If we write $\underline{P}(t) = \underline{P}(s(t))$ then the chain rule implies

$$\underline{P}'(t) = \frac{d\underline{P}}{ds}\frac{ds}{dt} \tag{10.9}$$

Comparing this result with (10.8) leads to the conclusion

$$\frac{d\underline{P}}{ds} = T(t). \tag{10.10}$$

ACCELERATION VECTOR

We define the acceleration vector for the curve Γ to be the derivative with respect to t of the velocity vector. That is $P''(t) = (x''(t), y''(t), z''(t))$. In view of (10.8) this can also be written in the form

$$P''(t) = \frac{d}{dt}\left(\frac{ds}{dt}\right) \underline{T}(t) + \frac{ds}{dt} \underline{T}'(t).$$

UNIT NORMAL TO A CURVE, CURVATURE

Since $\underline{T}(t)$ is a unit vector for all t, $\underline{T} \cdot \underline{T} = 1$, hence

$$\frac{d}{dt}(\underline{T} \cdot \underline{T}) = \underline{T}'(t) \cdot \underline{T}(t) + \underline{T}(t) \cdot \underline{T}'(t) = 2\,\underline{T}(t) \cdot \underline{T}'(t) = 0.$$

It follows that $\underline{T}'(t)$ is orthogonal to $\underline{T}(t)$. If we define $\underline{N}(t)$ to be a unit vector whose direction for each t is orthogonal to $\underline{T}(t)$ and whose sense is the same as that of P(t), then $\underline{T}'(t) = \alpha\, \underline{N}(t)$ where $\alpha = \|\underline{T}'(t)\|$. But

$$\|\underline{T}'(t)\| = \left\|\frac{d\underline{T}}{ds}\frac{ds}{dt}\right\| = \left\|\frac{d\underline{T}}{ds}\right\| \frac{ds}{dt} = \kappa \frac{ds}{dt}$$

where we define the parameter κ, the *curvature* of Γ, by

$$k = \left\|\frac{d\underline{T}}{ds}\right\|. \tag{10.11}$$

TANGENTIAL AND CENTRIFUGAL ACCELERATION

We have now

$$P''(t) = \frac{d^2s}{dt^2} \underline{T}(t) + \kappa \left(\frac{ds}{dt^2}\right)^2 \underline{N}(t). \tag{10.12}$$

We refer to

$$\frac{d^2s}{dt^2} \underline{T}(t) \quad \text{and} \quad \kappa \left(\frac{ds}{dt}\right)^2 \underline{N}(t)$$

respectively as the *tangential acceleration* and *centrifugal acceleration* on Γ.

Example 10.2 Velocity and Acceleration

10.1 (a) For α a positive constant and $t \geq 0$, consider the straight line,

$$P(t) = ((x_1 - x_0)\alpha t + \alpha x_0, (y_1 - y_0)\alpha t + \alpha y_0, (z_1 - z_0)\alpha t + \alpha z_0).$$

Then $P'(t) = ((x_1 - x_0)\alpha, (y_1 - y_0)\alpha, (z_i - z_0)\alpha) = \alpha(P(1) - P(0))$ and $P''(t) = (0, 0, 0) = \underline{0}$. Note that

$$\underline{T}(t) = \frac{P'(t)}{\| P'(t) \|} = \frac{P(1) - P(0)}{\| P(1) - P(0)\|}$$

Hence

$$P'(t) = \alpha(P(1) - P(0)) = \alpha\| P(1) - P(0)\| \frac{P(1) - P(0)}{\| P(1) - P(0) \|}$$

$$= \alpha\| P(1) - P(0) \| \underline{T}(t).$$

It follows from (10.8) that $ds/dt = \alpha\| P(1) - P(0)\|$. Note that since the tangent vector has constant direction the curvature equals zero (as you would expect for a straight line) so there is no centrifugal acceleration. In addition, ds/dt is constant so there is no tangential acceleration either.

10.2 (b) Consider the circular helix of Example 10.1(b). Then

$$P'(t) = (- R\Omega \text{ Sin } \Omega t, R\Omega \text{ Cos } \Omega t, k) \text{ and } \|P'(t) \| = \sqrt{R^2\Omega^2 + k^2}$$

$$P''(t) = (- R\Omega^2 \text{Cos } \Omega t, - R\Omega^2 \text{Sin } \Omega t, 0) \text{ and } \|P''(t)\| = R\Omega^2.$$

If $k = 0$ then Γ becomes a circle in the xy-plane. In this case

$$\underline{T}(t) = \frac{P'(t)}{\|P'(t)\|} = (-\text{Sin } \Omega t, \text{Cos } \Omega, 0) \quad \text{and} \frac{ds}{dt} = R\Omega.$$

Note further that for $k = 0$ we have $P''(t) \cdot \underline{T}(t) = 0$. Then it follows from (10.12) that

$$P''(t) = \kappa(ds/dt)^2\underline{N}(t).$$

Hence

$$\|P''(t)\| = R\Omega^2 = \kappa(ds/dt)^2 = \kappa R^2\Omega^2$$

It follows that the curvature κ for the circular path is equal to $1/R$ and the magnitude of the centrifugal acceleration equals $R\Omega^2$. The tangential component of acceleration is zero for the circular path.

Vector Differential Operators

SCALAR FIELDS, VECTOR FIELDS

A scalar valued function $F = F(x, y, z)$ defined over some domain Ω in $\mathbb{R}^3$ will be called a *scalar field*. Similarly a function defined on Ω with vector values

$$\underline{V}(x, y, z) = v_1(x, y, z)\underline{i} + v_2(x, y, z)\underline{j} + v_3(x, y, z)\underline{k}$$

will be referred to as a *vector field.* We shall suppose throughout this chapter that all scalar and vector fields are at least differentiable.

The Del Operator: Gradient, Divergence and Curl

We define a vector differential operator

$$\nabla = \frac{\partial}{\partial x}\underline{i} + \frac{\partial}{\partial y}\underline{j} + \frac{\partial}{\partial z}\underline{k}$$

which we will call the *del operator*. We can apply this operator to both scalar and vector fields in the following ways:

Gradient of a Scalar Field For scalar field $F = F(x, y, z)$

$$\nabla F = F_x\underline{i} + F_y\underline{j} + F_z\underline{k}$$

is a vector field called the *gradient* of F. It is denoted by ∇F or grad F.

Divergence of a Vector Field For vector field $\underline{V} = \underline{V}(x, y, z)$, the inner product

$$\begin{aligned}\text{div } \underline{V} &= \nabla \cdot \underline{V} \\ &= \left(\frac{\partial}{\partial x}\underline{i} + \frac{\partial}{\partial y}\underline{j} + \frac{\partial}{\partial z}\underline{k}\right) \cdot (v_1(x, y, z)\underline{i} + v_2(x, y, z)\underline{j} + v_3(x, y, z)\underline{k}) \\ &= \frac{\partial v_1}{\partial x} + \frac{\partial v_2}{\partial y} + \frac{\partial v_3}{\partial z}\end{aligned}$$

is a scalar field called the *divergence* of $\underline{V}(x, y, z)$. It is denoted by div $\underline{V}$ or $\nabla \cdot \underline{V}$.

Curl of a Vector Field For a vector field $\underline{V} = \underline{V}(x, y, z)$ the vector product

$$\nabla \times \underline{V}(x, y, z) = \text{Curl } \underline{V} = \begin{vmatrix} \underline{i} & \underline{j} & \underline{k} \\ \partial/\partial x & \partial/\partial y & \partial/\partial z \\ v_1 & v_2 & v_3 \end{vmatrix}$$

$$= \underline{i}\left(\frac{\partial v_3}{\partial y} - \frac{\partial v_2}{\partial z}\right) - \underline{j}\left(\frac{\partial v_3}{\partial x} - \frac{\partial v_1}{\partial z}\right) + \underline{k}\left(\frac{\partial v_2}{\partial x} - \frac{\partial v_1}{\partial y}\right)$$

is a vector field called the *curl* of $\underline{V}(x, y, z)$. It is denoted by Curl $\underline{V}$ or by $\nabla \times \underline{V}$.

Each of these fields has a variety of applications that will be described in this chapter and the next.

Example 10.3 Gradient, Divergence and

10.3 (a) For the scalar field $F(x, y, z) = (x^2 + y^2 + z^2)/2$ we have

$$\nabla F = F_x\underline{i} + F_y\underline{j} + F_z\underline{k} = x\underline{i} + y\underline{j} + z\underline{k}.$$

10.3 (b) For the vector fields $\underline{V}(x, y, z) = x\underline{i} + y\underline{j} + z\underline{k}$ and $\underline{W}(x, y, z) = f(x, y, z)\underline{C}$ where f denotes a scalar field and $\underline{C}$ is a constant vector

$$\text{div } \underline{V} = 1 + 1 + 1 = 3$$

and

$$\text{div } \underline{W} = f_x C_1 + f_y C_2 + f_z C_3 = \nabla f \cdot \underline{C}$$

10.3 (c) For vector fields $\underline{V}(x, y, z) = x\underline{i} + y\underline{j} + z\underline{k}$ and $\underline{W}(x, y, z) = \underline{\Omega} \times \underline{V}(x, y, z)$ for $\underline{\Omega}$ a constant vector, we compute

$$\text{Curl } \underline{V} = 0$$

and

$$\underline{W} = \underline{\Omega} \times \underline{V} = (\Omega_2 z - \Omega_3 y)\underline{i} + (\Omega_3 x - \Omega_1 z)\underline{j} + (\Omega_1 y - \Omega_2 x)\underline{k}$$

so

$$\text{Curl } \underline{W} = (\Omega_1 + \Omega_1)\underline{i} + (\Omega_2 + \Omega_2)\underline{j} + (\Omega_3 + \Omega_3)\underline{k} = 2\Omega$$

Vector Identities

The gradient, divergence and curl operators may be combined in various ways. If we suppose that $F = F(x, y, z)$ is a $\mathbb{C}^2$ scalar field and $V = V(x, y, z)$ is a $\mathbb{C}^2$ vector field then grad F and curl V are vector fields while div V is a scalar field. Then the following combinations of vector differential operators are defined:

$$\text{grad}(\text{div } \underline{V}) = \nabla(\nabla \cdot \underline{V})$$
$$\text{div}(\text{curl } \underline{V}) = \nabla \cdot (\nabla \times \underline{V})$$
$$\text{curl}(\text{curl } \underline{V}) = \nabla \times (\nabla \times \underline{V})$$
$$\text{curl}(\mathbf{grad}\ F) = \nabla \times (\nabla F)$$
$$\text{div}(\mathbf{grad}\ F) = \nabla \cdot \nabla F \text{ (also denoted by } \nabla^2 F)$$

We have the following results regarding these combinations of operators:

Theorem 10.1

Theorem 10.1 Let $F = F(x, y, z)$ denote a $\mathbb{C}^2$ scalar field and let $\underline{V} = \underline{V}(x, y, z)$ be a $\mathbb{C}^2$ vector field. Then at each point of the domain of F we have

$$\text{div}(\text{grad}F) = \nabla^2 F = F_{xx} + F_{yy} + F_{zz}$$

and at each point of the domain of $\underline{V}$ we have

$$\text{curl}(\text{curl}\underline{V}) = \mathbf{grad}\ (\text{div}\underline{V}) - (\nabla \cdot \nabla)\underline{V}.$$

Theorem 10.2

Theorem 10.2 Let $\underline{V} = \underline{V}(x, y, z)$ denote a $\mathbb{C}^2$ vector field. Then

(a) $\text{div}(\text{curl}\underline{V}) = 0$; i. e., if $\underline{W} = \text{curl } \underline{V}$ then $\text{div } \underline{W} = 0$.

(b) If $\text{div}\underline{W} = 0$ then $\underline{W} = \text{curl } \underline{V}$ for some smooth vector field $\underline{V}$.

Theorem 10.3

Theorem 10.3 Let $F = F(x, y, z)$ denote a $\mathbb{C}^2$ scalar field. Then

(a) $\text{curl}(\mathbf{grad}\ F) = 0$; i. e., if $\underline{V} = \mathbf{grad}\ F$ then $\text{curl } \underline{V} = \underline{0}$.

(b) If $\text{curl}\underline{V} = 0$ then $\underline{V} = \mathbf{grad}\ F$ for some smooth scalar field F.

In addition to these results there are a great number of identities relating to the application of vector differential operators to combinations of functions, separately and in combination with one another. We list several of these here.

ADDITIONAL OPERATOR IDENTITIES

Let $\underline{U}(x, y, z)$ and $\underline{V}(x, y, z)$ denote smooth vector fields and let $\varphi(x, y, z)$ and $\psi(x, y, z)$ denote smooth scalar fields.

1. $\nabla f(\varphi) = f'(\varphi)\nabla\varphi$ for $f \in \mathbb{C}^1(\mathbb{R}^1)$
2. $\nabla(\varphi\psi) = \varphi\nabla\psi + \psi\nabla\varphi$
3. $\nabla \cdot (\varphi\underline{V}) = \varphi\nabla \cdot \underline{V} + \underline{V} \cdot \nabla\varphi$
4. $\nabla \times (\varphi\underline{V}) = \varphi\nabla \times \underline{V} + \nabla\varphi \times \underline{V}$
5. $\nabla \cdot (\underline{U} \times \underline{V}) = \underline{V} \cdot (\nabla \times \underline{U}) - \underline{U} \cdot (\nabla \times \underline{V})$
6. $\nabla \times (\underline{U} \times \underline{V}) = (\underline{V} \cdot \nabla)\underline{U} - (\underline{U} \cdot \nabla)\underline{V} + (\nabla \cdot \underline{V})\underline{U} - (\nabla \cdot \underline{U})\underline{V}$

Maximum-Minimum Principles

Suppose that $f = f(x)$ is continuous on the closed interval $[a, b]$ and that f is $\mathbb{C}^2$ on the open interval (a, b). Let M denote the larger of the two values f(a) and f(b) and let m denote the smaller of these two values. Then we can draw the following conclusions based on the sign of $f''(x)$ on (a, b):

i) If $f''(x) \geq 0$ on (a, b) then $f(x) \leq M$ on [a, b]

ii) If $f''(x) = 0$ on (a, b) then $m \leq f(x) \leq M$ on [a, b]

iii) If $f''(x) \leq 0$ on (a, b) then $f(x) \geq m$ on [a, b]

It is evident from Figure 10.1 why these conclusions are justified. Figure 10.2 illustrates why it is necessary for f(x) to be continuous on the closed interval [a, b] and for $f''(x)$ to be continuous on the open interval (a, b). In Figure 10.2(a) we have $f''(x) > 0$ but f(x) exceeds M at a point of (a, b) since $f''(x)$ is not continuous on (a, b). A similar example (not shown) is possible in which conclusion iii) fails because $f''(x)$ is not continuous on (a, b). Figure 10.2(b) shows a function f(x) with $f''(x) = 0$ for which conclusion ii) fails because f(x) is not continuous on [a, b]. We refer to these results as *maximum-minimum principles*. Similar results hold for functions of several variables $F = F(x, y, z)$ with $\nabla^2 F$ playing the role of $f''(x)$.

Theorem 10.4

Theorem 10.4 Let Ω denote a bounded domain in $\mathbb{R}^3$ with boundary Γ and suppose $F = F(x, y, z)$ is continuous on the compact set $\Omega \cup \Gamma$ composed of Ω together with its boundary Γ. Suppose further that $\nabla^2 F$ is continuous on Ω and let m, M denote respectively the minimum and maximum values for F over the boundary, Γ.

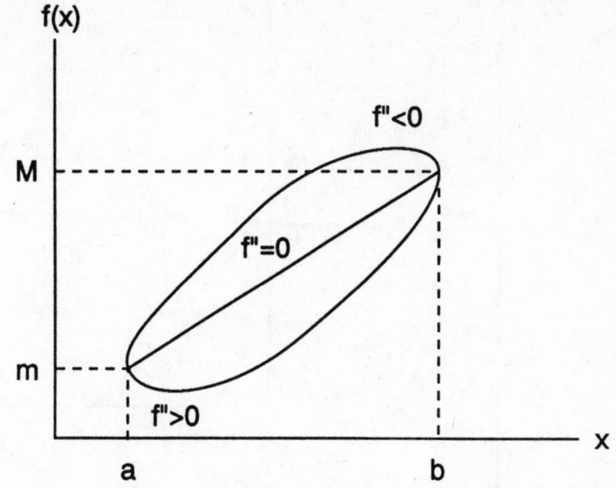

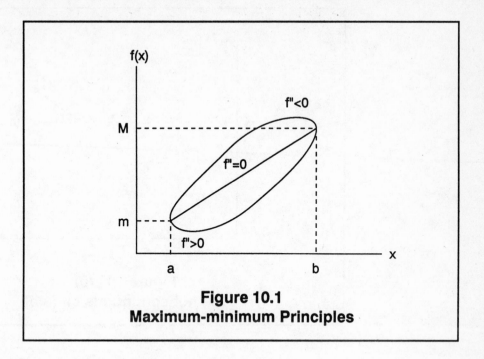

Figure 10.1
Maximum-minimum Principles

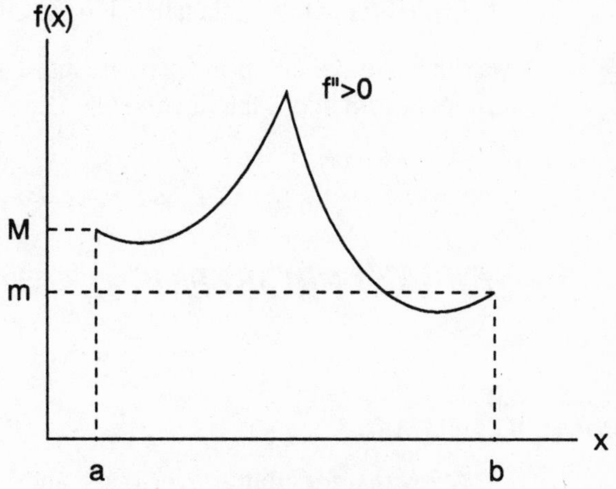

Figure 10.2(a)
f" not continuous on (a,b)

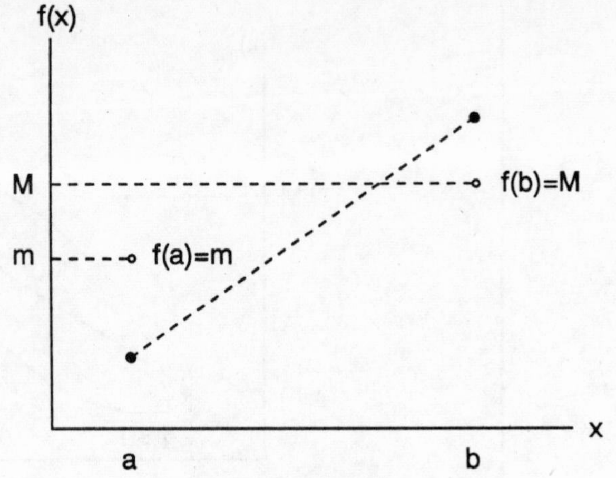

Figure 10.2(b)
f(x) not continuous on [a,b]

i) If $\nabla^2 F \geq 0$ in Ω then $M \geq F$ on $\Omega \cup \Gamma$

ii) If $\nabla^2 F = 0$ in Ω then $m \leq F \leq M$ on $\Omega \cup \Gamma$

iii) If $\nabla^2 F \leq 0$ in Ω then $m \leq F$ on $\Omega \cup \Gamma$.

Maximum-minimum principles are used extensively in the study of partial differential equations that involve the Laplace operator.

SOLVED PROBLEMS

Vector Algebra

PROBLEM 10.1

Show that for arbitrary vectors $\underline{V}$ and $\underline{W}$, $\underline{V} \cdot \underline{W} = \|\underline{V}\| \, \|\underline{W}\| \text{Cos } \vartheta$, where ϑ denotes the angle between the two vectors.

SOLUTION 10.1

The vectors $\underline{V}$ and $\underline{W}$ together with $\underline{W} - \underline{V}$ form a triangle with angle ϑ opposite the side $\underline{W} - \underline{V}$. Then the law of Cosines implies that

$$\|\underline{W} - \underline{V}\|^2 = \|\underline{W}\|^2 + \|\underline{V}\|^2 - 2\|\underline{W}\|\,\|\underline{V}\| \operatorname{Cos} \vartheta.$$

But

$$\begin{aligned}\|\underline{W} - \underline{V}\|^2 &= (w_1 - v_1)^2 + (w_2 - v_2)^2 + (w_3 - v_3)^2 \\ &= w_1{}^2 + w_2{}^2 + w_3{}^2 + v_1{}^2 + v_2{}^2 + v_3{}^2 - 2w_1v_1 - 2w_2v_2 - 2w_2v_2 \\ &= \|\underline{W}\|^2 + \|\underline{V}\|^2 - 2\underline{W} \cdot \underline{V}.\end{aligned}$$

Thus

$$\underline{W} \cdot \underline{V} = \|\underline{W}\|\,\|\underline{V}\| \operatorname{Cos} \vartheta.$$

PROBLEM 10.2

Show that for arbitrary vectors $\underline{V}$, $\underline{W}$ and $\underline{X}$

$$(\underline{V} \times \underline{W}) \times \underline{X} = (\underline{V} \cdot \underline{X})\underline{W} - (\underline{W} \cdot \underline{X})\underline{V} \tag{1}$$

SOLUTION 10.2

Note first that for arbitrary vectors $\underline{V}$, $\underline{W}$ and $\underline{X}$,

$$\underline{V} \cdot (\underline{W} \times \underline{X}) = \underline{W} \cdot (\underline{X} \times \underline{V}) = \underline{X} \cdot (\underline{V} \times \underline{W}) = \begin{vmatrix} v_1 & v_2 & v_3 \\ w_1 & w_2 & w_3 \\ x_1 & x_2 & x_3 \end{vmatrix} \tag{2}$$

It also follows by direct computation using the definition of the dot and cross products, that for arbitrary vectors $\underline{U}$, $\underline{V}$, $\underline{W}$ and $\underline{X}$,

$$(\underline{U} \times \underline{V}) \cdot (\underline{W} \times \underline{X}) = (\underline{U} \cdot \underline{W})(\underline{V} \cdot \underline{X}) - (\underline{U} \cdot \underline{X})(\underline{V} \cdot \underline{W}) \tag{3}$$

Equation (2) implies that for arbitrary vectors $\underline{U}$, $\underline{V}$, $\underline{W}$ and $\underline{X}$,

$$((\underline{V} \times \underline{W}) \times \underline{X}) \cdot \underline{U} = (\underline{V} \times \underline{W}) \cdot (\underline{X} \times \underline{U})$$

and then (3) leads to

$$((\underline{V} \times \underline{W}) \times \underline{X}) \cdot \underline{U} = (\underline{V} \cdot X)(\underline{W} \cdot \underline{U}) - (\underline{W} \cdot X)(\underline{V} \cdot \underline{U}) \tag{4}$$

Choosing $\underline{U}$ in (4) to be $\underline{i}$, $\underline{j}$ and $\underline{k}$ in turn then yields (1).

PROBLEM 10.3

A flat rectangular plate is photographed from three mutually pependicular directions. In each photograph the plate appears as a polygonal shape. If the areas of the polygonal shapes in the photographs are found to be A_1, A_2 and A_3 respectively, then what is the area of the plate?

SOLUTION 10.3

Let the rectangular plate be denoted by R and let the perpendicular edges of R be denoted by $\underline{U}$ and $\underline{V}$. Then $\|\underline{U}\| = L$ and $\|\underline{V}\| = W$ denote the length and

Space Curves

width of the plate. If we let $\underline{i}$, $\underline{j}$ and $\underline{k}$ denote unit vectors in the mutually perpendicular directions of the three photographs, then

$$\underline{U} = L(\cos \vartheta_1\underline{i} + \cos \vartheta_2\underline{j} + \cos \vartheta_3\underline{k}) \tag{1}$$

$$\underline{V} = W(\cos \varphi_1\underline{i} + \cos \varphi_2\underline{j} + \cos \varphi_3\underline{k}) \tag{2}$$

where ϑ_1, ϑ_2, ϑ_3, and φ_1, φ_2, φ_3 denote the angles between $\underline{U}$ and $\underline{V}$ and the unit vectors $\underline{i}$, $\underline{j}$ and $\underline{k}$. In addition, (10.5) implies

$$\underline{U} \times \underline{V} = LW\underline{N} \tag{3}$$

for $\underline{N}$ a unit vector normal to the plate R. Then $LW = \|\underline{U} \times \underline{V}\|$ and from (1) and (2) we have

$$\underline{U} \times \underline{V} = LW((\cos \vartheta_2 \cos \varphi_3 - \cos \vartheta_3 \cos \varphi_2)\,\underline{i}$$
$$+ (\cos \vartheta_3 \cos \varphi_1 - \cos \vartheta_1 \cos \varphi_3)\,\underline{j} + (\cos \vartheta_1 \cos \varphi_2 - \cos \vartheta_2 \cos \varphi_1)\underline{k})$$
$$= A_1\underline{i} + A_2\underline{j} + A_3\underline{k}.$$

Evidently the areas in the three photographs are the projections of $\underline{U} \times \underline{V}$ onto the planes having $\underline{i}$, $\underline{j}$ and $\underline{k}$ as normal vectors. Then

$$LW = \|\underline{U} \times \underline{V}\| = \sqrt{A_1^2 + A_2^2 + A_3^2}$$

This example is intended to illustrate the efficiency that results in using vector methods to solve geometric problems.

PROBLEM 10.4

Express the position, velocity and acceleration vectors in cylindrical coordinates.

SOLUTION 10.4

The transformation from cylindrical to Cartesian coordinates is described by

$$x = r \cos \vartheta, \quad y = r \sin \vartheta, \quad z = z;$$
$$r = \sqrt{x^2 + y^2}, \quad \vartheta = \operatorname{ArcTan}(y/x), \quad z = z. \tag{1}$$

Then as shown in Figure 10.3,

$$\underline{e}_r = \cos \vartheta\,\underline{i} + \sin \vartheta\,\underline{j} = \text{unit vector in the radial direction}$$

$$\underline{e}_\vartheta = \sin \vartheta\,\underline{i} - \cos \vartheta\,\underline{j} = \text{unit vector in the } \vartheta \text{ direction}$$

$$\underline{e}_z = \underline{k} = \text{unit vector in the z direction.}$$

Along any path we have

$$\underline{P}(t) = x(t)\underline{i} + y(t)\underline{j} + z(t)\underline{k} = r(t)\underline{e}_r + z(t)\underline{k} \tag{2}$$

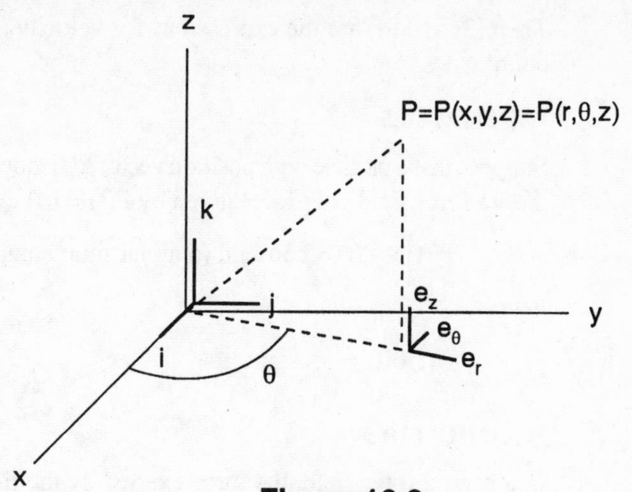

Figure 10.3
Unit Vectors in the Coordinate Directions

Note that the direction of the unit vectors $\underline{e}_r$ and $\underline{e}_\vartheta$ varies with t. Thus

$$\underline{P}'(t) = r'(t)\underline{e}_r + r(t)(d/dt)\underline{e}_r(t) + z'(t)\underline{k}.$$

But

$$(d/dt)\underline{e}_r = -\operatorname{Sin}\vartheta\vartheta'(t)\underline{i} + \operatorname{Cos}\vartheta\vartheta'(t)\,\underline{j} = -\vartheta'(t)\underline{e}_\vartheta.$$

Thus

$$\underline{P}'(t) = r'(t)\underline{e}_r - r\vartheta'(t)\underline{e}_\vartheta + z'(t)\underline{k} \qquad (3)$$

Similarly, by differentiating (3) with respect to t,

$$\underline{P}''(t) = r''(t)\underline{e}_r + r'(t)(d/dt)\underline{e}_r(t) - (d/dt)(r\vartheta')\underline{e}_\vartheta$$
$$- r\vartheta'(t)(d/dt)\underline{e}_\vartheta(t) + z''(t)k.$$

Since

$$(d/dt)(r\vartheta') = r'(t)\vartheta'(t) + r(t)\vartheta''(t)$$

and

$$(d/dt)\underline{e}_\vartheta(t) = \vartheta'(t)\underline{e}_r(t)$$

it follows that

$$\underline{P}''(t) = (r''(t) - r(\vartheta'(t))^2)\underline{e}_r - (2r'(t)\vartheta'(t) + r\vartheta''(t))\underline{e}_\vartheta + z''(t)\underline{k} \tag{4}$$

Then (3) and (4) are the expressions for velocity and acceleration in cylindrical coordinates.

PROBLEM 10.5

Suppose that a particle with position vector $\underline{P}(t)$ moves in a plane central force field; i. e., a force field with force given by $F(t) = f(t)\ \underline{e}_r$. Then show that

$$\underline{P}'(t) \times \underline{P}(t) = \text{constant (angular momentum is conserved)}$$

hence

$$r^2(t)\vartheta(t) = \text{constant.}$$

SOLUTION 10.5

In a central force field, the force exerted by the field on a particle in the field is always in the radial direction. Then Newton's second law states that for a particle of mass m

$$m\underline{P}''(t) = f(t)\underline{e}_r(t).$$

It follows that

$$m\underline{P}''(t) \times \underline{P}(t) = (f(t)\underline{e}_r) \times (r\underline{e}_r) = 0 \tag{1}$$

Now (1) implies that

$$d/dt(\underline{P}' \times \underline{P}) = P'' \times \underline{P} + \underline{P}' \times \underline{P}' = 0$$

thus

$$\underline{P}' \times \underline{P} = C \text{ (constant).} \tag{2}$$

The quantity $\underline{P}' \times \underline{P}$ is referred to as *angular momentum* and (2) is the statement that angular momentum is conserved in a central force field. Using the results of the previous problem, we see that (2) becomes

$$(r'(t)\underline{e}_r - r\vartheta'(t)\underline{e}_\vartheta) \times (r(t)\ \underline{e}_r) = C$$

or

$$-\,r^2\vartheta'(t)(\underline{e}_\vartheta \times \underline{e}_r) = r^2\vartheta'\underline{k} = \underline{C}.$$

Then $r^2\vartheta'$ is equal to the constant $\underline{C} \cdot \underline{k} = \| C \|$. Note that if this constant is zero, then the angular momentum is constantly zero and $\vartheta'(t) = 0$; ϑ is constant.

PROBLEM 10.6

Consider the path described by $P(t) = e^{-t}\text{Cos}\Omega t\,\underline{i} + e^{-t}\text{Sin}\Omega t\,\underline{j} + 4\,k$. Compute ds/dt as well as the tangential and centrifugal components of acceleration for this path.

SOLUTION 10.6

We have

$$P'(t) = -e^{-t}(\text{Cos}\,\Omega t + \Omega\text{Sin}\Omega t)\,\underline{i} + e^{-t}(\Omega\text{Cos}\Omega t - \text{Sin}\Omega t)\,\underline{j}$$

hence

$$\| P'(t) \| = e^{-t}\sqrt{1+\Omega^2} = \frac{ds}{dt}$$

and

$$\underline{T}(t) = \frac{-(\text{Cos}\Omega t + \Omega\text{Sin}\Omega t)}{\sqrt{1+\Omega^2}}\,\underline{i} + \frac{(\Omega\text{Cos}\Omega t - \text{Sin}\Omega t)}{\sqrt{1+\Omega^2}}\,j$$

We differentiate $\underline{P}'(t)$ to get the acceleration vector. After simplifying the result we have

$$\underline{P}''(t) = e^{-t}((1-\Omega^2)\text{Cos}\Omega t + 2\Omega\text{Sin}\Omega t)\underline{i} + e^{-t}((1-\Omega^2)\text{Sin}\Omega t - 2\Omega\text{Cos}\Omega t)\underline{j}$$

The tangential component of acceleration equals $\underline{P}'' \cdot \underline{T}$ which eventually simplifies to

$$\underline{P}'' \cdot \underline{T} = e^{-t}\sqrt{1+\Omega^2}.$$

Since

$$\| \underline{P}''(t) \| = e^{-t}\sqrt{(1-\Omega^2)^2 + 4\Omega^2}$$

it follows that the normal component of acceleration is given by

$$\sqrt{\| \underline{P}''(t) \| - (\underline{P}''(t) \cdot \underline{T}(t))^2} = e^{-t}\sqrt{(1-\Omega^2)^2 + 4\Omega^2 - (1+\Omega^2)}$$
$$= e^{-t}\Omega\sqrt{1+\Omega^2}.$$

Vector Differential Operators

PROBLEM 10.7

In Cartesian coordinates the gradient of the scalar field $F = F(x, y, z)$ is given by

$$\nabla F = F_x\underline{i} + F_y\underline{j} + F_z\underline{k}. \qquad (1)$$

Compute the gradient in terms of the variables u, v, w which are related to x, y and z by the following nonsingular transformation

$$x = x(u, v, w)$$
$$y = y(u, v, w) \tag{2}$$
$$z = z(u, v, w).$$

SOLUTION 10.7

We use the chain rule to compute:

$$F_x = F_u u_x + F_v v_x + F_w w_x$$
$$F_y = F_u u_y + F_v v_y + F_w w_y \tag{3}$$
$$F_z = F_u u_z + F_v v_z + F_w w_z.$$

We substitute (3) into (1) and collect terms to obtain

$$\nabla F = F_u \underline{e}_u + F_v \underline{e}_v + F_w \underline{e}_w \tag{4}$$

where

$$\underline{e}_u = u_x i + u_y j + u_z k = \nabla u$$

$$\underline{e}_v = v_x i + v_y j + v_z k = \nabla v$$

$$\underline{e}_w = w_x i + w_y j + w_z k = \nabla w.$$

Note that while the $\underline{e}$'s are not necessarily unit vectors, $\underline{e}_u$ = **grad** u is normal to the level surfaces of u and that $\underline{e}_v$ and $\underline{e}_w$ are normal to the level surfaces of v and w, respectively. Similarly $\underline{i}$ = **grad** x is normal to level surfaces of x, that is to planes parallel to the yz-plane, just as $\underline{j}$ = **grad** y, $\underline{k}$ = **grad** z are normal to level surfaces of y and z.

PROBLEM 10.8

Transform the gradient $\nabla F = F_x \underline{i} + F_y \underline{j} + F_z \underline{k}$ from Cartesian coordinates to cylindrical coordinates,

$$x = u \operatorname{Cos} v$$
$$y = u \operatorname{Sin} v \tag{1}$$
$$z = w.$$

SOLUTION 10.8

Differentiating (1) with respect to x yields

$$1 = \operatorname{Cos} v\, u_x - u \operatorname{Sin} v\, v_x$$

$$0 = \operatorname{Sin} v\, u_x + u \operatorname{Cos} v\, v_x$$

$$0 = w_x$$

This leads to: $u_x = \text{Cos } v$, $v_x = -(\text{Sin } v)/u$, $w_x = 0$.
Similarly, $u_y = \text{Sin } v$, $v_y = (\text{Cos } v)/u$, $w_y = 0$
and $u_z = v_z = 0$, $w_z = 1$.
Then

$$\underline{e}_u = \textbf{grad } u = \text{Cos } v\, \underline{i} + \text{Sin } v\, \underline{j}$$

$$\underline{e}_v = \textbf{grad } v = \frac{1}{u}(-\text{Sin } v\, \underline{i} + \text{Cos } v\, \underline{j}) \quad (2)$$

$$\underline{e}_w = \textbf{grad } w = \underline{k}$$

Then $\nabla F(u, v, w) = F_u\underline{e}_u + F_v\underline{e}_v + F_w\underline{e}_w$. The vectors $\underline{e}_u$, $\underline{e}_v$ and $\underline{e}_w$ in the u, v and w coordinate directions are shown if Figure 10.4. These vectors are similar to the vectors $\underline{e}_r$, $\underline{e}_\theta$ and $\underline{e}_z$ in Figure 10.3. Note that while the vectors $\underline{i}$, $\underline{j}$, $\underline{k}$ associated with Cartesian coordinates are all unit vectors with constant direction, the vectors $\underline{e}_u$, $\underline{e}_v$ and $\underline{e}_w$ associated with cylindrical coordinates are not all unit vectors nor do they all have constant direction. However, the vectors $\underline{e}_u$, $\underline{e}_v$ and $\underline{e}_w$ are mutually orthogonal at all points. This implies that the level surfaces of the coordinate u, v and w are mutually orthogonal surfaces at all points since two surfaces are orthogonal at a point of intersection if and only if their normal vectors are orthogonal at the point of intersection. A coordinate system with this property is called an *orthogonal coordinate system*. Cartesian and cylindrical coordinates are examples of orthogonal coordinate systems.

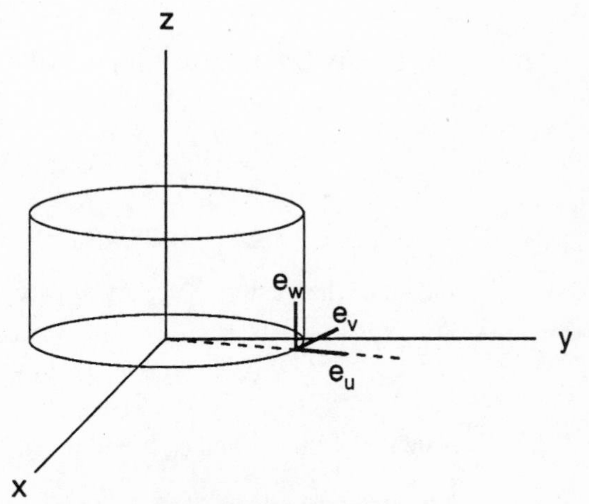

Figure 10.4

PROBLEM 10.9

Compute the divergence of a $\mathbb{C}^1$ vector field $\underline{G}$ in cylindrical coordinates.

SOLUTION 10.9

It follows from the results of the previous problem that the del operator can be expressed in cylindrical coordinates as

$$\nabla = \underline{e}_u \partial/\partial u + \underline{e}_v \partial/\partial v + \underline{e}_w \partial/\partial w.$$

Then for $\underline{G}(u, v, w) = g_1\underline{e}_u + g_2\underline{e}_v + g_3\underline{e}_w$ where $g_k = g_k\,(u, v, w)$ for $k = 1, 2, 3$ we have

$$\begin{aligned}\text{div}\,\underline{G} &= (\underline{e}_u \partial/\partial u + \underline{e}_v \partial/\partial v + \underline{e}_w \partial/\partial w)\cdot(g_1\underline{e}_u + g_2\underline{e}_v + g_3\underline{e}_w)\\ &= \underline{e}_u \partial/\partial u(g_1\underline{e}_u) + \underline{e}_u \partial/\partial u(g_2\underline{e}_v) + \underline{e}_u \partial/\partial u(g_3\underline{e}_w)\\ &\quad + \underline{e}_v \partial/\partial v(g_1\underline{e}_u) + \underline{e}_v \partial/\partial v(g_2\underline{e}_v) + \underline{e}_v \partial/\partial v(g_3\underline{e}_w)\\ &\quad + \underline{e}_w \partial/\partial w(g_1\underline{e}_u) + \underline{e}_w \partial/\partial w(g_2\underline{e}_v) + \underline{e}_w \partial/\partial w(g_3\underline{e}_w)\end{aligned}$$

For $\underline{e}_u$, $\underline{e}_v$ and $\underline{e}_w$ as in (2) of the previous problem, we compute the following derivatives

$$(\partial/\partial u)\underline{e}_u = 0\,, \qquad (\partial/\partial v)\underline{e}_u = u\underline{e}_v\,, \qquad (\partial/\partial w)\underline{e}_u = \underline{0}$$

$$(\partial/\partial u)\underline{e}_v = \frac{-1}{u^2}\,\underline{e}_v\,, \qquad (\partial/\partial v)\underline{e}_v = \frac{-1}{u}\,\underline{e}_u, \qquad (\partial/\partial w)\underline{e}_v = \underline{0}$$

$$(\partial/\partial u)\underline{e}_w = \underline{0}\,, \qquad (\partial/\partial v)\underline{e}_w = \underline{0}, \qquad (\partial/\partial w)\underline{e}_w = \underline{0}$$

Then

$$\begin{aligned}\text{div}\,\underline{G} &= \underline{e}_u\cdot\left((\partial/\partial u)g_1\underline{e}_u + (\partial/\partial u)g_2\underline{e}_v + g_2\left(\frac{-1}{u^2}\,\underline{e}_v\right) + (\partial/\partial u)g_3\underline{e}_w\right)\\ &\quad + \underline{e}_v\cdot\left((\partial/\partial u)g_1\underline{e}_u + g_1(u\underline{e}_v) + (\partial/\partial v)g_2\underline{e}_v + g_2\left(\frac{-1}{u^2}\,\underline{e}_u\right) + (\partial/\partial v)g_3\underline{e}_w\right)\\ &\quad + \underline{e}_w\cdot((\partial/\partial w)g_1\underline{e}_u + (\partial/\partial w)g_2\underline{e}_v + (\partial/\partial w)g_3\underline{e}_w)\end{aligned}$$

and since the vectors $\underline{e}_u$, $\underline{e}_v$ and $\underline{e}_w$ are mutually orthogonal, this reduces to

$$\text{div}\,\underline{G} = (\partial/\partial u)g_1 + \frac{g_1}{u} + (\partial/\partial v)g_2 + (\partial/\partial w)g_3.$$

Note that while $\underline{e}_u$ and $\underline{e}_w$ are unit vectors, $\underline{e}_v \cdot \underline{e}_v = u^{-2}$.

PROBLEM 10.10

Suppose that $F = F(x, y, z)$ is a $\mathbb{C}^2$ scalar field. Then show that

$$\text{div}\,\textbf{grad}\,F = F_{xx} + F_{yy} + F_{zz}.$$

SOLUTION 10.10

Since F = F(x, y, z) we are using Cartesian coordinates. Then

$$\text{div } \mathbf{grad}\, F = (\underline{i}(\partial/\partial x) + \underline{j}(\partial/\partial y) + \underline{k})\partial/\partial z)) \cdot (F_x\underline{i} + F_y\underline{j} + F_z\underline{k})$$

Since the vector $\underline{i}$, $\underline{j}$ and $\underline{k}$ are mutually orthogonal, constant, unit vectors this reduces to

$$\begin{aligned}\text{div } \mathbf{grad}\, F &= (\partial/\partial x)F_x + (\partial/\partial y)F_y + (\partial/\partial z)F_z\\ &= F_{xx} + F_{yy} + F_{zz}.\end{aligned}$$

Sometimes div **grad** F is denoted by $\nabla^2 F$. This is referred to as the *Laplacian* of F and ∇^2 is called the *Laplacian operator.*

PROBLEM 10.11

Suppose that $\underline{G} = \underline{G}(x, y, z)$ is a C^2 vector field. Then show that curl curl $\underline{G}$ = **grad** div$\underline{G}$ – $(\nabla \cdot \nabla)\, \underline{G}$

SOLUTION 10.11

By definition of the curl,

$$\begin{aligned}\text{curl curl } \underline{G} &= \nabla \times (\nabla \times \underline{G})\\
&= \nabla \times ((\partial_y g_3 - \partial_z g_2)\underline{i} + (\partial_z g_1 - \partial_x g_3)\underline{j} + (\partial_x g_2 - \partial_y g_1)\underline{k})\\
&= (\partial_y(\partial_x g_2 - \partial_y g_1) - \partial_z(\partial_z g_1 - \partial_x g_3))\underline{i}\\
&\quad + (\partial_z(\partial_y g_3 - \partial_z g_2) - \partial_x(\partial_x g_2 - \partial_y g_1))\underline{j}\\
&\quad + (\partial_x(\partial_z g_1 - \partial_x g_3) - \partial_y(\partial_y g_3 - \partial_z g_2))\underline{k}\\
&= (\partial_{yx} g_2 + \partial_{zx} g_3 - \partial_{yy} g_1 - \partial_{zz} g_1)\underline{i}\\
&\quad + (\partial_{zy} g_3 + \partial_{xy} g_1 - \partial_{xx} g_2 - \partial_{zz} g_2)\underline{j}\\
&\quad + (\partial_{xz} g_1 + \partial_{yz} g_2 - \partial_{xx} g_3 - \partial_{yy} g_3)\underline{k}\\
&= (\partial_x(\partial_x g_1 + \partial_y g_2 + \partial_z g_3) - \nabla^2 g_1)\underline{i}\\
&\quad + (\partial_y(\partial_x g_1 + \partial_y g_2 + \partial_z g_3) - \nabla^2 g_2)\underline{j}\\
&\quad + (\partial_z(\partial_x g_1 + \partial_y g_2 + \partial_z g_3) - \nabla^2 g_3)\underline{k}\end{aligned}$$

But $\partial_x g_1 + \partial_y g_2 + \partial_z g_3$ = div $\underline{G}$ hence

$$\begin{aligned}\text{curl curl } \underline{G} &= (\underline{i}\, \partial_x + \underline{j}\, \partial_y + \underline{k}\, \partial_z) \text{ div } \underline{G} - \nabla^2(\, g_1\underline{i} + g_2\underline{j} + g_3\underline{k})\\
&= \mathbf{grad} \text{ div } \underline{G} - (\nabla \cdot \nabla)\underline{G}\end{aligned}$$

Here we have introduced the more conveniently written notation ∂_x for $\partial/\partial x$ etc.

PROBLEM 10.12

Show that if $\underline{G} = \underline{G}(x, y, z)$ is a $\mathbb{C}^2$ vector field then div curl $\underline{G} = 0$.

SOLUTION 10.12

By definition of the divergence and curl operators,

$$\begin{aligned}\text{div curl } \underline{G} &= (\underline{i}\partial_x + \underline{j}\partial_y + \underline{k}\partial_z) \cdot ((\partial_y g_3 - \partial_z g_2)\underline{i} + (\partial_z g_1 - \partial_x g_3)\underline{j} + (\partial_x g_2 - \partial_y g_1)\underline{k}) \\ &= \partial_x(\partial_y g_3 - \partial_z g_2) + \partial_y(\partial_z g_1 - \partial_x g_3) + \partial_z(\partial_x g_2 - \partial_y g_1) \\ &= \partial_{xy} g_3 - \partial_{yx} g_3 + \partial_{zx} g_2 - \partial_{xz} g_2 + \partial_{yz} g_1 - \partial_{zy} g_1.\end{aligned}$$

Since g_1, g_2 and g_3 are all $\mathbb{C}^2$ functions their mixed partial derivatives are equal and it follows that div curl $\underline{G} = 0$. That is, if $\underline{W}$ = curl $\underline{G}$ for some $\mathbb{C}^2$ vector field $\underline{G}$ then the divergence of $\underline{W}$ must vanish.

The converse result is also true. That is, if div $\underline{W} = 0$ then it follows that $\underline{W}$ = curl $\underline{G}$ for some $\mathbb{C}^2$ vector field $\underline{G}$. We delay the proof of this result until the next chapter.

PROBLEM 10.13

Show that if $F = F(x, y, z)$ is a $\mathbb{C}^2$ scalar field, then curl **grad** $F = 0$.

SOLUTION 10.13

By definition of the curl and gradient operators,

$$\text{curl } \mathbf{grad}\, F = \begin{vmatrix} \underline{i} & \underline{j} & \underline{k} \\ \partial_x & \partial_y & \partial_z \\ F_x & F_y & F_z \end{vmatrix} = (F_{yz} - F_{zy})\underline{i} + (F_{zx} - F_{xz})\underline{j} + (F_{xy} - F_{yx})\underline{k}.$$

For F a $\mathbb{C}^2$ scalar field, the mixed partials are all equal and curl **grad** $F = 0$. Thus if $\underline{W}$ is a so-called *gradient field,* i. e., if $\underline{W}$ = **grad** F for some $\mathbb{C}^2$ scalar field F, then curl $\underline{W}$ vanishes.

The converse of this result is also true. If curl $\underline{W} = \underline{0}$, then $\underline{W}$ = **grad** F for some $\mathbb{C}^2$ scalar field F. A vector field whose curl vanishes is said to be a *conservative field* or *irrotational field.* Then every conservative field is a gradient field. This result will be proved in the next chapter.

PROBLEM 10.14

For smooth vector field $\underline{V}$ and scalar field φ show that

$$\nabla \cdot (\varphi \underline{V}) = \varphi \nabla \cdot \underline{V} + \underline{V} \cdot \nabla \varphi.$$

SOLUTION 10.14

Write

$$\nabla \cdot (\varphi \underline{V}) = \partial_x(\varphi v_1) + \partial_y(\varphi v_2) + \partial_z(\varphi v_3).$$

Then

$$\nabla \cdot (\varphi \underline{V}) = \varphi_x v_1 + \varphi \partial_x v_1 + \varphi_y v_2 + \varphi \partial_y v_2 + \varphi_z v_3 + \varphi \partial_z v_3$$

$$= \nabla \varphi \cdot \underline{V} + \varphi \operatorname{div} \underline{V}.$$

PROBLEM 10.15

For smooth vector fields $\underline{U}$ and $\underline{V}$ show that

$$\nabla \cdot (\underline{U} \times \underline{V}) = \underline{V} \cdot (\nabla \times \underline{U}) - \underline{U} \cdot (\nabla \times \underline{V})$$

SOLUTION 10.15

We can proceed as in the previous problem to write out all the terms of $\underline{U} \times \underline{V}$ and compute the required derivatives. However a more formal approach saves a lot of this effort. Write

$$\nabla \cdot (\underline{U} \times \underline{V}) = \nabla_u \cdot (\underline{U} \times \underline{V}) + \nabla_v \cdot (\underline{U} \times \underline{V})$$

where $\nabla_u \cdot (\underline{U} \times \underline{V})$ means that ∇ operates on $\underline{U}$ while $\underline{V}$ is treated as a constant vector. Similarly $\nabla_v \cdot (\underline{U} \times \underline{V})$ means ∇ operates on $\underline{V}$ while $\underline{U}$ is treated as a constant vector. Then, using the vector identity (2) from Problem 10.2,

$$\nabla_u \cdot (\underline{U} \times \underline{V}) = \underline{V} \cdot \nabla \times \underline{U} \quad \text{and} \quad \nabla_v \cdot (\underline{U} \times \underline{V}) = -\underline{U} \cdot \nabla \times \underline{V}.$$

Hence

$$\nabla \cdot (\underline{U} \times \underline{V}) = \underline{V} \cdot \nabla \times \underline{U} - \underline{U} \cdot \nabla \times \underline{V}.$$

PROBLEM 10.16

Let $\underline{R} = x\underline{i} + y\underline{j} + z\underline{k}$ and $r = \| \underline{R} \| = \sqrt{x^2 + y^2 + z^2}$. Then show that

$$\nabla r = \frac{1}{r} \underline{R} = \underline{e}_r = \text{unit vector in the radial direction.}$$

SOLUTION 10.16

Since

$$r = (x^2 + y^2 + z^2)^{1/2} \quad \text{we have } r_x = \frac{1}{2}(x^2 + y^2 + z^2)^{-1/2}\, 2x = \frac{x}{r}$$

Similarly,

$$r_y = \frac{y}{r} \quad \text{and} \quad r_z = \frac{z}{r}.$$

Then

$$\nabla r = r_x\underline{i} + r_y\underline{j} + r_y\underline{k} = \frac{1}{r}(x\underline{i} + y\underline{j} + z\underline{k}) = \frac{1}{r}\underline{R} = \underline{e}_r$$

PROBLEM 10.17

Let r and $\underline{R}$ be as in the previous problem. Then find **grad** r^n for n a positive integer and **grad** 1/r.

SOLUTION 10.17

Let $f(r) = r^n$. Then by identity 1 listed under the additional operator identities,

$$\nabla f(r) = f'(r)\nabla r = nr^{n-1}\nabla r = nr^{n-1}\underline{e}_r = nr^{n-2}\underline{R}$$

and

$$\nabla(1/r) = -\frac{1}{r^2}\underline{e}_r = -\frac{1}{r^3}\underline{R}$$

Maximum-minimum Principles

PROBLEM 10.18

Let Ω denote a bounded domain in R with boundary Γ and suppose $F = F(x, y, z)$ is continuous on the compact set W∪Γ composed of W together with its boundary Γ. Suppose further that $\nabla^2 F$ is continuous on Ω and let m, M denote respectively the minimum and maximum values for F over the boundary, Γ. Show that if $\nabla^2 F = 0$ in Ω then $m \leq F \leq M$ on Ω∪Γ.

SOLUTION 10.18

Suppose the conclusion fails. In particular, suppose there exists a point $\underline{X}_0 = (x_0, y_0, z_0)$ inside Ω where $F(x_0, y_0, z_0) > M$. Then

$$F(\underline{X}_0) = F(x_0, y_0, z_0) \geq M + \rho$$

for $\rho > 0$ sufficiently small and the function $G(\underline{X}) = F(\underline{X}) + \in \|\underline{X} - \underline{X}_0\|$ satisfies

$$G(\underline{X}_0) = F(\underline{X}_0) \geq M + \rho \geq M + \in \|\underline{X} - \underline{X}_0\|^2 \geq \max_{\Gamma} G(\underline{X})$$

provided $\in > 0$ is sufficiently small that $\rho \geq \in \|\underline{X} - \underline{X}_0\|^2$. This implies that G does not attain its maximum value on Γ. Hence the maximum value for G on Ω∪Γ must occur at a point $Y = (x_1, y_1, z_1)$ inside Ω where we then have

$$\textbf{grad } G(\underline{Y}) = \underline{0} \quad \text{and} \quad G_{xx}(\underline{Y}) \leq 0,\ G_{yy}(\underline{Y}) \leq 0,\ G_{zz}(\underline{Y}) \leq 0\,.$$

But this implies

$$\nabla^2 G(\underline{Y}) \leq 0$$

which is contradictory to the statement

$$\nabla^2 G(\underline{X}) = \nabla^2 F(\underline{X}) + \nabla^2(\in \|\underline{X} - \underline{X}_0\|^2) = 6\in > 0 \quad \text{for all } \underline{X} \text{ in } \Omega.$$

Then there can exist no point $\underline{X}_0$ inside Ω where $F(\underline{X}_0) = F(x_0, y_0, z_0) > M$. Similarly there can exist no point $\underline{X}_0$ in Ω where $F(\underline{X}_0) < m$.

PROBLEM 10.19

Let Ω denote a bounded domain in $\mathbb{R}^3$ with boundary Γ. For given functions F continuous on Ω and f continuous on Γ, we say that $u = u(x, y, z)$ is a *classical solution of the Dirichlet problem* in Ω if u is $\mathbb{C}^2$ in Ω, continuous on $\Omega \cup \Gamma$ and satisfies

$$\nabla^2 u(x, y, z) = F(x, y, z) \quad \text{in } \Omega$$

$$u = f \quad \text{on } \Gamma.$$

Show that if a classical solution of the Dirichlet problem exists, then it is unique; i. e., if u and v are two such solutions then $u = v$.

SOLUTION 10.19

Suppose $u = u(x, y, z)$ and $v = v(x, y, z)$ are each classical solutions of the Dirichlet problem above. Then $w(x, y, z) = u(x, y, z) - v(x, y, z)$ satisfies,

$$\nabla^2 w(x, y, z) = 0 \quad \text{in } \Omega \tag{1}$$

$$w = 0 \text{ on } \Gamma. \tag{2}$$

It follows from (2) that the maximum and minimum values for w over Γ are both zero. Then (1) together with Theorem 10.4(ii) implies

$$0 \le w(x, y, z) \le 0 \quad \text{on } \Omega \cup \Gamma.$$

That is $w = 0$ and $u = v$ on $\Omega \cup \Gamma$.

PROBLEM 10.20

Let Ω denote a bounded domain in $\mathbb{R}^3$ with boundary Γ. For given functions F continuous on Ω and f continuous on Γ, suppose $u = u(x, y, z)$ is a classical solution of the Dirichlet problem in Ω. Show that u satisfies

$$|u(x, y, z)| \le M_f + \frac{R^2}{6} M_F \tag{1}$$

where M_f and M_F denote the maximum values of f and F on Γ and on Ω respectively and $R > 0$ is such that $x^2 + y^2 + z^2 \le R^2$ for all $\underline{X}$ in Ω.

SOLUTION 10.20

Let

$$v(x, y, z) = u(x, y, z) + \frac{1}{6} M_F (x^2 + y^2 + z^2)$$

Then the maximum value assumed by $v(x, y, z)$ on Γ is less than or equal to $M_f + R^2 M_F/6$ where R^2 denotes the maximum value of $(x^2 + y^2 + z^2)$ for (x, y, z) on Γ. Since Ω is assumed to be bounded, R is a finite constant. Now computing the Laplacian of v(x, y, z) we find that

$$\nabla^2 v = \nabla^2 u + M_F = F + M_F \geq 0 \quad \text{in } \Omega.$$

Then Theorem 10.4(i) applied to the function v = v(x, y, z), implies

$$v = u(x, y, z) + \frac{1}{6} M_F(x^2 + y^2 + z^2) < M_f + \frac{R^2}{6} M_F \quad \text{on } \Omega \cup \Gamma.$$

By applying the same argument to –u, we obtain (1).

The estimate (1) implies that a classical solution of the Dirichlet problem depends continuously on the data. That is, if the data F and f in the problem is changed slightly, then the corresponding new solution v is only slightly different from the original solution u since (1) implies

$$|u - v| \leq \Delta f + \frac{R^2}{6} \Delta F \quad \text{on } \Omega \cup \Gamma,$$

where Δf and ΔF denote respectively the maximum differences in the data on Γ and the data on Ω.

In this chapter we understand a vector to mean an ordered triple of real numbers. In addition to the standard operations of vector addition and scalar multiplication we define the inner product via (10.1) and the cross product via (10.3). These products satisfy the identities (10.2) and (10.5) respectively. We have the following rules for differentiating vector functions

1. $\frac{d}{dt}(\underline{U}(t) + \underline{V}(t)) = \underline{U}'(t) + \underline{V}'(t)$

2. $\frac{d}{dt}(f(t)\, \underline{U}(t)) = f'(t)\underline{U}(t) + f(t)\, \underline{U}'(t)$

3. $\frac{d}{dt}(\underline{U}(t) \cdot V(t)) = \underline{U}'(t) \cdot \underline{V}(t) + \underline{U}(t) \cdot \underline{V}'(t)$

4. $\frac{d}{dt}(\underline{U}(t) \times \underline{V}(t)) = \underline{U}'(t) \times \underline{V}(t) + \underline{U}(t) \times \underline{V}'(t)$

A curve in $\mathbb{R}^3$ can be viewed as a vector valued function of one real variable. The vector $P(t) = (x(t), y(t), z(t))$ is referred to as the position vector for the curve $\Gamma = (x, y, z){:}\, x = x(t), y = y(t), z = z(t)$. Then the velocity vector $P'(t)$ is tangent to Γ at each point and its magnitude, $s'(t)$, can be interpreted as the speed at which $P(t)$ moves along Γ. The acceleration vector $P''(t)$ has a component equal to $s''(t)$ in the direction tangent to Γ and a component equal to $\kappa s'(t)^2$ in the direction normal to Γ. These are referred to as the tangential and centrifugal components of acceleration. Here κ defined by (10.11) can be interpreted as the *curvature of the path Γ. For example when Γ is a circle of radius R, κ turns out to equal 1/R.*

A real valued function F = F(x, y, z) is referred to as a scalar field while a vector valued function $\underline{V}(x, y, z) = v_1(x, y, z)\underline{i} + v_2(x, y, z)\underline{j} + v_3(x, y, z)\underline{k}$ *is called a vector field. To such fields we can apply the vector differential operator*

$$\nabla = \partial_x\underline{i} + \partial_y\underline{j} + \partial_z\underline{k}$$

to form:

$$\textbf{grad}\, F = \nabla F = F_x\underline{i} + F_y\underline{j} + F_z\underline{k} \quad \textit{(a vector field)}$$

$$\text{div}\, \underline{V} = \nabla \cdot V = \partial_x v_1 + \partial_y v_2 + \partial_z v_3 \quad \textit{(a scalar field)}$$

$$\text{curl}\, \underline{V} = \nabla \times V = \quad \textit{(a vector field)}$$

$$= (\partial_y v_3 - \partial_z v_2)\underline{i} + (\partial_z v_1 - \partial_x v_3)\underline{j} + (\partial_x v_2 - \partial_y v_1)\underline{k}$$

These operations may be combined in various ways leading to the following identities:

$$\text{div}\, \textbf{grad}\, F = F_{xx} + F_{yy} + F_{zz}$$

$$\text{curl curl}\, \underline{V} = \textbf{grad}(\text{div}\, \underline{V}) - (\nabla \cdot \nabla)\underline{V}$$

$$\text{div}\, \underline{V} = 0 \quad \textit{if and only if } \underline{V} = \text{curl}\, \underline{W} \textit{ for some } \underline{W}$$

$$\text{curl}\, \underline{V} = 0 \quad \textit{if and only if } \underline{V} = \textbf{grad}\, F \textit{ for some } F$$

Here F and V are assumed to be $\mathbb{C}^2$ *fields. In addition, we have*

1. $\nabla f(\varphi) = f'(\varphi)\nabla\varphi \quad \textit{for } f \in \mathbb{C}^1(\mathbb{R}^1)$
2. $\nabla(\varphi\psi) = \varphi\nabla\psi + \psi\nabla\varphi$
3. $\nabla \cdot (\varphi\underline{V}) = \varphi\nabla \cdot \underline{V} + \underline{V} \cdot \nabla\varphi$
4. $\nabla \times (\varphi\underline{V}) = \varphi\nabla \times \underline{V} + \nabla\varphi \times \underline{V}$
5. $\nabla \cdot (\underline{U} \times \underline{V}) = \underline{V} \cdot (\nabla \times \underline{U}) - \underline{U} \cdot (\nabla \times \underline{V})$
6. $\nabla \times (\underline{U} \times \underline{V}) = (\underline{V} \cdot \nabla)\underline{U} - (\underline{U} \cdot \nabla)\underline{V} + (\nabla \cdot \underline{V})\underline{U} - (\nabla \cdot \underline{U})\underline{V}$

Finally, for a smooth scalar field F whose Laplacian is positive (negative) inside a bounded region Ω we have the so-called maximum-minimum principles which state that F assumes its maximum (minimum) value on the boundary Γ of Ω.

11

Vector Integral Calculus

In Chapter 5 we introduced the notion of the integral of a function over an interval of the real axis. In this chapter we extend that concept. We begin by defining a space curve as the continuous image in $\mathbb{R}^3$ of an interval of the real axis. Then we define what is meant by the integral of a function along a curve. Such an integral is called a line integral.

Using the parametric description of the curve C, a line integral along C can be reduced to a standard integral over an interval of the real axis. Line integrals have properties analogous to the properties of integrals discussed in Chapter 5. In addition there are various identities for line integrals. The simplest of these, called Green's theorem asserts the equivalence of a double integral over a bounded region in the plane to an associated line integral. This theorem is, in fact, a generalization of the fundamental theorem of calculus. Consequences of this theorem include the so-called Green's identities in the plane.

Line integrals are the natural extensions of single variable integrals on intervals. Similarly we can extend the concept of a double integral over a region in the plane to permit integration over surfaces in $\mathbb{R}^3$. This extension introduces numerous technical difficulties but in the simplest cases the results can be conveniently stated. Important results include the higher dimensional analogues of the Green's theorem, namely the divergence theorem of Gauss and the Stokes' theorem. We also have higher dimensional versions of the Green's identities.

LINE INTEGRALS

Space Curves

PARAMETRIC DESCRIPTION OF A CURVE

Let C denote the set of points $\{\underline{P}(t) \in \mathbb{R}^3 : \underline{P}(t) = (x(t), y(t), z(t)),\ a \le t \le b\}$. Then we say that C is a *curve* in $\mathbb{R}^3$ described *parametrically* in terms of the parameter t. As t varies from a to b, $\underline{P}(t)$ traces out the points of C. The functions x(t), y(t), z(t) in the parametric description are not unique. In fact, for any given curve C, there are infinitely many different parameterizations. Curves may be described in other ways as well; e.g., as the intersection of two surfaces.

ARCS

If $\underline{P}(a)$ and $\underline{P}(b)$ are distinct points then C is said to be an arc and if there are no values $t_1, t_2, a \le t_1 < t_2 \le b$, such that $\underline{P}(t_1) = \underline{P}(t_2)$ then we say that C is a simple arc; i.e., C does not cross over itself at any point.

SIMPLE CLOSED CURVES

If $\underline{P}(a) = \underline{P}(b)$ then we say that C is a closed curve; i.e., the initial and final points of C coincide. A closed curve for which $\underline{P}(t_1) \neq \underline{P}(t_2)$ for $a < t_1 < t_2 < b$, is called a *simple closed curve.*

SMOOTH CURVES

If the functions x(t), y(t) and z(t) belong to $\mathbb{C}^1[a, b]$ we say that C is a *smooth* curve. We will always assume that there is no t in [a, b] where the derivatives x′(t), y′(t), z′(t) vanish simultaneously. This ensures that C has a continuously turning tangent line at each point.

PIECEWISE SMOOTH CURVES

More generally we say that C is *piecewise smooth* if there exists a partition $\{a = t_0, t_1, \ldots, t_n = b\}$ for [a, b] such that x, y, z are $\mathbb{C}^1$ on $[t_{k-1}, t_k]$ for $k = 1, \ldots, n$. Then each of the subarcs $C_k = \{\underline{P}(t) \in R : \underline{P}(t) = (x(t), y(t), z(t)), t_{k-1} \le t \le t_k\}$ is a smooth arc.

ORIENTATION

Once a parameterization has been chosen for a curve

$$C = \{\underline{P}(t) \in R : \underline{P}(t) = (x(t), y(t), z(t)), a \le t \le b\}$$

then an orientation is induced on C. We say that C is traced in the *positive sense* if C is traversed from $\underline{P}(a)$ to $\underline{P}(b)$ and is traced in the *negative* sense from $\underline{P}(b)$ to $\underline{P}(a)$. We refer to $\underline{P}(a)$ and $\underline{P}(b)$ respectively as the *initial* endpoint and *final* endpoint for C.

If C is a simple closed curve with $\underline{P}(a) = \underline{P}(b)$ then C may be traversed in either the clockwise or the counter clockwise sense. Either of these may be designated as the positive sense.

Example 11.1 Curves

11.1 (a) The curve $C = \underline{P} \in \mathbb{R}^3$: $\underline{P}(t) = (2\,\text{Cos}t, 2\text{Sin}\,t, 2\text{Sin}\,t)$, $0 < t < 2\pi$ is easily shown to be a circle of radius 2 lying in the plane z = y. This is an example of a curve that is simple, closed and smooth. Alternatively, C may be defined as the intersection of the spherical surface $S = \{x^2 + y^2 + z^2 = 4\}$ and the plane $M = \{z = y\}$. Any curve that is the intersection of a surface and a plane is called a *plane curve.*

11.1 (b) The curve $C = \{\underline{P} \in \mathbb{R}^3: \underline{P}(t) = (f(t), g(t), 0), 0 \le t \le 2\}$ where

$$f(t) = \begin{cases} 1 - t & \text{if } 0 \le t \le 1 \\ 0 & \text{if } 1 \le t \le 2 \end{cases} \qquad g(t) = \begin{cases} 0 & \text{if } 0 \le t \le 1 \\ t - 1 & \text{if } 1 \le t \le 2 \end{cases}$$

is an L-shaped path from the point (1, 0, 0) to the origin and then to (0, 1, 0). This is an example of piecewise smooth, simple arc. Note that

$$f(t) = \begin{cases} 1 - t^2 & \text{if } 0 \le t \le 1 \\ 0 & \text{if } 1 \le t \le \sqrt{2} \end{cases} \qquad g(t) = \begin{cases} 0 & \text{if } 0 \le t \le 1 \\ t^2 - 1 & \text{if } 1 \le t \le \sqrt{2} \end{cases}$$

is a second parametric description of the same curve.

Line Integrals of Scalar Fields

Let $C = \{\underline{P}(t) \in \mathbb{R}^3: \underline{P}(t) = (x(t), y(t), z(t)), a \le t \le b\}$ be a piecewise smooth curve and let the real valued function $F = F(x, y, z)$ be defined and bounded at each point of C. We may define a partition $\Delta = \{a = t_0, t_1, \ldots, t_n = b\}$ for [a, b] and consider the following sum based on this partition and the evaluation points $\tau_j \in [t_{j-1}, t_j]$, $j = 1, \ldots, n$

$$S_\Delta = \sum_{j=1}^{n} F(x(\tau_j), y(\tau_j), z(\tau_j))(x(t_j) - x(t_{j-1}))$$

We define the *line integral* of F over C with respect to x by

$$\int_C F\,dx = \lim_{\|\Delta\| \to 0} S_\Delta$$

if the limit exists. We may similarly define the line integrals

$$\int_C F\,dy, \quad \int_C F\,dz, \quad \text{and} \int_C F\,ds.$$

In particular the last integral is defined to be the limit of sums of the form

$$S_\Delta = \sum_{j=1}^{n} F(x(\tau_j), y(\tau_j), z(\tau_j))\sqrt{\Delta x_j^2 + \Delta y_j^2 + \Delta z_j^2}\,(t_j - t_{j-1})$$

where $\Delta x_j = x(t_j) - x(t_{j-1})$ etc.

Theorem 11.1

Theorem 11.1 Let $C = \{\underline{P}(t) \in \mathbb{R} : \underline{P}(t) = (x(t), y(t), z(t)), a \le t \le b\}$ be a piecewise smooth curve and let $F = F(x, y, z)$ be continuous at each point of C. Then

$$\int_C F\,dx = \int_a^b F(x(t), y(t), z(t))x'(t)dt$$

$$\int_C F\,dy = \int_a^b F(x(t), y(t), z(t))y'(t)dt$$

$$\int_C F\,dz = \int_a^b F(x(t), y(t), z(t))z'(t)dt$$

$$\int_C F\,ds = \int_a^b F(x(t), y(t), z(t))\sqrt{x'(t)^2 + y'(t)^2 + z'(t)^2}\,dt$$

In particular, each of the line integrals listed above exists.

Theorem 11.1 expresses the line integrals as single variable integrals over an interval. This permits the line integrals to be evaluated by means of the fundamental theorem of calculus.

ARCLENGTH

The length of a simple arc C is defined to be the limit of the sums

$$S_\Delta = \sum_{j=1}^{n} \sqrt{\Delta x_j^2 + \Delta y_j^2 + \Delta z_j^2}\,(t_j - t_{j-1})$$

as the mesh size of the partition tends to zero. If the limit exists the curve C is said to be *rectifiable*. If the curve C is piecewise smooth then we can show that C is rectifiable with length L_C given by

$$L_C = \int_C 1\,ds = \int_a^b \sqrt{x'(t)^2 + y'(t)^2 + z'(t)^2}\,dt$$

where $C = \{\underline{P}(t) \in \mathbb{R} : \underline{P}(t) = (x(t), y(t), z(t)),\ a \le t \le b\}$.

Line Integrals of Vector Fields

Let $\underline{V} = \underline{V}(x, y, z)$ denote a vector field which is continuous on the piecewise smooth curve C. Then we may consider the line integral

$$\int_C \underline{V} \cdot dr = \int_C \underline{V}(x(t), y(t), z(t)) \cdot \underline{P}'(t)dt = \int_C \underline{V} \cdot \underline{T}\,ds$$

where

$$dr = dx\,\underline{i} + dy\,\underline{j} + dz\,\underline{k} = \underline{P}'(t)\,dt = \underline{T}(t)\frac{ds}{dt}\,dt.$$

WORK

If the vector field $\underline{V} = \underline{V}(x, y, z)$ is interpreted as a force field then the integral

$$W = \int_C \underline{V} \cdot dr$$

is interpreted as the *work* done by the field as a particle moves along the path C. For $\underline{V} = v_1(x, y, z)\underline{i} + v_2(x, y, z)\underline{j} + v_3(x, y, z)\underline{k}$ we have

$$W = \int_C v_1dx + v_2dy + v_3dz.$$

Then the work can be evaluated by means of Theorem 11.1.

CIRCULATION OF A VECTOR FIELD

In the case of a piecewise smooth, simple closed curve C and a smooth vector field $\underline{V} = v_1(x, y, z)\underline{i} + v_2(x, y, z)\underline{j} + v_3(x, y, z)\underline{k}$ we define the *circulation* of $\underline{V}$ around C to be

$$\Gamma[\underline{V}; C] = \int_C \underline{V} \cdot \underline{T}\,ds.$$

Then

$$\Gamma[\underline{V}; C] = \int_C v_1dx + v_2dy + v_3dz.$$

Thus work and circulation are different interpretations of the same line integral (although circulation is defined only for *closed* curves C).

Example 11.2 Line Integrals

11.2 (a) Let C denote the curve $\{\underline{P}(t) = (R\,\mathrm{Cos}\,t, R\,\mathrm{Sin}\,t, 0), 0 \le t \le 2\pi\}$; i.e., a circle of radius R in the xy-plane. Then

$$L = \int_0^{2\pi} \sqrt{R^2\mathrm{Sin}^2 t + R^2\mathrm{Cos}^2 t + 0}\, dt = R\int_0^{2\pi} dt = 2\pi R$$

11.2 (b) For the vector field $\underline{V} = a\underline{i} + b\underline{j} + c\underline{k}$, a, b, c = constants, we compute

$$\Gamma[\underline{V}; C] = \int_0^{2\pi} (ax'(t) + by'(t) + 0)dt$$

$$= \int_0^{2\pi} (-aR\,\mathrm{Sin}\,t + bR\,\mathrm{Cos}\,t)dt = 0$$

11.2(c) Consider the vector field

$$\underline{V} = \underline{\Omega} \times \underline{r} = (\omega\underline{k}) \times (x\underline{i} + y\underline{j} + z\underline{k}) = \omega\,(-y\underline{i} + x\underline{j}).$$

This vector field can be interpreted as the velocity field associated with a body that is rotating about the z-axis with angular speed equal to ω. For this choice of $\underline{V}$ and C, we compute

$$\Gamma[\underline{V}; C] = \int_0^{2\pi} (-\omega y x'(t) + \omega x y'(t) + 0)dt$$

$$= \int_0^{2\pi} (\omega R\,\mathrm{Sin}^2 t + \omega R\,\mathrm{Cos}^2 t)dt = 2\pi R\omega.$$

Thus the circulation of the vector field $\underline{V}$ is proportional to the angular speed associated with $\underline{V}$. In general the circulation of a vector field is a measure of the "rotational tendency" of the vector field.

11.2 (d) Let $\underline{V}(x, y, z) =$ **grad** $F(x, y, z)$ for some smooth scalar field $F(x, y, z)$. Then for the piecewise smooth path $C = \{\underline{P}(t) = (x(t), y(t), z(t)), a \le t \le b\}$ we compute

$$W = \int_C \underline{V} \cdot dr = \int_C \partial_x F\,dx + \partial_y F\,dy + \partial_z F\,dz$$

$$= \int_C dF$$

$$= F(x(b), y(b), z(b)) - F(x(a), y(a), z(a))$$

Then the work done by this force field $\underline{V} =$ **grad** F depends only on the endpoints of the path C and not on the path itself.

Properties of Line Integrals

Suppose simple piecewise smooth arcs C_1 and C_2 have been parameterized and that the terminal endpoint of C_1 coincides with the initial endpoint of C_2. Let C denote the simple arc $C_1 \cup C_2$. Then for any line integral over C we have

$$\int_C = \int_{C_1} + \int_{C_2}$$

If the piecewise smooth, simple arc C has been parameterized and – C denotes the same curve but with the opposite orientation, then for any line integral on C

$$\int_{-C} = -\int_{C_1}$$

Finally, line integrals are linear. For example, if $\underline{P} = \underline{P}(x, y, z)$ and $\underline{Q} = \underline{Q}(x, y, z)$ are smooth vector fields in domain Ω in $\mathbb{R}^3$ and C is a piecewise smooth simple arc in Ω then for arbitrary constants a and b,

$$\int_C (a\underline{P} + b\underline{Q}) \cdot \underline{dr} = a\int_C \underline{P} \cdot \underline{dr} + b\int_C \underline{Q} \cdot \underline{dr}$$

Integral Identities in the Plane

We state now a result which asserts the equivalence of a double integral over a bounded region in $\mathbb{R}^2$ to an associated line integral over the boundary of the region. This result is known variously as Green's Theorem or as Gauss' Theorem and has several important corollaries. Most of these results generalize to higher dimensional settings but it is worthwhile stating the plane versions separately. Since we will be restricting our attention in this section to plane vector fields, $\underline{V}(x, y, z) = v_1(x, y)\underline{i} + v_2(x, y)\underline{j} + 0\underline{k}$, we will write these more compactly as $\underline{V} = \underline{V}(x, y) = v_1(x, y)\underline{i} + v_2(x, y)\underline{j}$.

ORIENTATION OF THE BOUNDARY OF A PLANE REGION

Let Ω denote a bounded region in $\mathbb{R}^2$ and suppose that the boundary consists of one or more simple closed curves. Then we assign the positive sense on each of these curves as the one for which the region Ω lies to the left as the boundary curve is traced out. For example if Ω is the annular region between an inner circle C_1 and outer circle C_2 then the clockwise sense is the positive sense on C_1 while counterclockwise is the positive sense on C_2. We shall refer to this convention as the *usual orientation* for boundaries of plane regions.

FUNDAMENTAL DOMAINS

A bounded region Ω in $\mathbb{R}^2$ will be called an R_x if there exist real numbers $a < b$, and functions $f, g \in \mathbb{C}[a, b]$ such that $\Omega = \{(x, y): a \le x \le b, f(x) \le y \le g(x)\}$. In the same way, we say that Ω is an R_y if there exist real numbers $c < d$, and functions $p, q \in \mathbb{C}[c, d]$ such that $\Omega = \{(x, y): c \le y \le d, p(y) \le x \le q(y))$. If Ω is both an R_x and R_y then we say that Ω is a *fundamental domain*. Rectangles, circles and ellipses are fundamental domains.

Theorem 11.2 Green's Theorem

Theorem 11.2 Let W be a closed and bounded fundamental domain in $\mathbb{R}^2$ with boundary Γ consisting of finitely many piecewise smooth, simple closed curves with the usual orientation. Suppose $\underline{V}(x, y) = v_1(x, y)\underline{i} + v_2(x, y)\underline{j}$ is a plane smooth vector field in Ω. Then

$$\int_\Gamma \underline{V} \cdot \underline{dr} = \int_\Gamma v_1(x, y)dx + v_2(x, y)dy = \iint_\Omega (\partial_x v_2(x, y) - \partial_y v_1(x, y))dx\,dy$$

The result extends to the case where Ω can be expressed as the union of finitely many bounded nonoverlapping fundamental domains Ω_1 , ..., Ω_n whose boundaries $G_1, \ldots, G_n$ consist of finitely many piecewise smooth simple closed curves with the usual orientation. Green's theorem has a number of useful corollaries.

Corollary 11.3 Let Ω in $\mathbb{R}^2$ be a domain in which Theorem 11.2 applies and let $\underline{V}(x, y)$ be a plane vector field, defined and smooth in Ω. Let $\underline{N}(x, y)$ denote the outward pointing unit normal vector to the boundary Γ at $(x, y) \in \Gamma$. Then

$$\int_\Gamma \underline{V} \cdot \underline{N}\, ds = \iint_\Omega \text{div}\, \underline{V}\, dx\, dy$$

Corollary 11.4 Let Ω in $\mathbb{R}^2$ be a domain in which Theorem 11.2 applies and let $\varphi \in C^2(\Omega)$. Let $\underline{N}(x, y)$ denote the outward pointing unit normal vector to the boundary Γ of Ω. Then

$$\iint_\Omega \nabla^2\varphi\, dxdy = \int_\Gamma \nabla\varphi \cdot \underline{N}\, ds$$

The next two corollaries are referred to as Green's first and second identities, respectively.

Corollary 11.5 Let Ω in $\mathbb{R}^2$ be a domain in which Theorem 11.2 applies and let $\psi, \varphi \in C^2(\Omega)$. Let $\underline{N}(x, y)$ denote the outward pointing unit normal vector to the boundary Γ of Ω. Then

$$\iint_\Omega \psi\nabla^2\varphi\, dxdy = \int_\Gamma \psi\nabla\varphi \cdot \underline{N}\, ds - \iint_\Omega \nabla\varphi \cdot \nabla\psi\, dx\, dy$$

Corollary 11.6 Let Ω in $\mathbb{R}$ be a domain in which Theorem 11.2 applies and let $\psi, \varphi \in C^2(\Omega)$. Let $\underline{N}(x, y)$ denote the outward pointing unit normal vector to the boundary Γ of Ω. Then

$$\iint_\Omega (\psi\nabla^2\varphi - \varphi\nabla^2\psi)dx\, dy = \int_\Gamma (\psi\nabla\varphi \cdot \underline{N} - \varphi\nabla\psi \cdot \underline{N})ds$$

Corollary 11.7 Let Ω in $\mathbb{R}^2$ be a domain in which Theorem 11.2 applies and let $\underline{V}(x, y)$ be a plane vector field, defined and smooth in Ω. Let $\underline{T}(x, y)$ denote the unit tangent vector to the boundary Γ at $(x, y) \in \Gamma$. Then

$$\int_\Gamma \underline{V} \cdot \underline{T}\, ds = \iint_\Omega \text{curl}\underline{V} \cdot \underline{k}\, dx\, dy$$

Equivalent Properties of Smooth Vector Fields

SIMPLY CONNECTED DOMAINS

We return now to consideration of vector fields in $\mathbb{R}^3$. Recall that a connected open set Ω in $\mathbb{R}^3$ is called a *domain*. If, for every simple closed curve C in the domain Ω, there is a surface in Ω having C as its boundary, then we say that Ω is *simply connected*. More plainly, a simply connected domain is a domain without "holes" bored completely through it. For example, a sphere is a simply connected domain and the annular region between concentric spheres is simply connected but the annular region between two infinitely long coaxial cylinders is not simply connected.

PATH INDEPENDENT VECTOR FIELDS

A continuous vector field $\underline{V}(x, y, z)$ on Ω in $\mathbb{R}^3$ is said to be *path independent* if for any two curves C_1 and C_2 in Ω having the same initial endpoint and the same final endpoint, we have

$$\int_{C_1} \underline{V} \cdot dr = \int_{C_2} \underline{V} \cdot dr$$

CONSERVATIVE VECTOR FIELDS

A continuous vector field $\underline{V}(x, y, z)$ on Ω in $\mathbb{R}^3$ is said to be *conservative* if for any simple closed curve C in Ω we have

$$\int_{C} \underline{V} \cdot dr = 0$$

POTENTIAL FIELDS

A continuous vector field $\underline{V}(x, y, z)$ on Ω in $\mathbb{R}^3$ is said to be a *potential field* if $\underline{V}(x, y, z) = \textbf{grad}\ F$ for some scalar field $F = F(x, y, z)$ in $\mathbb{C}^1(\Omega)$.

IRROTATIONAL FIELDS

A $\mathbb{C}^1$ vector field $\underline{V}(x, y, z)$ on Ω in $\mathbb{R}$ is said to be *irrotational* if curl $\underline{V} = 0$ throughout W.

Theorem 11.8

Theorem 11.8 Let $\underline{V} = \underline{V}(x, y, z)$ be a continuous vector field in domain Ω in $\mathbb{R}^3$. Then the following are equivalent:

1. $\underline{V}$ is path independent in Ω
2. $\underline{V}$ is conservative in Ω
3. $\underline{V}$ is a potential field

If $\underline{V} \in \mathbb{C}^1(\Omega)$ and the domain Ω is simply connected, then each of the above is equivalent to

4. $\underline{V}$ is irrotational in Ω

The following results were stated previously as parts of Theorems 9.2 and 9.3. At that stage we were not in a position to prove these results.

Theorem 11.9

Theorem 11.9 Let $\Omega = \{(x, y, z) \in \mathbb{R}^3 : x_1 \le x \le x_2, y_1 \le y \le y_2, z_1 \le z \le z_2\}$ and suppose $\underline{V}(x, y, z)$ is $\mathbb{C}^1$ in Ω. Then

(a) curl $\underline{V} = 0$ in Ω implies $\underline{V} = \textbf{grad}\ F$ for some scalar field $F \in \mathbb{C}^1(\Omega)$

(b) div $\underline{V} = 0$ in Ω implies $\underline{V} = \text{curl}\ \underline{G}$ for some vector field $\underline{G} \in \mathbb{C}^1(\Omega)$

Theorem 11.9 holds for more general domains than a rectangular box. The precise description of the most general domain for which the theorem is valid involves technicalities that are beyond the scope of the text.

DIFFERENTIAL FORMS

Sometimes the results of the previous two theorems are stated in terms of *differential forms* instead of vector fields. For example

$$w = v_1(x, y, z)dx + v_2(x, y, z)dy + v_3(x, y, z)dz$$

is called a differential form of order one or a differential 1-form. The association between differential forms and vector fields is apparent. If the vector field $\underline{V}$ is a potential field we say the associated 1-form is *exact* and if the vector field is irrotational then we say the associated 1-form is *closed.*

SURFACE INTEGRALS

Surfaces

REPRESENTATIONS OF SURFACES

While a curve in $\mathbb{R}^3$ is a set of points having a 1-dimensional character, a surface is a point set where the points in the set have two degrees of freedom. There are several ways to represent a surface:

Explicit Representation S is the set of points $\{(x, y, z)\}$ where

$$z = f(x, y) \text{ for smooth function } f \text{ with domain } U_{xy} \text{ in } \mathbb{R}^2$$
or $$y = g(x, z) \text{ for smooth function } g \text{ with domain } V_{xz} \text{ in } \mathbb{R}^2$$
or $$x = h(y, z) \text{ for smooth function } h \text{ with domain } W_{yz} \text{ in } \mathbb{R}^2$$

Implicit Representation S is the set of points $\{(x, y, z): F(x, y, z) = 0\}$ for F a smooth function on domain Ω in $\mathbb{R}^3$.

Parametric Representation S is the set of points (x, y, z) where

$$\begin{aligned} x &= x(s, t) \\ y &= y(s, t) \quad \text{for } a \le s \le b, \quad c \le t \le d, \\ z &= z(s, t) \end{aligned}$$

for x, y, z smooth functions on the rectangle $[a, b] \times [c, d]$.

Example 11.3

11.3 (a) The plane through the points (4, 0, 0), (0, 2, 0), (0, 0, 3) has the representations

$$\frac{x}{4} + \frac{y}{2} + \frac{z}{3} - 1 = 0 \text{ (implicit)}$$

$$z = 3\left(1 - \frac{x}{4} - \frac{y}{2}\right) \text{ (explicit)}$$

and

$$x = 4s$$
$$y = 2t \qquad s, t \in \mathbb{R}^1 \text{ (parametric)}$$
$$z = 3(1 - s - t)$$

11.3 (b) The ellipsoidal surface described implicitly by

$$\frac{x^2}{4} + \frac{y^2}{9} + \frac{z^2}{16} - 1 = 0$$

has the parametric description

$$x = 2 \text{ Sin } s \text{ Sin } t$$
$$y = 3 \text{ Cos } s \text{ Sin } t \quad 0 \le s \le 2\pi, 0 \le t \le \pi$$
$$z = 4 \text{ Cos } t$$

The explicit representations

$$z = \sqrt{1 - x^2/4 - y^2/9} \quad \text{and} \quad z = -\sqrt{1 - x^2/4 - y^2/9}$$

represent, respectively, the upper and lower portions of the ellipsoidal surface. There is no single explicit representation that describes this entire surface.

NORMAL VECTOR TO A SMOOTH SURFACE

Suppose S is given explicitly in the form

$$z = f(x, y) \text{ for smooth function f with domain } U_{xy} \text{ in } \mathbb{R}^2$$

Then

$$\underline{P} = x\underline{i} + y\underline{j} + f(x, y)\underline{k}$$

is a vector that sweeps out the surface S as (x, y) varies over U_{xy}. It is not hard to show that the vectors

$$\partial_x\underline{P} = \underline{i} + 0\underline{j} + f_x\underline{k} \quad \text{and} \quad \partial_y\underline{P} = 0\underline{i} + \underline{j} + f_y\underline{k}$$

are independent vectors, each of which is tangent to S. It follows that

$$n = \partial_y\underline{P} \times \partial_x\underline{P} = -f_x\underline{i} - f_y\underline{j} + \underline{k}$$

is normal to S. If S has a continuously varying normal vector $\underline{n}$ we say that S is a *smooth surface*. If S is the union of finitely many nonoverlapping smooth surfaces then we say S is *piecewise smooth*. The surface of a cube is piecewise smooth but not smooth.

Surface Area

We can define the surface area of a surface S as a limit of appropriate Riemann sums. When the limit exists, we write

$$\iint_S dA = \text{Lim} \sum_{i,j} \Delta A_{ij}$$

We can show that if S is piecewise smooth the increment of surface area ΔA_{ij} is approximated by the element of surface area dA defined as follows:

$$dA = \| \partial_x \underline{P} \times \partial_y \underline{P} \| \, dx\, dy = \sqrt{f_x^2 + f_y^2 + 1}\, dx\, dy$$

Then the surface area of S can be computed from the double integral

$$\iint_S dA = \iint_{U_{xy}} \sqrt{f_x^2 + f_y^2 + 1}\, dx\, dy.$$

Similarly for S given by :

$$y = g(x, z) \text{ for smooth function g with domain } V_{xz} \text{ in } \mathbb{R}^2$$

we have $\underline{P} = x\underline{i} + g(x, z)\underline{j} + z\underline{k}$ and

$$dA = \| \partial_x \underline{P} \times \partial_z \underline{P} \| \, dx\, dz = \sqrt{g_x^2 + g_z^2 + 1}\, dx\, dz$$

and for $x = h(y, z)$ for smooth function h with domain W_{yz} in $\mathbb{R}^2$

$$\underline{P} = h(y, z)\underline{i} + y\underline{j} + z\underline{k}$$

$$dA = \| \partial_y \underline{P} \times \partial_z \underline{P} \| \, dy\, dz = \sqrt{h_y^2 + h_z^2 + 1}\, dy\, dz$$

Then the surface area of S is equal to:

$$\iint_S dA = \iint_{U_{xy}} \sqrt{f_x^2 + f_y^2 + 1}\, dx\, dy$$

$$\iint_{V_{xz}} \sqrt{g_x^2 + g_z^2 + 1}\, dx\, dz$$

$$\iint_{W_{yz}} \sqrt{h_y^2 + h_z^2 + 1}\, dy\, dz$$

SURFACE INTEGRALS

The integral of a smooth scalar field $F = F(x, y, z)$ over a smooth surface S can also be defined in terms of limits of appropriate Riemann sums. It can then be shown that

$$\iint_S F\, dA = \iint_{U_{xy}} F(x, y, f(x, y))\sqrt{f_x^2 + f_y^2 + 1}\, dx\, dy$$

$$\iint_{V_{xz}} F(x, g(x, y), z)\sqrt{g_x^2 + g_z^2 + 1}\, dx\, dz$$

$$\iint_{W_{yz}} F(h(x, y), y, z)\sqrt{h_y^2 + h_z^2 + 1}\, dy\, dz$$

ALTERNATE REPRESENTATIONS FOR N AND dA

In the case that S is expressed explicitly as a function $z = f(x, y)$ over U_{xy}, we can reduce the normal vector n to a unit vector by dividing by its length,

$$\underline{N} = \frac{\underline{n}}{\|\underline{n}\|} = \frac{-f_x\underline{i} - f_y\underline{j} + \underline{k}}{\sqrt{f_x^2 + f_y^2 + 1}}$$

Then we can define the angle φ by the relation $\text{Cos}\varphi = \underline{N} \cdot \underline{k}$; i.e., φ is the angle between the z-axis and the normal vector $\underline{N}$. Alternatively, when S is specified in terms of the functions g(x, z) or h(y, z) we have

$$\underline{N} = \frac{-g_x\underline{i} + \underline{j} - g_z\underline{k}}{\sqrt{g_x^2 + g_z^2 + 1}} \quad \text{and } \text{Cos}\,\beta = \underline{N} \cdot \underline{j}$$

and

$$\underline{N} = \frac{\underline{i} + h_y\underline{j} - h_z\underline{k}}{\sqrt{h_x^2 + h_z^2 + 1}} \quad \text{and } \text{Cos}\,\alpha = \underline{N} \cdot \underline{i}$$

It follows that

$$\underline{N} = \text{Cos}\,\alpha\underline{i} + \text{Cos}\,\beta\underline{j} + \text{Cos}\,\varphi\underline{k}$$

and

$$\text{Sec}\,\alpha = \sqrt{h_y^2 + h_z^2 + 1}$$

$$\text{Sec}\,\beta = \sqrt{g_x^2 + g_z^2 + 1}$$

$$\text{Sec}\,\varphi = \sqrt{f_x^2 + f_y^2 + 1}$$

This leads to the result

$$dA = \text{Sec}\,\alpha\, dy\, dz = \text{Sec}\,\beta\, dx\, dz = \text{Sec}\,\varphi\, dx\, dy.$$

SURFACE INTEGRAL OF A VECTOR FIELD

The analogue of the line integral of a vector field $\underline{V}(x, y, z)$ on a curve C is the surface integral of $\underline{V}(x, y, z)$ on a surface S. For S a smooth surface in $\mathbb{R}^3$ and vector field $\underline{V} = \underline{V}(x, y, z) = v_1\underline{i} + v_2\underline{j} + v_3\underline{k}$ continuous on S, define the integral of $\underline{V}$ over S to equal

$$\iint_S \underline{V} \cdot \underline{N}\, dA = \iint_S v_1 dy\, dz + v_2 dx\, dz + v_3 dx\, dy$$

$$= \iint_S (v_1\text{Cos}\alpha + v_2\text{Cos}\beta + v_3\text{Cos}\varphi)dA$$

whenever the double integrals on the right exist.

Example 11.4 Surface Integral of a Vector Field

If S is given explicitly in the form $S = \{(x, y, z) : z = f(x, y), (x, y) \in U_{xy}\}$ then

$$N = \frac{-f_x}{M}\underline{i} + \frac{-f_y}{M}\underline{j} + \frac{1}{M}\underline{k} \quad \text{for } M = \sqrt{1 + f_x^2 + f_y^2}$$

and

$$dA = \sec\varphi\, dx\, dy = \sqrt{1 + f_x^2 + f_y^2}\, dx\, dy.$$

Thus

$$\iint_S \underline{V} \cdot \underline{N}\, dA = \iint_{U_{xy}} v_1(x, y, f(x, y))(-f_x(x, y))dxdy +$$

$$+ \iint_{U_{xy}} v_2(x, y, f(x, y))(-f_y(x, y))dx\, dy + \iint_{U_{xy}} v_3(x, y, f(x, y))dx\, dy$$

In particular, for $\underline{V} = z\underline{i} - y\underline{j} + x\underline{k}$ and S equal to the planar surface,

$$S = \{(x, y, z) : z = 1 - x/2 - y, 0 \le x \le 2, 0 \le y \le 1\},$$

we have $f_x = - 1/2$, $f_y - 1$ and $v_1(x, y, f(x, y)) = z(x, y) = 1 - x/2 - y$, $v_2 = - y$, $v_3 = x$. Then

$$\iint_S \underline{V} \cdot \underline{N}\, dA = \int_0^1 \int_0^2 ((1 - x/2 - y)(1/2) + (-y)(+1) + x)dx\, dy = 1/2$$

PROPERTIES OF SURFACE INTEGRALS

Properties analogous to those stated previously for line integrals are valid for surface integrals. In particular surface integrals are linear and for a surface S that is the union of nonoverlapping surfaces S_1 and S_2

$$\iint_S = \iint_{S_1} + \iint_{S_2}$$

for all surface integrals over S.

Integral Identities For Surface Integrals

Theorem 11.2 expresses the equivalence between a double integral and an associated line integral. This theorem has extensions to higher dimensions. In particular we will state a theorem expressing the equivalence of a triple integral to an associated surface integral and we will state a second extension asserting the equivalence of a surface integral to an associated line integral over a space curve.

FUNDAMENTAL DOMAINS IN $\mathbb{R}^3$

A solid region Ω in $\mathbb{R}^3$ will be said to be xy-fundamental if there exists a fundamental domain U_{xy} in $\mathbb{R}^2$ and functions f(x, y), g(x, y) in $\mathbb{C}(U_{xy})$ such that

$$\Omega = (x, y, z) \in \mathbb{R}^3: f(x, y) \le z \le g(x, y), (x, y) \in U_{xy}\}.$$

We have similar definitions for xz – fundamental and yz – fundamental. We say that Ω is fundamental if Ω is xy- , yz- and xz-fundamental. Spheres, ellipsoids and cubes are examples of such solids.

OUTWARD NORMAL TO A BOUNDARY SURFACE

If S is the boundary surface for a solid region Ω in $\mathbb{R}^3$ then S is a *closed surface*. If S is smooth then S has a well defined unit normal vector N at each point $\underline{P} = (x_0, y_0, z_0) \in S$. We say that $\underline{N}$ is the *outward normal* to S if for each $t < 0$ sufficiently small, $\underline{P} + t\underline{N}$ belongs to Ω and for each $t > 0$ sufficiently small, $\underline{P} + t\underline{N}$ lies outside Ω.

THE DIVERGENCE THEOREM

We have the following generalization of Corollary 11.3. This theorem is known as Gauss' theorem or the divergence theorem and is frequently applied in the derivation of the equations of mathematical physics.

Theorem 11.10

Theorem 11.10 Let Ω denote a closed bounded domain in $\mathbb{R}^3$ that can be written as the union of finitely many nonoverlapping fundamental domains. Suppose the boundary S of Ω is a piecewise smooth closed surface with outward unit normal $\underline{N}$ and suppose $\underline{V}(x, y, z)$ is a vector field that is $\mathbb{C}^1$ in Ω. Then

$$\iint_S v_1 dy\, dz + v_2 dx\, dz + v_3 dx\, dy = \iiint_\Omega ((\partial_x v_1 + \partial_y v_2 + \partial_z v_3)\, dx\, dy\, dz$$

i.e.,

$$\iint_S \underline{V} \cdot \underline{N}\, dA = \iiint_\Omega \text{div}\underline{V}\, dx\, dy\, dz.$$

Corollary 11.11 Let Ω denote a closed bounded domain in $\mathbb{R}^3$ as in the theorem and let F(x, y, z) a scalar field that is $\mathbb{C}^1$ in Ω. Then

$$\iint_S F\underline{N}\, dA = \iiint_\Omega \textbf{grad}\, F\, dx\, dy\, dz$$

GREEN'S IDENTITIES

Corollary 11.12 Let Ω denote a closed bounded domain in $\mathbb{R}^3$ as in the theorem and let $\varphi, \psi \in \mathbb{C}^2(\Omega)$ and let $\underline{N}$ denote the outward unit normal to S the boundary of Ω. Then

$$\iiint_\Omega \nabla^2\varphi\, dx\, dy\, dz = \iint_S \nabla\varphi \cdot \underline{N}\, dA$$

$$\iiint_\Omega \psi\nabla^2\varphi\, dx\, dy\, dz = \iint_S \nabla\varphi \cdot \underline{N} dA - \iiint_\Omega \nabla\varphi \cdot \nabla\psi\, dx\, dy\, dz$$

$$\iiint_\Omega (\psi\nabla^2\varphi - \varphi\nabla^2\psi) dx\, dy\, dz = \iint_S (\psi\nabla\varphi \cdot \underline{N} - \varphi\nabla\psi \cdot \underline{N}) dA$$

The last two results in this corollary are known respectively as Green's first and second identities in $\mathbb{R}^3$.

ORIENTATION ON THE BOUNDARY OF A SURFACE

We have also a generalization of Corollary 11.7. This theorem applies to a two sided surface S that is not the boundary of a solid region in $\mathbb{R}^3$. If S is given parametrically by $S = \{(x, y, z): x = x(s, t), y = y(s, t), z = z(s, t), (s, t) \in R\}$ where R denotes a compact set in $\mathbb{R}^2$, then the image in $\mathbb{R}^3$ of the boundary of R is referred to as the boundary Γ of S. If the boundary γ of R is described parametrically in terms of parameter σ by $\gamma = \{(s(\sigma), t(\sigma)) \in \mathbb{R}^2: \sigma_1 < \sigma < \sigma_2\}$ then Γ, the boundary of S, is described parametrically in terms of σ by

$$\Gamma = \{(x, y, z) = (x(\sigma), y(\sigma), z(\sigma)): \sigma_1 \leq \sigma \leq \sigma_2\}$$

where $x(\sigma) = x(s(\sigma), t(\sigma))$, etc. If the simple closed plane curve g is given the positive orientation then this induces an orientation on Γ which we refer to as the positive orientation for Γ.

Theorem 11.13 STOKES' THEOREM

Theorem 11.13 Let S denote a smooth bounded but not closed surface in $\mathbb{R}^3$ with positively oriented boundary Γ. Let $\underline{V}(x, y, z)$ denote a vector field that is continuous on S. Then

$$\int_\Gamma v_1 dx + v_2 dy + v_3 dz = \iint_S (\partial_y v_3 - \partial_z v_2) dy\,dz + \iint_S (\partial_z v_1 - \partial_x v_3) dx\,dz$$

$$+ \iint_S (\partial_x v_2 - \partial_y v_1) dx\,dy$$

i.e.,

$$\int_\Gamma \underline{V} \cdot \underline{T}\,ds = \iint_S \text{curl}\underline{V} \cdot \underline{N}\,dA$$

where $\underline{N}$ denotes the unit normal to S.

Example 11.5 An Illustration of Stokes' Theorem

Let S denote the lateral surface of the cone $x^2 + y^2 = (1 - z)^2$, $z > 0$, and let

$$\underline{V} = x^3\underline{j} - (z + 1)\underline{k}.$$

Then Γ is the circle $x^2 + y^2 = 1$, which we can parameterize by

$$\Gamma: \begin{cases} x = \text{Cos } t \\ y = \text{Sin } t \end{cases} \quad 0 \le t \le 2\pi$$

and

$$\int_\Gamma \underline{V} \cdot \underline{T}\,ds = \int_0^{2\pi} (\text{Cos}^3 t\,\underline{j} - k) \cdot (-\text{Sin } t\,\underline{i} + \text{Cos } t\,\underline{j} + 0\underline{k}) dt$$

$$= \int_0^{2\pi} \text{Cos}^4 t\,dt = \frac{3\pi}{4}$$

The surface S can be represented parametrically as

$$S: \begin{cases} x = s\,\text{Cos } t \\ y = s\,\text{Sin } t \\ z = 1 - s \end{cases} \quad 0 \le t \le 2\pi, 0 \le s \le 1.$$

Then
$$E = (s\,\text{Cos } t)^2 + (-s\,\text{Sin } t)^2 = s^2$$
$$G = (\text{Cos } t)^2 + (\text{Sin } t)^2 + (-1)^2 = 2$$
$$F = s\,\text{Sin } t\,\text{Cos } t - s\,\text{Sin } t\,\text{Cos } t + 0 = 0$$

so

$$dA = \sqrt{EG - F^2}\,ds\,dt = s\sqrt{2}\,ds\,dt.$$

If we write S in the form $z = 1 - \sqrt{x^2 + y^2}$ then

$$z_x = \frac{-x}{\sqrt{x^2+y^2}} \quad \text{and } z_y = \frac{-y}{\sqrt{x^2+y^2}}$$

so

$$\sqrt{1+z_x{}^2+z_y{}^2} = \sqrt{2} \quad \text{and} \quad N = \frac{x}{M}\underline{i} + \frac{y}{M}\underline{j} + \frac{1}{\sqrt{2}}\underline{k}$$

Then since curl $\underline{V} = -\,3x^2\underline{k}$ we find

$$\iint_S \text{curl } \underline{V} \cdot N \, dA = \int_0^1\int_0^{2\pi} -\frac{3}{\sqrt{2}} s^2\text{Cos}^2 t \, s\sqrt{2} \, ds \, dt$$

$$= \pi\int_0^1 3s^3 ds = \frac{3\pi}{4} = \int_\Gamma \underline{V} \cdot \underline{T} \, ds$$

Theorem 11.11 and 11.13 are true under more general conditions than have been stated here. However finding the most general hypotheses under which these two theorems hold is a matter that is far beyond the level of this text.

SOLVED PROBLEMS

Space Curves

PROBLEM 11.1

Find a parametric description for the curve C formed by the intersection in the first octant of the two surfaces:

$$F(x, y, z) = x^2 + y^2 + z^2 - 9R^2 = 0$$

and $\quad G(x, y, z) = x^2 + (y - 2R)^2 - R^2 = 0.$

SOLUTION 11.1

The surface F = 0 is a sphere of radius 3R centered at the origin. The surface G = 0 is a circular cylinder with axis of symmetry passing through x = 0, y = 2R parallel to the z-axis. The first octant portion of C begins at the point (0, 3R, 0) on the y-axis and moves along the spherical surface to the point $(0, R, 2R\sqrt{2})$ where it passes into another octant. We will use z as the parameter in this description with z varying over the interval $0 \le z \le 2R\sqrt{2}$.

Subtracting the equation G = 0 from F = 0 eliminates x. Then solving for y in terms of z leads to

$$y = 3R - \frac{z^2}{4R}$$

Using this in the equation G = 0 then allows x to be expressed in terms of z

$$x = \frac{z}{4R} \sqrt{8R^2 - z^2}$$

Then $C = \{(x(z), y(z), z): 0 \le z \le 2R\sqrt{2}\}$ is a parametric description of the curve in terms of the parameter z.

PROBLEM 11.2

Let C denote the circular helix $\{\underline{P}(t) = (4\text{Cos } t, 4\text{Sin } t, 3t) : 0 \le t \le 6\pi\}$. Compute the arc length along C between $\underline{P}(0)$ and $\underline{P}(t)$, then give a parametric description of C using the arc length s as the parameter.

SOLUTION 11.2

Since the component functions in $\underline{P}(t)$ are smooth functions of t, C is a smooth and therefore rectifiable arc. Then

$$L = \int_0^t \sqrt{x'(\tau)^2 + y'(\tau)^2 + z'(\tau)^2}\, d\tau = \int_0^t \sqrt{16\text{Sin}^2\tau + 16\text{Cos}^2\tau + 9}\, d\tau = 5\tau$$

If we denote the arc length L_C by s, then s(t) = 5t and t(s) = s/5. Then

$$C = \{\underline{P}(s) = (\text{Cos } s/5 , \text{Sin } s/5 , 3s/5): 0 \le s \le 30\pi\}$$

is the parametric description of C in terms of the parameter s(arclength). Note that in this case $\| \underline{P}'(s) \| = 1$. We can show that for a curve C given in the form

$$C = \{\underline{P}(t) \in \mathbb{R} : \underline{P}(t) = (x(t), y(t), z(t)), a \le t \le b\}$$

the parameter t is the parameter of arclength if and only if $\| \underline{P}'(t) \| = 1$.

Line Integrals

PROBLEM 11.3

For F(x, y, z) = x + 2y + 3z compute the line integral $\int_C F\, dx$ with:

(a) $C = \{x = t, y = t, z = 4t^2: 0 \le t \le 1\}$

(b) $C = \{x = t^2, y = t^2, z = 4t^2: 0 \le t \le 1\}$

(c) $C = \{x = 1 - t, y = 1 - t, z = 4(1 - t)^2: 0 \le t \le 1\}$

SOLUTION 11.3

The path C is in each of the three cases here, the parabolic curve obtained when the plane x = y intersects the paraboloid $z = 2x^2 + 2y^2$. In case (a) the curve is oriented to begin at the origin and end at the point (1, 1, 4). Then

$$\int_C F\,dx = \int_0^1 (t + 2t + 3(4t^2))(dt) = \frac{11}{2}$$

In case (b) we have a different parameterization but the curve and the orientation are the same as in case (a). Thus

$$\int_C F\,dx = \int_0^1 (t^2 + 2t^2 + 3(4t^2))(2t\,dt) =$$

$$= \int_0^1 (6t^3 + 24t^5)dt = \frac{11}{2}$$

i.e., this is an illustration of the fact that the value of the line integral is independent of the parameterization chosen for the path.

The parameterization in (c) produces the same curve but with the opposite orientation from cases (a) and (b). Thus

$$\int_C F\,dx = \int_0^1 ((1-t) + 2(1-t) + 3(4(1-t)^2))(dt)$$

$$= \int_0^1 (15 - 27t + 12t^2)dt = -\frac{11}{2}$$

As expected, reversing the orientation changes the sign of the line integral.

PROBLEM 11.4

For $\underline{V}(x, y, z) = 2xy\underline{i} + x^2\underline{j}$ compute $\int_C \underline{V} \cdot \underline{dr}$ if C denotes the piecewise smooth path composed of the straight line from (–1, 0, 0) to (1, 0, 0) followed by the circular arc from (1, 0, 0) to (0, 1, 0).

SOLUTION 11.4

Let the two parts of C be denoted by C_1 and C_2 respectively. Then we can parameterize these curves as follows: $C_1 = \{x = t, y = 0, z = 0, -1 \le t \le 1\}$ and $C_2 = \{x = \text{Cos}\,t, y = \text{Sin}\,t, z = 0, 0 \le t \le \pi/2\}$. Then on C_1 we have $x = t$, $dx = dt$ and $y = 0$, $dy = 0$. On C_2, $x = \text{Cos}\,t$, $dx = -\text{Sin}\,t\,dt$ and $y = \text{Sin}\,t$, $dy = \text{Cos}\,t\,dt$. It follows that

$$\int_C \underline{V} \cdot \underline{dr} = \int_{C_1} \underline{V} \cdot \underline{dr} + \int_{C_2} \underline{V} \cdot \underline{dr}$$

$$= \int_{-1}^1 (2t)(0)dt + t^2\,0\,dt + \int_0^{\pi/2} 2\text{Cos}\,t\,\text{Sin}\,t(-\text{Sin}\,t\,dt)$$

$$+ \int_0^{\pi/2} \text{Cos}^2 t\,(\text{Cos}\,t\,dt) = 0$$

PROBLEM 11.5

Show that the work done in moving a particle along path C in a force field F is equal to the change in the kinetic energy of the particle. Show that if F is a potential field, then the total energy of a particle in the field is conserved.

SOLUTION 11.5

The work done in moving a particle along path C in force field F is defined to be

$$W = \int_C \underline{F} \cdot \underline{dr}$$

If the field exerts force $\underline{F}$ on the particle, then the position vector $\underline{P}(t)$ of the particle must satisfy Newton's second law,

$$F = m\underline{P}''(t) = m \frac{d}{dt} \underline{V}(t)$$

where m denotes the mass of the particle and $\underline{V}(t) = \underline{P}'(t)$ denotes the velocity vector. Then since $dr = \underline{P}'(t)dt = \underline{V}(t)dt$

$$W = \int_C m\underline{V}'(t) \cdot dr = \int_C m\underline{V}'(t) \cdot \underline{P}'(t)dt = \int_C m\underline{V}'(t) \cdot \underline{V}(t)dt$$

But

$$\underline{V}'(t) \cdot \underline{V}(t) = -\frac{1}{2}\frac{d}{dt} (\underline{V} \cdot \underline{V})$$

hence

$$W = \frac{1}{2}\int_C m \frac{d}{dt} (\underline{V} \cdot \underline{V})\, dt = \frac{1}{2} m \,|\, \underline{V} \,|^2 (t_2) - \frac{1}{2} m \,|\, \underline{V} \,|^2 (t_1)$$

where t_1 and t_2 denote the parameter value at the initial endpoint and final endpoint of C, respectively. If we define the *kinetic energy* of the particle by

$$KE(t) = \frac{1}{2} m \,\|\, \underline{V} \,\|^2(t)$$

then

$$W = KE(t_2) - KE(t_1);$$

i.e., the work done by the field equals the change in the kinetic energy of the particle.

If $\underline{F} = -\textbf{grad}\, f$ for some smooth scalar field f, then

$$\underline{F} \cdot \underline{dr} = -f_x dx - f_y dy - f_z dz$$

$$= -(f_x x' + f_y y' + f_z z')dt = -\frac{df}{dt}\, dt.$$

Hence

$$W = \int_C \underline{F} \cdot \underline{dr} = \int_{t_1}^{t_2} -\, df = f(t_1) - f(t_2)$$

and by the previous result, $f(t_1) - f(t_2) = KE(t_2) - KE(t_1)$. That is,

$$f(t_1) + KE(t_1) = f(t_2) + KE(t_2)$$

which is to say, the sum of the *potential energy*, f(t), and the kinetic energy, KE(t), is constant as the particle moves along the path. A vector field F such that F = – **grad** f for some smooth scalar field f is referred to as a *potential* field. We have just shown that for motion in a potential field, the total energy is conserved. The negative sign before the gradient is just a sign convention that causes the potential energy to be positive.

Integral Identities in the Plane

PROBLEM 11.6

Let Ω denote a closed bounded fundamental domain in $\mathbb{R}^2$. Let the boundary Γ of Ω consist of a single smooth simple closed curve with the usual orientation. Then for $\underline{V} = \underline{V}(x, y)$ a smooth vector field in Ω, show that

$$\int_\Gamma v_1 dx + v_2 dy = \iint_\Omega (\partial_x v_2 - \partial_y v_1) dx\, dy \tag{1}$$

This statement is a somewhat simplified version of Theorem 11.2.

SOLUTION 11.6

Note that even for $\underline{V}$ smooth in Ω it may not be possible to define the partial derivatives $\partial_x v_2 - \partial_y v_1$ at points on Γ by means of the limit definition of the partial derivative. This is because the shape of Ω may make it impossible to approach certain points of Γ along a line parallel to one of the coordinate axes. This technicality can be overcome if we agree to say that for f defined in Ω, f_x is continuous in Ω if there exists a function $g \in \mathbb{C}(\Omega)$ such that $f_x = g$ at all interior points of Ω. Here the function g is obtained by computing f_x by the usual rules of differentiation inside Ω. A similar definition permits the extension of f_y to Γ.

For simplicity suppose that Ω is as shown in Figure 11.1. That is, Ω is contained in and is tangent to the sides of the rectangle $R = \{a \le x \le b, c \le y \le d\}$. The case in which Ω is tangent to the sides of R on an interval can be handled without significant additional difficulty.

Since we have assumed that Ω is a fundamental domain, we may suppose that for functions $f_1(x), f_2(x) \in \mathbb{C}[a, b]$,

$$\Omega = \{(x, y): f_1(x) \le y \le f_2(x), a \le x \le b\}$$

and for functions $g_1(y), g_2(y) \in \mathbb{C}[c, d]$,

$$\Omega = \{(x, y): g_1(y) \le x \le g_2(y), c \le y \le d\}.$$

Then

$$\int_\Gamma v_1 dx = \int_{C_1} v_1 dx + \int_{-C_2} v_1 dx = \int_{C_1} v_1 dx - \int_{C_2} v_1 dx$$

where

$$C_1 = \{x = t, y = f_1(t): a \le t \le b\}$$
$$C_2 = \{x = t, y = f_2(t): a \le t \le b\}.$$

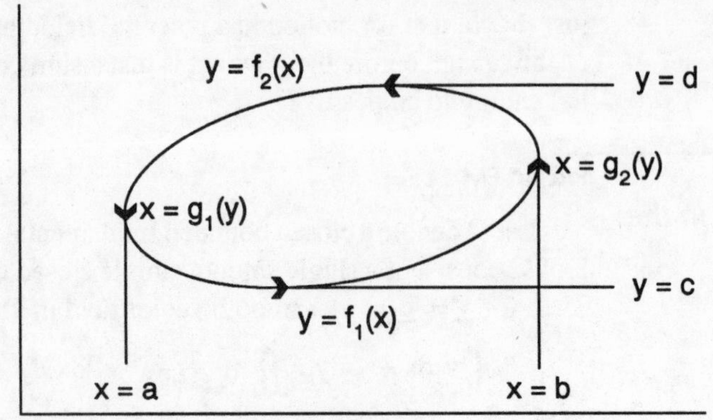

Figure 11.1
Problem 11.6

Note that because of the orientation of C_1 and C_2, Γ consists of $C_1 + (-C_2)$. Thus

$$\int_\Gamma v_1 dx = \int_a^b v_1(t, f_1(t))dt - \int_a^b v_1(t, f_2(t))dt.$$

In addition, evaluating the following double integral in the usual way as an iterated integral leads to

$$\iint_\Omega \partial_y v_1 dx\, dy = \int_a^b \int_{f_1(x)}^{f_2(x)} \partial_y v_1(x, y)dy\, dx$$

$$= \int_a^b (v_1(x, f_2(x)) - v_1(x, f_1(x)))dx.$$

That is,

$$\iint_\Omega \partial_y v_1 dxdy = -\int_\Gamma v_1 dx. \qquad (2)$$

In much the same way we can show

$$\iint_\Omega \partial_x v_2 dxdy = \int_\Gamma v_2 dy. \qquad (3)$$

Then subtracting (2) from (3) leads to (1).

We have proved Green's Theorem for a domain Ω whose boundary is a single simple closed curve. Then any simple closed curve inside Ω contains only points of Ω in its interior. Any domain Ω in $\mathbb{R}^2$ having this property is said to be *simply connected*. We can, however, prove Green's Theorem for domains Ω that are not simply connected.

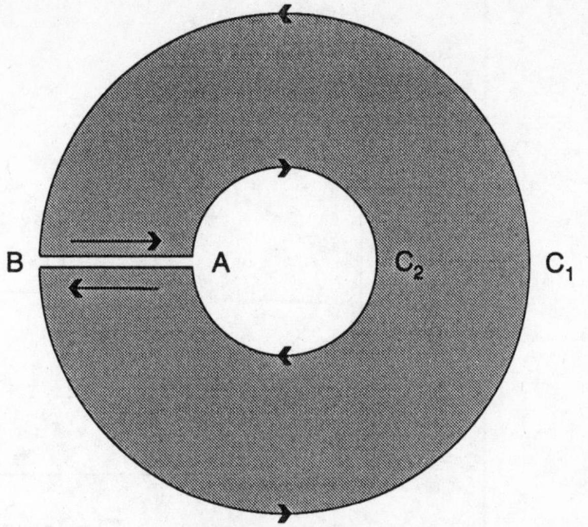

Figure 11.2
Cutting a Multiply Connected Domain

For example, consider the annular domain W contained between concentric circles C_1 and C_2 (see Figure 11.2). The boundary Γ of Ω consists of the two simple closed curves C_1 and C_2. Clearly any circle C_0 concentric with C_1 and C_2 with radius between r_1 and r_2 lies in Ω but contains points in its interior that are not points of Ω. Thus Ω is not simply connected. Let A and B denote points of C_1 and C_2, respectively, colinear with the center O of the circles and let C_+ denote the straight line joining A to B. Similarly let C_- denote the same straight line path with the opposite orientation; i.e., C_- is the line from B to A. Finally let Ω' denote the bounded open domain whose boundary Γ' is $C_1 \cup C_+ \cup C_2 \cup C_-$. Then Ω' is simply connected (e.g., C_0 is not a curve inside Ω') and for any line integral over the boundary of Ω'

$$\int_{\Gamma'} = \int_{C_1 \cup C_+ \cup C_2 \cup C_-} = \int_{C_1} + \int_{C_+} + \int_{C_2} + \int_{C_-}$$
$$= \int_{C_1} + \int_{C_2} + \int_{C_+} - \int_{C_+} = \int_{C_1} + \int_{C_2} = \int_{\Gamma}$$

To apply Green's Theorem in the domain Ω we just cut Ω along the line AB to form the simply connected domain Ω' and note that

$$\iint_{\Omega} \partial_x v_2 - \partial_y v_1 = \iint_{\Omega'} \partial_x v_2 - \partial_y v_1 = \int_{\Gamma'} v_1 dx + v_2 dy = \int_{\Gamma} v_1 dx + v_2 dy$$

We can also use the method of cuts to extend Green's Theorem to domains Ω that are not fundamental but are the union of finitely many nonoverlapping fundamental domains.

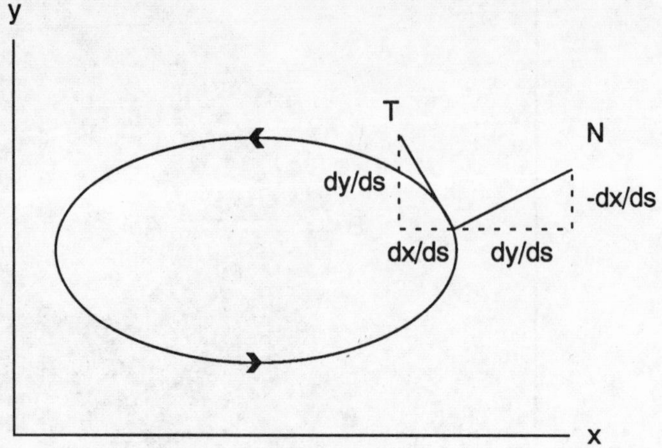

Figure 11.3
Unit Tangent and Normal to a Plane Curve

PROBLEM 11.7

Let Ω denote a closed bounded fundamental domain in $\mathbb{R}$. Let the boundary Γ of Ω consist of a smooth simple closed curve with the usual orientation and outward unit normal denoted by N. Then for $\underline{V} = \underline{V}(x, y)$ a smooth vector field in Ω, show that

$$\int_\Gamma \underline{V} \cdot \underline{N}\, ds = \iint_\Omega \text{div}\, \underline{V}\, dx\, dy \tag{1}$$

SOLUTION 11.7

Suppose Γ is given parametrically in terms of the arc length s by

$$\Gamma = \{x = x(s), y = y(s): a \le s \le b\}.$$

Then $\underline{T}(s) = x'(s)\underline{i} + y'(s)\underline{j}$ is the unit tangent to Γ then the unit vector $\underline{N}(s) = y'(s)\underline{i} - x'(s)\underline{j}$ is clearly normal to $\underline{T}$ and it is evident from Figure 11.3 that $\underline{N}$ is the outward normal to Γ. Then

$$\int_\Gamma \underline{V} \cdot \underline{N}\, ds = \int_\Gamma (v_1 y'(s) - v_2 x'(s))ds = \int_\Gamma v_1 dy - v_2 dx$$

and it follows from Green's Theorem that

$$\int_\Gamma v_1 dy - v_2 dx = \iint_\Omega (\partial_x v_1 + \partial_y v_2)dx\, dy.$$

But this is (1).

PROBLEM 11.8

Let $\underline{V}(x, y)$ denote a smooth vector field in $\mathbb{R}^2$. Show that at each point $\underline{P} \in \mathbb{R}^2$ the divergence of $\underline{V}$ satisfies

$$\text{div } \underline{V}(\underline{P}) = \lim_{\rho \to 0} \frac{1}{\pi\rho^2} \int_{\Gamma_\rho} \underline{V} \cdot \underline{N} \, ds = \text{"normalized flux at } \underline{P}\text{"} \tag{1}$$

SOLUTION 11.8

For an arbitrary point $\underline{P} \in \mathbb{R}^2$ let Ω_ρ denote the circular disc of radius $\rho > 0$ centered at $\underline{P}$ and let Γ_ρ denote the circular boundary of this disc. Then the previous result implies that

$$\iint_{\Omega_\rho} \text{div}\underline{V} \, dx \, dy = \int_{\Gamma_\rho} \underline{V} \cdot N \, ds$$

Now the mean value theorem for integrals implies that for some $\underline{Q} \in \Omega_\rho$,

$$\iint_{\Omega_\rho} \text{div}\underline{V} \, dx \, dy = \text{div } \underline{V}(\underline{Q}) \iint_{\Omega_\rho} dx \, dy = \text{div } \underline{V}(\underline{Q})\pi\rho^2$$

Then

$$\text{div } \underline{V}(\underline{Q}) = \frac{1}{\pi\rho^2} \int_{\Gamma_\rho} \underline{V} \cdot \underline{N} \, ds$$

As ρ tends to zero, $\underline{Q}$ must approach $\underline{P}$ and we obtain (1). The integral

$$\int_{\Gamma_\rho} \underline{V} \cdot N \, ds$$

is often referred to as the *flux* of the vector field $\underline{V}$ through Γ_ρ. The flux can be interpreted physically as the total net outflow of the vector field $\underline{V}$ through Γ_ρ. Dividing the flux by the area of the region inside Γ_ρ produces a "normalized" flux. Thus the divergence of $\underline{V}$ at a point $\underline{P}$ can be interpreted as the flow rate out of $\underline{P}$. If div $\underline{V}(\underline{P})$ is positive we say $\underline{P}$ is a *source* and if the divergence at $\underline{P}$ is negative we refer to $\underline{P}$ as a *sink.*

We can in fact define the divergence of $\underline{V}$ by

$$\text{div}\underline{V} = \lim_{|\Omega| \to 0} \frac{1}{|\Omega|} \int_{\Gamma_\rho} \underline{V} \cdot \underline{N} \, ds$$

where Ω denotes a domain of area $|\Omega|$ with simple closed boundary Γ. In order for the limit to exist we specify that as Ω shrinks to a point $\underline{P}$, this must occur in such a way that $\underline{P}$ is constantly inside Ω. This integral definition of divergence is independent of any coordinate system.

PROBLEM 11.9

Let Ω denote a closed bounded fundamental domain in $\mathbb{R}^2$. Let the boundary Γ of Ω consist of a smooth simple closed curve with the usual orientation and unit tangent vector denoted by T. Then for $\underline{V} = \underline{V}(x, y)$ a smooth vector field in Ω, show that

$$\int_\Gamma \underline{V} \cdot \underline{T} \, ds = \iint_\Omega \text{curl } \underline{V} \cdot \underline{k} \, dx \, dy \tag{1}$$

SOLUTION 11.9

By definition

$$\int_\Gamma \underline{V}\cdot\underline{T}\,ds = \int_\Gamma v_1dx + v_2dy \tag{2}$$

and by Green's theorem,

$$\int_\Gamma v_1dx + v_2dy = \iint_\Omega (\partial_x v_2 - \partial_y v_1)dx\,dy \tag{3}$$

But

$$\text{curl}\underline{V} = \begin{vmatrix} \underline{i} & \underline{j} & \underline{k} \\ \partial_x & \partial_y & \partial_z \\ v_1 & v_2 & 0 \end{vmatrix} = (\partial_x v_2 - \partial_y v_1)\underline{k} \tag{4}$$

and (4) together with (2) and (3) leads to (1).

PROBLEM 11.10

Let $\underline{V}(x, y)$ denote a smooth vector field in $\mathbb{R}^2$. Show that at each point $\underline{P} \in \mathbb{R}^2$ the curl of $\underline{V}$ satisfies

$$\text{curl }\underline{V}(\underline{P})\cdot k = \lim_{\rho\to 0}\frac{1}{\pi\rho^2}\int_{\Gamma_\rho}\underline{V}\cdot\underline{T}\,ds = \text{"normalized circulation at }\underline{P}\text{"} \tag{1}$$

SOLUTION 11.10

For an arbitrary point $\underline{P} \in \mathbb{R}^2$ let Ω_ρ denote the circular disc of radius $\rho > 0$ centered at $\underline{P}$ and let Γ_ρ denote the circular boundary of this disc. Then the previous result implies that

$$\iint_{\Omega_\rho} \text{curl }\underline{V}\cdot\underline{k}\,dx\,dy = \int_{\Gamma_\rho}\underline{V}\cdot\underline{T}\,ds$$

Now the mean value theorem for integrals implies that for some $\underline{Q} \in \Omega_\rho$,

$$\iint_{\Omega_\rho} \text{curl }\underline{V}\cdot\underline{k}\,dx\,dy = \text{curl}\underline{V}(\underline{Q})\cdot\underline{k}\iint_{\Omega_\rho} dx\,dy = \text{curl}\underline{V}(\underline{Q})\cdot\underline{k}\,\pi\rho^2$$

Then

$$\text{curl}\underline{V}(\underline{Q})\cdot\underline{k} = \frac{1}{\pi\rho^2}\int_{\Gamma_\rho}\underline{V}\cdot\underline{T}\,ds$$

As ρ tends to zero, $\underline{Q}$ must approach $\underline{P}$ and we obtain (1). The integral

$$\int_{\Gamma_\rho}\underline{V}\cdot\underline{T}\,ds$$

is equal to the circulation of the vector field $\underline{V}$ around the path Γ_ρ. Dividing by the area contained in Γ_ρ "normalizes" this quantity. Then (1) asserts that as ρ tends to zero the normalised circulation tends to the component of curl$\underline{V}$ in the direction of the normal to the plane in which Γ_ρ lies. Thus the value of curl$\underline{V}\cdot\underline{k}$ at a point is a measure of the intensity of circulation in the xy-plane at that point. The circulation is in some sense a measure of the rotational tendency of the vector field. We saw this earlier in Example 11.2(c).

PROBLEM 11.11

Let Ω in $\mathbb{R}$ be a domain in which Theorem 11.2 applies and let $\varphi, \psi \in \mathbb{C}^2(\Omega)$. Let N(x, y) denote the outward pointing unit normal vector to the boundary Γ of Ω. Then show

(a) $$\iint_\Omega \nabla^2\varphi \, dx\, dy = \int_\Gamma \nabla\varphi \cdot \underline{N} \, ds$$

(b) $$\iint_\Omega \psi\nabla^2\varphi \, dxdy = \int_\Gamma \psi\nabla\varphi \cdot \underline{N} \, ds - \iint_\Omega \nabla\varphi \cdot \nabla\psi \, dx\, dy$$

(c) $$\iint_\Omega (\psi\nabla^2\varphi - \varphi\nabla^2\psi) dx\, dy = \int_\Gamma (\psi\nabla\varphi \cdot \underline{N} - \varphi\nabla\psi \cdot \underline{N}) ds$$

SOLUTION 11.11

Part (a) We apply the result (1) in problem 11.7 in the case $\underline{V} = \nabla\varphi$ for $\varphi \in \mathbb{C}^1(\Omega)$. Then div$\underline{V} = \nabla^2\varphi$ and

$$\iint_\Omega \text{div}\underline{V} \, dx\, dy = \iint_\Omega \nabla^2\varphi \, dx\, dy = \int_\Gamma \underline{V} \cdot \underline{N} \, ds = \int_\Gamma \nabla\varphi \cdot \underline{N} \, ds.$$

Part (b) For $\varphi, \psi \in \mathbb{C}^1(\Omega)$ let $\underline{V} = \psi\nabla\varphi$. Then by the rules for differentiation of vector functions,

$$\text{div}\underline{V} = \nabla \cdot (\psi\nabla\varphi) = \nabla\psi \cdot \nabla\varphi + \psi\nabla^2\varphi.$$

Now apply the result (1) from Problem 11.7 to obtain (b)

$$\iint_\Omega (\nabla\psi \cdot \nabla\varphi + \psi\nabla^2\varphi) \, dx\, dy = \int_\Gamma \psi\nabla\varphi \cdot \underline{N} \, ds.$$

Part (c) Repeat the steps of part (b) with $\underline{V} = \varphi\nabla\psi$ to obtain

$$\iint_\Omega (\nabla\psi \cdot \nabla\varphi + \varphi\nabla^2\psi) \, dx\, dy = \int_\Gamma \varphi\nabla\psi \cdot \underline{N} \, ds.$$

Subtracting this equation from the equation in (b) yields (c).

PROBLEM 11.12

Let Ω denote a bounded domain in $\mathbb{R}^2$ having as its boundary the smooth, simple closed curve Γ. Suppose $\varphi \in \mathbb{C}^2(\Omega) \cap \mathbb{C}^1(\Omega + \Gamma)$ satisfies

$$\nabla^2\varphi = F \quad \text{in } \Omega \text{ and } \quad \nabla^2\varphi \cdot \underline{N} = g \quad \text{on } \Gamma \tag{1}$$

for given functions $F \in \mathbb{C}(\Omega)$, $g \in \mathbb{C}(\Gamma)$ and N the unit outward normal to G. Show that in order for this problem to have a solution it is necessary that the data F, g satisfy the following condition of compatibility

$$\iint_\Omega F \, dxdy = \int_\Gamma g \, ds. \tag{2}$$

SOLUTION 11.12

Suppose that there exists a function f satisfying the conditions of the problem (1). Such a function is said to be a classical solution of the Neumann problem for Laplace's equation. Apply Green's first identity, the result (b) of the previous problem, for this function φ and $\psi = 1$. Then $\nabla\psi = 0$ and (b) reduces to

$$\iint_{\Omega} \nabla^2\varphi \, dx\, dy = \int_{\Gamma} \nabla\varphi \cdot \underline{N}\, ds - 0$$

Since φ satisfies the conditions of (1) this is precisely the result (2).

PROBLEM 11.13

Let Ω denote a bounded domain in $\mathbb{R}^2$ having as its boundary the smooth, simple closed curve Γ composed of nonoverlapping arcs Γ_1 and Γ_2. Show that if there exists a function $\varphi \in \mathbb{C}^2(\Omega) \cap \mathbb{C}^1(\Omega + \Gamma)$ satisfying

$$\nabla^2\varphi = F \quad \text{in } \Omega \quad \text{and } \varphi = f \text{ on } \Gamma_1, \quad \nabla\varphi \cdot \underline{N} = g \text{ on } \Gamma_2 \tag{1}$$

for given functions $F \in \mathbb{C}(\Omega)$, $f, g \in \mathbb{C}(\Gamma)$ and $\underline{N}$ the unit outward normal to Γ, then φ must be unique.

SOLUTION 11.13

Suppose there exist two functions u and v satisfying the conditions of problem (1). Such functions are said to be classical solutions of the mixed boundary value problem for Laplace's equation. Let w = u – v and note that it follows from (1) that

$$\nabla^2 w = 0 \text{ in } \Omega \text{ and } w = 0 \text{ on } \Gamma_1,\ \nabla w \cdot \underline{N} = 0 \text{ on } \Gamma_2. \tag{2}$$

Apply Green's first identity for the choice φ = ψ = w to obtain

$$\iint_{\Omega} w\nabla^2 w \, dx\, dy = \int_{\Gamma_1} w\nabla w \cdot \underline{N}\, ds + \int_{\Gamma_2} w\nabla w \cdot \underline{N}\, ds - \iint_{\Omega} \nabla w \cdot \nabla w \, dx\, dy$$

i.e.,

$$0 = 0 + 0 - \iint_{\Omega} \nabla w \cdot \nabla w \, dx\, dy \tag{3}$$

Since $\nabla w \cdot \nabla w = |\nabla w| \geq 0$, it follows from (3) that $\nabla w = \underline{0}$ on Ω. This implies that w is constant on Ω + Γ and since w = 0 on Γ_1, it follows that w = 0 on Ω + Γ. But then u = v and if (1) has any solution, that solution is unique.

Surfaces

PROBLEM 11.14

Suppose the surface S is given parametrically in the form:

$$S: \begin{cases} x = x(s, t) \\ y = y(s, t) \\ z = z(s, t) \end{cases} \qquad \begin{matrix} a \leq s \leq b \\ c \leq t \leq d \end{matrix}$$

Find an expression for a normal vector to surface S in terms of the parametric description.

SOLUTION 11.14

For S is given parametrically then $\underline{P}(s, t) = x(s, t)\underline{i} + y(s, t)\underline{j} + z(s, t)\underline{k}$ is the position vector for a typical point on S. Then each of the following vectors is tangent to S

$$\partial_s\underline{P} = \frac{\partial x}{\partial s}\underline{i} + \frac{\partial y}{\partial s}\underline{j} + \frac{\partial z}{\partial s}\underline{k} \quad \text{and} \quad \partial_t\underline{P} = \frac{\partial x}{\partial t}\underline{i} + \frac{\partial y}{\partial t}\underline{j} + \frac{\partial z}{\partial t}\underline{k};$$

i.e., the set of points C_t in S where t is held constant is a curve with tangent vector $\partial_s \underline{P}$. Similarly, $\partial_t \underline{P}$ is tangent to the curve C_s in S where s is held constant. Then

$$\underline{n} = \partial_s \underline{P} \times \partial_t \underline{P} = \begin{vmatrix} \underline{i} & \underline{j} & \underline{k} \\ x_s & y_s & z_s \\ x_t & y_t & z_t \end{vmatrix}$$

$$= (y_s z_t - z_s y_t)\underline{i} - (x_s z_t - x_t z_s)\underline{j} + (x_s y_t - x_t y_s)\underline{k}$$

is normal to both $\partial_s \underline{P}$ and to $\partial_t \underline{P}$, hence it is normal to S. Note that $\underline{n}$ can be expressed in terms of Jacobian determinants

$$n = \frac{\partial(y, z)}{\partial(s, t)} \underline{i} + \frac{\partial(z, x)}{\partial(s, t)} \underline{j} + \frac{\partial(x, y)}{\partial(s, t)} \underline{k}.$$

PROBLEM 11.15

Find an expression for the surface integrals

$$\iint_S dA \quad \text{and} \quad \iint_S \underline{V} \cdot \underline{N}\, dA$$

when S is given parametrically as in the previous problem.

SOLUTION 11.15

Let C_s (respectively C_t) denote the curve mentioned in the previous problem lying in the surface S along which s (respectively t) is constant. If S is covered by a uniform grid of these curves then S will be divided into finitely many small parallelograms with sides of the form $\partial_s \underline{P}\, \Delta s$ and $\partial_t \underline{P}\, \Delta t$. The area of one of these parallelograms is equal to $\| \partial_s \underline{P}\, \Delta s \times \partial_t \underline{P}\, \Delta t \|$. Then the area of S can be approximated by the sum

$$\sum \| \partial_s \underline{P}\, \Delta s \times \partial_t \underline{P}\, \Delta t \|$$

where the sum runs over all the parallelograms on S. If the sum tends to a limit as the mesh size of the grid tends to zero, we define the limit as the area of S. This motivates the definition

$$\iint_S dA = \iint_{R_{st}} \| \partial_s \underline{P} \times \partial_t \underline{P} \|\, ds\, dt$$

where R_{st} denotes the rectangle $\{(s, t){:}\ a \le s \le b,\ c \le t \le d\}$. Then

$$dA = \| \partial_s \underline{P} \times \partial_t \underline{P} \|\, ds\, dt$$

Recalling the vector identity

$$\| A \times B \| = \sqrt{(A \cdot A)(B \cdot B) - (A \cdot B)^2}$$

we have

$$\| \partial_s \underline{P} \times \partial_t \underline{P} \| = \sqrt{EG - F^2} \quad \text{where}$$

$$E = \partial_s \underline{P} \times \partial_s \underline{P}, \quad G = \partial_t \underline{P} \times \partial_t \underline{P} \quad \text{and} \quad F = \partial_s \underline{P} \times \partial_t \underline{P}.$$

Then the *unit* normal to S is given by

$$N = \frac{n}{\sqrt{EG - F^2}},$$

the element of surface area can be written $dA = \sqrt{EG - F^2}\, ds\, dt$ and

$$\iint_S dA = \iint_{R_{st}} \sqrt{EG - F^2}\, ds\, dt.$$

Finally, the surface integral of a vector field $\underline{V}$ can be written

$$\iint_S \underline{V} \cdot \underline{N}\, dA = \iint_S \underline{V} \cdot \underline{N} \,\|\, \partial_s \underline{P} \times \partial_t \underline{P} \,\|\, ds\, dt$$

$$= \iint_{R_{st}} \underline{V} \cdot \underline{n}\, ds\, dt$$

$$= \iint_{R_{st}} \left(v_1 \frac{\partial(y, z)}{\partial(s, t)} + v_2 \frac{\partial(z, x)}{\partial(s, t)} + v_3 \frac{\partial(x, y)}{\partial(s, t)} \right) ds\, dt$$

PROBLEM 11.16

Let S denote the plane triangle with vertices A = (1, 0, 0), B = (0, 2, 0) and C = (0, 0, 3). Represent S in explicit form and evaluate the integral

$$I = \iint_S x\, dy\, dz + y\, dx\, dz + z\, dx\, dy.$$

SOLUTION 11.16

The vectors $\underline{AB} = (-1, 2, 0)$ and $\underline{BC} = (0, -2, 3)$ form two sides of the triangle S. Then $\underline{AB} \times \underline{BC} = \underline{n} = 6\underline{i} + 3\underline{j} + 2\underline{k}$ is normal to S, pointing away from the origin. It follows that the equation of the plane with this normal and containing the point A is given by $6(x - 1) + 3y + 2z = 0$. Then solving for z in terms of x and y yields

$$z(x, y) = 3 - 3x - \frac{3}{2}y \quad 0 \le x \le 1, \quad 0 \le y \le 2 - 2x \tag{1}$$

This is the explicit representation for S in terms of $z = z(x, y)$ for which the corresponding expression for dA is

$$dA = \sqrt{1 + z_x^2 + z_y^2}\, dxdy = \sqrt{1 + 9 + 9/4}\, dx\, dy = \frac{7}{2} dx\, dy$$

Then we have

$$I = \iint_S x\, dy\, dz + y\, dx\, dz + z\, dx\, dy = \iint_S \underline{V} \cdot \underline{N}\, dA$$

for

$$\underline{V} = x\underline{i} + y\underline{j} + z\underline{k} \quad \text{and} \quad \underline{N} = \frac{6}{7}\underline{i} + \frac{3}{7}\underline{j} + \frac{2}{7}\underline{k}$$

Then

$$\underline{V} \cdot \underline{N} = (x\underline{i} + y\underline{j} + z\underline{k}) \cdot \left(\frac{6}{7}\underline{i} + \frac{3}{7}\underline{j} + \frac{2}{7}\underline{k} \right) = \frac{6x + 3y + 2z}{7} = \frac{6}{7}$$

and

$$I = \int_0^1 \int_0^{2-2x} \frac{6}{7}\frac{7}{2}\, dy\, dx = 3\int_0^1 (2 - 2x)dx = 3$$

PROBLEM 11.17

Give a parametric description of the surface S in the previous problem. Express the integral I in terms of the parametric variables and evaluate.

SOLUTION 11.17

Note that the vector

$$\underline{P}(s, t) = \underline{OA} + s\underline{AB} + t\underline{BC} = \begin{bmatrix} 1 \\ 0 \\ 0 \end{bmatrix} + s\begin{bmatrix} -1 \\ 2 \\ 0 \end{bmatrix} + t\begin{bmatrix} 0 \\ -2 \\ 3 \end{bmatrix} = \begin{bmatrix} 1-s \\ 2s-2t \\ 3t \end{bmatrix}$$

is the position vector of a typical point on S. As t varies between 0 and s and s varies between 0 and 1, $\underline{P}(s, t)$ sweeps out the points of S. Then

$$\begin{aligned} x &= 1 - s \\ y &= 2s - 2t \quad 0 \le s \le 1, \quad 0 \le t \le s \\ z &= 3t \end{aligned}$$

provides a parametric description of S. According to the result of Problem 11.15 we have

$$I = \iint_S \underline{V} \cdot N\, dA = \iint_{R_{st}} \left(v_1 \frac{\partial(y, z)}{\partial(s, t)} + v_2 \frac{\partial(z, x)}{\partial(s, t)}\, v_3 \frac{\partial(x, y)}{\partial(s, t)} \right) ds\, dt$$

and since

$$\frac{\partial(y, z)}{\partial(s, t)} = 6, \quad \frac{\partial(z, x)}{\partial(s, t)} = 3, \quad \frac{\partial(x, y)}{\partial(s, t)} = 2,$$

$$v_1 = x(s, t) = 1 - s, \quad v_2 = y(s, t) = 2s - 2t \quad \text{and} \quad v_3 = z(s, t) = 3t$$

we find

$$I = \int_0^1 \int_0^s 6(1 - s) + 3(2s - 2t) + 2(2t)\, ds\, dt = 6 \int_0^1 \int_0^s ds\, dt = 3.$$

PROBLEM 11.18

Let S denote the lateral surface of the cylinder $\{x^2 + y^2 = 1, 0 \le z \le 1\}$. Give a parametric description of S and evaluate the integral

$$I = \iint_S x^2 z\, dA.$$

SOLUTION 11.18

The vector

$$\underline{P}(s, t) = \text{Cos } s\underline{i} + \text{Sin } s\underline{j} + t\underline{k} \quad 0 \le s \le 2\pi, 0 \le t \le 1,$$

is the position vector for a typical point on S. The corresponding parametric description of S is then

$$x = \text{Cos } s,\ y = \text{Sin } s,\ z = t,\ 0 \le s \le 2\pi,\ 0 \le t \le 1.$$

We can compute $\partial_s \underline{P} = -\text{Sin } s\underline{i} + \text{Cos } s\underline{j}$ and $\partial_t \underline{P} = \underline{k}$ and thus

$$E = \partial_s\underline{P} \cdot \partial_s\underline{P} = 1,\ F = \partial_s\underline{P} \cdot \partial_t\underline{P} = 0,\ G = \partial_t\underline{P} \cdot \partial_t\underline{P} = 1.$$

Then using the results of Problem 11.15,

$$I = \int_0^1 \int_0^{2\pi} x(s, t)^2 z(s, t)\sqrt{EG - F^2}\, ds dt$$

$$= \int_0^1 \int_0^{2\pi} \text{Cos}^2 s\ t\ 1\ ds dt = \int_0^1 \pi t\, dt = \pi/2.$$

PROBLEM 11.19

Let S denote the surface of a sphere in $\mathbb{R}^3$ and let $\underline{N}$ denote the unit outward normal to S. Show that

$$\text{Volume of Sphere} = \frac{1}{3}\iint_S x\, dy\, dz + y\, dx\, dz + z\, dx\, dy$$

Does this formula hold for the boundary S of any other solid in $\mathbb{R}^3$?

SOLUTION 11.19

Write

$$\frac{1}{3}\iint_S x\, dy\, dz + y\, dx\, dz + z\, dx\, dy = \iint_S \underline{V} \cdot \underline{N}\, dA$$

where

$$\underline{V} = \frac{1}{3}(x\underline{i} + y\underline{j} + z\underline{k}). \tag{1}$$

Then by Theorem 11.10,

$$\iint_S \underline{V} \cdot \underline{N}\, dA = \iiint_\Omega \text{div } \underline{V}\, dx\, dy\, dz$$

where Ω denotes the spherical domain inside S. But div$\underline{V}$ = (1 + 1 + 1)/3 = 1. Thus

$$\iint_S \underline{V} \cdot \underline{N}\, dA = \iiint_\Omega 1\, dx\, dy\, dz = \text{volume of } \Omega. \tag{2}$$

For $\underline{V}$ given by (1), the result (2) holds for any solid region Ω for which Theorem 11.10 is valid.

Properties of Vector Fields

PROBLEM 11.20

Prove that the continuous vector field $\underline{V}(x, y, z) = v_1\underline{i} + v_2\underline{j} + v_3\underline{k}$ is path independent in a domain $\Omega \subset \mathbb{R}^3$ if and only if $\underline{V}$ is conservative in Ω.

SOLUTION 11.20

Suppose first that $\underline{V}$ is such that

$$\text{for any smooth simple closed curve C in } \Omega, \int_C \underline{V} \cdot \underline{dr} = 0 \qquad (1)$$

Let A and B denote arbitrary points of Ω. Since Ω is connected, there exists a smooth curve C_1 in Ω joining A to B. Let C_2 denote another smooth curve in Ω that goes from A to B. Without loss of generality we may suppose that C_1 and C_2 have only their endpoints in common. Let C denote the curve that goes from A to B along C_1 and then goes from B back to A along the curve C_2. Then C consists of C_1 followed by $-C_2$. Then C is a simple closed curve in Ω and it follows from our hypothesis (1) that

$$0 = \int_C \underline{V} \cdot \underline{dr} = \int_{C_1} \underline{V} \cdot \underline{dr} + \int_{-C_2} \underline{V} \cdot \underline{dr} = \int_{C_1} \underline{V} \cdot \underline{dr} - \int_{C_2} \underline{V} \cdot \underline{dr}$$

This proves for C_1 and C_2 two curves having the same initial endpoint and the same final endpoint then

$$\int_{C_1} \underline{V} \cdot \underline{dr} = \int_{C_2} \underline{V} \cdot \underline{dr} \qquad (2)$$

i.e. (1) implies (2).

To show that (2) implies (1), suppose C is a simple closed curve in Ω and let A and B denote distinct points on C. Let C_1 denote the curve obtained when C is traced from A to B and let C_2 denote the remainder of C, the part that goes from B back to A. Let C_2 denote this arc but with orientation opposite to that of C. Then C_2 is a smooth curve from A to B and C is composed of C_1 and $-C_2$. Therefore

$$\int_C \underline{V} \cdot \underline{dr} = \int_{C_1} \underline{V} \cdot \underline{dr} + \int_{-C_2} \underline{V} \cdot \underline{dr} = \int_{C_1} \underline{V} \cdot \underline{dr} - \int_{C_2} \underline{V} \cdot \underline{dr} = 0$$

since C_1 and C_2 have the same initial endpoints and the same final endpoints. Then (2) implies (1) and we have proved that these conditions are equivalent.

PROBLEM 11.21

Show that for vector field $\underline{V}(x, y, z)$ continuous in domain Ω in $\mathbb{R}^3$, the following conditions are equivalent:

1. $\underline{V}$ is path independent in Ω
2. $\underline{V}$ is conservative in Ω
3. $\underline{V} = \textbf{grad}\, F$ for some scalar field $F \in \mathbb{C}^1(\Omega)$

SOLUTION 11.21

We showed in the previous problem that 1. is equivalent to 2. Suppose then that 3. holds. Then for any simple closed curve C in Ω

$$\int_C \underline{V} \cdot \underline{dr} = \int_C F_x dx + F_y dy + F_z dz$$

$$= \int_C dF = F_1 - F_2$$

where F_1 and F_2 denote the respective values for F at the initial endpoint and final endpoint of C. But for a closed path C these are the same and thus the value of the integral is zero. Since this must be true for any simple closed path C, it follows that 3 implies 2.

Now suppose that 1 holds. Fix a point A in Ω and for arbitrary point B = (x, y, z) in Ω define a function

$$F(x, y, z) = \int_C \underline{V} \cdot \underline{dr}$$

where C denotes a smooth curve in Ω having A as its initial endpoint and B as it final endpoint. Since 1 holds and A is held fixed, the value of the integral depends only on the endpoint B. Thus the function F is well defined for each B in Ω. Now we can show that **grad** F = $\underline{V}$. This will prove that 1 implies 3. Since we already have 1 = 2 this will prove that 1 = 2 = 3.

It remains to show that **grad** F = $\underline{V}$. For arbitrary B = (x, y, z) in Ω let h > 0 be chosen sufficiently small that B′ = (x + h, y, z) is also in Ω and the line segment joining B to B′ is also inside Ω. Since Ω is an open set in $\mathbb{R}^3$, this is always possible. Now

$$F_x(x, y, z) = \lim_{h \to 0} \frac{F(x + h, y, z) - F(x, y, z)}{h}$$

But

$$F(x, y, z) = \int_C \underline{V} \cdot \underline{dr}$$

and

$$F(x + h, y, z) = \int_C \underline{V} \cdot \underline{dr} + \int_L \underline{V} \cdot \underline{dr}$$

where L denotes the oriented line segment joining B = (x, y, z) to B′ = (x + h, y, z). Let L be described parametrically by

$$L = \{x(t) = x(1 - t) + t(x + h),\ y(t) = y,\ z(t) = z;\ 0 \le t \le 1\}.$$

Then

$$F(x + h, y, z) - F(x, y, z) = \int_L \underline{V} \cdot \underline{dr} = \int_0^1 v_1(x(t), y, z)(h\,dt)$$

and by the mean value theorem for integrals,

$$F(x + h, y, z) - F(x, y, z) = v_1(x(\xi), y, z)\,h \quad \text{for some } \xi,\ x \le \xi \le x + h.$$

Thus

$$F_x(x, y, z) = \lim_{h \to 0} \frac{F(x + h, y, z) - F(x, y, z)}{h} = v_1(x, y, z).$$

In the same way we can show

$$F_y(x, y, z) = v_2(x, y, z) \text{ and } F_z(x, y, z) = v_3(x, y, z).$$

Then **grad** F = $\underline{V}$ and 1 implies 3.

PROBLEM 11.22

Suppose $\underline{V} \in \mathbb{C}^1(\Omega)$ for Ω a simply connected domain in $\mathbb{R}^3$. Then show that the condition

4. $\underline{V}$ is irrotational in Ω

is equivalent to the conditions 1, 2 and 3 of the previous problem.

SOLUTION 11.22

We show first that 3 implies 4. Suppose that $\underline{V} = \textbf{grad}\ F$ for some $F \in \mathbb{C}^1(\Omega)$. For any $F \in \mathbb{C}^2(\Omega)$, curl **grad** F = 0, and thus 4 follows from 3.

Now we will show 4 implies 2. Let C denote any simple closed curve in Ω. Since Ω is simply connected, there exists a smooth surface S in Ω having C as its boundary. Then Theorem 11.13, Stokes' theorem, implies that

$$\int_C \underline{V} \cdot \underline{dr} = \iint_S \text{curl}\ \underline{V} \cdot \underline{N}\ dA$$

where $\underline{N}$ denotes the unit normal to S. If curl $\underline{V} = \underline{0}$ then the integrals vanish and since C is an arbitrary simple closed curve in Ω, 2 follows from 4.

Note that we have proved part (a) of Theorem 11.9 which was stated previously as part (b) of Theorem 10.3.

PROBLEM 11.23

Show by example that simple connectedness is a necessary condition on Ω in order for 4 to imply 2.

SOLUTION 11.23

Let Ω denote the annular region between two coaxial infinite cylinders whose axis of symmetry is the z-axis. Then Ω is not simply connected. The vector field

$$\underline{V} = \frac{-y}{x^2 + y^2}\underline{i} + \frac{x}{x^2 + y^2}\underline{j}$$

is $\mathbb{C}^1$ in Ω (note that the origin is excluded from Ω) and satisfies curl$\underline{V} = \underline{0}$ throughout Ω. However let C denote the simple closed curve x(t) = R Cos t, y(t) = R Sin t, z(t) = 0, $0 \le t \le 2\pi$ where R is chosen so that C lies in Ω. Then

$$\int_C \underline{V} \cdot \underline{dr} = \int_0^1 (-R \text{ Sin } t\underline{i} + R \text{ Cos } t\underline{j}) \cdot (-R \text{ Sin } t\underline{i} + R \text{ Cos } t\underline{j})dt = 2\pi R \ne 0.$$

Thus $\underline{V}$ is irrotational in Ω but not conservative. As this example illustrates the properties 1 through 4 of Theorem 11.8 depend not just on $\underline{V}$ but on Ω as well.

PROBLEM 11.24

Show that for $\Omega = \mathbb{R}^3\underline{V}(x, y, z) = yz\underline{i} + xz\underline{j} + xy\underline{k}$ is conservative and find a scalar field F(x, y, z) such that $\underline{V} = \textbf{grad}\ F$.

SOLUTION 11.24

It is simple to compute curl$\underline{V}$ = (z – z)i + (y – y)j + (x – x)k = 0. Then since $\underline{V}$ is irrotational and W is simply connected it follows that $\underline{V}$ = grad F for some F ∈ $\mathbb{C}^2$(W). To find F note that

$$F_x = yz \text{ implies } F = xyz + f_1(y, z)$$

$$F_y = xz \text{ implies } F = xyz + f_2(x, z)$$

$$F_z = xy \text{ implies } F = xyz + f_3(x, y).$$

Here f_1, f_2 and f_3 denote arbitrary functions of (y, z), (x, z) and (x, y) respectively. Differentiating the first expression for F with respect to y leads to

$$F_y = xz + \partial_y f_1 = xz \text{ i.e., } \partial_y f_1 = 0.$$

Similarly

$$F_z = xy + \partial_z f_1 = xz \text{ i.e., } \partial_z f_1 = 0.$$

We conclude that f_1 = constant. In the same way we can show $f_2 = f_3$ = constant. Then F(x, y, z) = xyz + constant is a scalar field whose gradient equals $\underline{V}$. The constant here is arbitrary which is to say, F is unique only up to an aribtrary additive constant.

PROBLEM 11.25

Let $\underline{V} = x\underline{i} - y\underline{j}$. Then show that div$\underline{V}$ = 0 and find $\underline{U}$ such that curl $\underline{U} = \underline{V}$.

SOLUTION 11.25

We compute div$\underline{V} = \partial_x x - \partial_y y = 0$. Now curl $\underline{U} = \underline{V}$ if

$$\partial_y u_3 - \partial_z u_2 = x, \quad \partial_z u_1 - \partial_x u_3 = -y, \quad \text{and} \quad \partial_x u_2 - \partial_y u_1 = 0.$$

It is evident that we can find infinitely many solutions to this system of equations. One possible solution is obtained if we arbitrarily choose $u_1 = u_2 = 0$. Then

$$\partial_y u_3 = x \quad \text{and} \quad \partial_x u_3 = y$$

i.e.

$$u_3 = xy + f_1(x) \quad \text{and} \quad u_3 = xy + f_2(y)$$

for arbitrary functions f_1 and f_2 of one variable. Then

$$\partial_x u_3 = y + f_1'(x) = y \quad \text{and} \quad \partial_y u_3 = x + f_2'(y) = x$$

which implies that $f_1(x) = f_2(y)$ = constant. Then $\underline{U}$ = (xy + C)k is one vector field whose curl equals $\underline{V}$.

A second choice for $\underline{U}$ is obtained if we arbitrarily choose $u_3 = 0$. Then

$$\partial_z u_2 = -x \quad \text{leads to } u_2 = -xz + C_1$$

and

$$\partial_z u_1 = -y \text{ leads to } u_1 = -yz + C_2$$

Thus $\underline{U} = -yz\underline{i} - xz\,\underline{j}$. It should not be surprising that there are many choices for $\underline{U}$ since if $\underline{U}$ is any vector field such that curl $\underline{U} = \underline{V}$ and F is a smooth scalar field then $\Omega = \underline{U} + \mathbf{grad}\, F$ satisfies, $\text{curl}W = \text{curl}(\underline{U} + \nabla F) = \text{curl}\, \underline{U} = \underline{V}$.

PROBLEM 11.26

Let Ω denote a fundamental domain in $\mathbb{R}^3$ with smooth boundary S and suppose the vector field $\underline{V}$ is $\mathbb{C}^1$ in Ω. Then show that

$$\iint_S \underline{V} \cdot \underline{N}\, dA = \iiint_\Omega \text{div}\, \underline{V}\, dx\, dy\, dz \tag{1}$$

where $\underline{N}$ denotes the outward unit normal to S.

SOLUTION 11.26

The equation (1) is equivalent to

$$\iint_S v_1 dy\, dz + v_2 dx\, dz + v_3 dx\, dy = \iiint_\Omega (\partial_x v_1 + \partial_y v_2 + \partial_z v_3) dx\, dy\, dz$$

Consider just the one term

$$\iiint_\Omega \partial_z v_3 dx\, dy\, dz = I_z$$

Since we have assumed that Ω is fundamental, there exists a fundamental domain U_{xy} in $\mathbb{R}^2$ and functions $f = f(x, y)$, $g = g(x, y)$ in $\mathbb{C}(U_{xy})$ such that

$$\Omega = \{(x, y, z): f(x, y) \le z \le g(x, y), (x, y) \in U_{xy}\}.$$

Then

$$\iiint_\Omega \partial_z v_3 dx\, dy\, dz = \iint_{U_{xy}} \int_{f(x,y)}^{g(x,y)} \partial_z v_3 dz\, dx\, dy$$

$$= \iint_{U_{xy}} (v_3(x, y, g(x, y)) - v_3(x, y, f(x, y)))\, dx\, dy$$

$$= \iint_{U_{xy}} v_3(x, y, g(x, y)) dx\, dy - \iint_{U_{xy}} v_3(x, y, f(x, y))\, dx\, dy$$

But

$$\iint_{U_{xy}} v_3(x, y, g(x, y)) dx\, dy = \iint_{U_{xy}} v_3(x, y, g(x, y)) \text{Cos}\, \varphi\, \text{Sec}\, \varphi\, dx\, dy$$

$$= \iint_{S_1} v_3 \text{Cos}\, \varphi\, dA$$

and

$$-\iint_{U_{xy}} v_3(x, y, f(x, y))\, dx\, dy = \iint_{U_{xy}} v_3(x, y, f(x, y)) \operatorname{Cos} \varphi\, (-\operatorname{Sec} \varphi) dx\, dy$$

$$= \iint_{S_2} v_3 \operatorname{Cos} \varphi\, dA$$

The surface S that bounds Ω consists of two parts. The upper part S_1, given in the form $z = g(x, y)$, has unit outward normal

$$N = \frac{-g_x}{M} \underline{i} + \frac{-g_y}{M} \underline{j} + \frac{1}{M} \underline{k} \quad \text{where } M = \sqrt{1 + g_x{}^2 + g_y{}^2} = \operatorname{Sec} \varphi$$

On the lower part of S, denoted by S_2 and given by $z = f(x, y)$, the vector

$$N = \frac{-f_x}{M} \underline{i} + \frac{-f_y}{M} \underline{j} + \frac{1}{M} \underline{k} \quad \text{where } M = \sqrt{1 + f_x{}^2 + f_y{}^2} = \operatorname{Sec} \varphi$$

points upward and hence is the *inward* normal to S. The vector $-\underline{N}$ is the outward normal and thus $dA = \operatorname{Sec}(\pi - \varphi) dx\, dy = -\operatorname{Sec} \varphi\, dx\, dy$ on S_2. It follows that

$$\iiint_\Omega \partial_z v_3 dx\, dy\, dz = \iint_{S_1} v_3 \operatorname{Cos} \varphi\, dA + \iint_{S_2} v_3 \operatorname{Cos} \varphi\, dA$$

$$= \iint_S v_3 \operatorname{Cos} \varphi\, dA = \iint_S v_3 dx\, dy$$

In the same way, we can show

$$\iiint_\Omega \partial_y v_2 dx\, dy\, dz = \iint_S v_2 \operatorname{Cos} \beta\, dA = \iint_S v_2 dx\, dz$$

and

$$\iiint_\Omega \partial_x v_1 dx dy dz = \iint_S v_1 \operatorname{Cos} \alpha\, dA = \iint_S v_1 dy\, dz$$

This proves Theorem 11.10 in the simplest case that Ω is a fundamental domain. Clearly the theorem extends to any domain Ω that is the union of finitely many nonoverlapping fundamental domains. The theorem may be extended to still more general domains than this by more sophisticated arguments.

PROBLEM 11.27

Let Ω denote a bounded domain in $\mathbb{R}^3$ for which the divergence theorem holds and let F denote a $\mathbb{C}^1$ scalar field in Ω. Then

$$\iint_S F\underline{N}\, dA = \iiint_\Omega \mathbf{grad}\, F\, dx\, dy\, dz \tag{1}$$

where S denotes the boundary of Ω and $\underline{N}$ denotes the outward unit normal to S.

SOLUTION 11.27

Let $\underline{V}$ denote the $\mathbb{C}^1$ vector field defined by $\underline{V} = F\, \underline{C}$ where $\underline{C}$ denotes an arbitrary constant vector. Then we can apply the divergence theorem to $\underline{V}$ to obtain

$$\iint_S \underline{V} \cdot \underline{N}\, dA = \iiint_\Omega \operatorname{div} \underline{V}\, dx\, dy\, dz$$

But

$$\text{div}\underline{V} = \nabla\cdot(F\underline{C}) = \text{grad } F\cdot\underline{C} + F\,\text{div }\underline{C} = \textbf{grad } F\cdot\underline{C}$$

and thus

$$\left(\iint_S F\underline{N}\,dA\right)\underline{C} = \left(\iiint_\Omega \textbf{grad } F\,dx\,dy\,dz\right)\underline{C} \tag{2}$$

Since the vector $\underline{C}$ is arbitrary, (2) implies (1).

PROBLEM 11.28

Let S denote a smooth bounded but not closed surface in $\mathbb{R}^3$ with positively oriented boundary G. Suppose S is described parametrically by

$$x = x(s, t),\ y = y(s, t),\ z = z(s, t) \quad (s, t) \in A$$

i.e.,

$$\underline{R}(s, t) = x(s, t)\underline{i} + y(s, t)\underline{j} + z(s, t)\underline{k}$$

is the position vector for a typical point on S. Then show that

$$\partial/\partial s = \partial_s\underline{R}\cdot\nabla \text{ and } \partial/\partial t = \partial_t\underline{R}\cdot\nabla \tag{1}$$

$$(\partial_s\underline{R}\times\partial_t\underline{R})\times\nabla = \partial_t\underline{R}(\partial_s\underline{R}\cdot\nabla) - \partial_s\underline{R}(\partial_t\underline{R}\cdot\nabla) \tag{2}$$

SOLUTION 11.28

For a smooth scalar field f(x, y, z) defined on S, we have by the chain rule

$$\begin{aligned}\partial_s f &= \partial_x f\,\partial_s x + \partial_y f\,\partial_s y + \partial_z f\,\partial_s z\\ &= (\partial_s x\underline{i} + \partial_s y\underline{j} + \partial_s z\underline{k})\cdot\nabla f = (\partial_s\underline{R}\cdot\nabla)f.\end{aligned}$$

Similarly, $\partial_t f = (\partial_t\underline{R}\cdot\nabla)f$ and since f is any smooth scalar field, (1) follows. By the vector identity (9.4)(b), we have

$$(\partial_s\underline{R}\times\partial_t\underline{R})\times\nabla = (\partial_s\underline{R}(\partial_t\underline{R}\cdot\nabla) - \partial_t\underline{R}(\partial_s\underline{R}\cdot\nabla)$$

which is (2).

PROBLEM 11.29

Let S denote a smooth bounded but not closed surface in $\mathbb{R}^3$ with positively oriented boundary Γ and let $\underline{V}$ denote a $\mathbb{C}^1$ vector field on S. Then show that

$$\int_\Gamma \underline{V}\cdot\underline{T}\,ds = \iint_S \text{curl }\underline{V}\cdot\underline{N}\,dA$$

where $\underline{N}$ denotes the unit normal to S.

SOLUTION 11.29

Let S be parameterized as in the previous problem. Then as s and t range over the set A in the uv-plane, (x(s, t), y(s, t), z(s, t)) ranges over S. Also as s and t range over the simple closed curve C that is the boundary of A, (x(s, t), y(s, t), z(s, t)) ranges over Γ. Then

$$\int_\Gamma \underline{V}\cdot\underline{T}\,ds = \int_\Gamma \underline{V}\cdot \underline{dR} = \int_C \underline{V}\cdot(\partial_s\underline{R}\,ds + \partial_t\underline{R}\,dt)$$

$$= \int_C (\underline{V}\cdot\partial_s\underline{R}\,ds + \underline{V}\cdot\partial_t\underline{R}\,dt)$$

We apply Green's theorem to this last integral to obtain

$$\int_\Gamma \underline{V}\cdot\underline{T}\,ds = \iint_\Sigma (\partial_s(\underline{V}\cdot\partial_t\underline{R}) - \partial_t(\underline{V}\cdot\partial_s\underline{R}))ds\,dt.$$

Then

$$\int_\Gamma \underline{V}\cdot\underline{T}\,ds = \iint_\Sigma (\partial_s(\underline{V}\cdot\partial_t\underline{R} + \underline{V}\cdot\partial_s\partial_t\underline{R} - \partial_t\underline{V}\cdot\partial_s\underline{R} - \underline{V}\cdot\partial_t\partial_s\underline{R})\,ds\,dt$$

$$= \iint_\Sigma (\partial_s\underline{V}\cdot\partial_t\underline{R} - \partial_t\underline{V}\cdot\partial_s\underline{R})\,ds\,dt$$

$$= \iint_\Sigma (\partial_t\underline{R}\,\partial_s - \partial_s\underline{R}\,\partial_t)\cdot\underline{V}\,ds\,dt$$

$$\int_\Gamma \underline{V}\cdot\underline{T}\,ds = \iint_\Sigma (\partial_t\underline{R}(\partial_s\underline{R}\cdot\nabla) - \partial_s\underline{R}(\partial_t\underline{R}\cdot\nabla))\cdot\underline{V}\,ds\,dt$$

where in the last step we used the identity (1) from the previous problem. We use also the other identity from the previous problem to obtain

$$\int_\Gamma \underline{V}\cdot\underline{T}\,ds = \iint_\Sigma ((\partial_s\underline{R}\times\partial_t\underline{R})\times\nabla)\cdot\underline{V}\,ds\,dt$$

$$= \iint_\Sigma (\partial_s\underline{R}\times\partial_t\underline{R})\cdot\nabla\times\underline{V}\,ds\,dt$$

$$= \iint_\Sigma \text{curl}\,\underline{V}\cdot(\partial_s\underline{R}\times\partial_t\underline{R})\,ds\,dt = \iint_S \text{curl}\underline{V}\cdot\underline{N}\,dA.$$

This proves Stokes' theorem.

PROBLEM 11.30

Let C denote a simple closed curve lying in the plane whose (unit) normal vector is $N = a\underline{i} + b\underline{j} + c\underline{k}$. Then show that the area enclosed by C is equal to

$$A = \frac{1}{2}\int_S (bz - cy)dx + (cx - az)dy + (ay - bx)dz$$

To what does this expression reduce if C lies in the xy-plane?

SOLUTION 11.30

Write

$$A = \frac{1}{2}\int_S (bz - cy)dx + (cx - az)dy + (ay - bx)dz$$

$$= \int_C \underline{V}\cdot\underline{T}\,ds$$

where

$$\underline{V} = \frac{1}{2}((bz - cy)\underline{i} + (cx - az)\underline{j} + (ay - bx)\underline{k})$$

Then by Stokes' theorem

$$\int_C \underline{V} \cdot \underline{T}\, ds = \iint_S \text{curl}\underline{V} \cdot \underline{N}\, dA$$

But curl$\underline{V}$ = a i + bj + ck = N and thus curl$\underline{V} \cdot \underline{N} = \underline{N} \cdot \underline{N} = 1$. Then

$$\int_C \underline{V} \cdot \underline{T}\, ds = \iint_S dA = \text{Area enclosed by C}$$

If C lies in the xy-plane then $\underline{N}$ = k (a = b = 0, c = 1) so

$$A = \frac{1}{2}\int_C x\, dy - y\, dx$$

If C is a piecewise smooth curve in $\mathbb{R}^3$ described parametrically by {x = x(t), y = y(t), z = z(t):a ≤ t ≤ b} then the length of C is equal to

$$L = \int_C 1\, ds = \int_a^b \sqrt{x'(t)^2 + y'(t)^2 + z'(t)^2}\, dt$$

In addition for F(x, y, z) a smooth scalar field, each of the following line integrals exists,

$$\int_C F dx = \int_a^b F(x(t), y(t), z(t))x'(t)dt$$

$$\int_C F dy = \int_a^b F(x(t), y(t), z(t))y'(t)dt$$

$$\int_C F dz = \int_a^b F(x(t), y(t), z(t))z'(t)dt$$

$$\int_C F ds = \int_a^b F(x(t), y(t), z(t))\sqrt{x'(t)^2 + y'(t)^2 + z'(t)^2}\, dt$$

For $\underline{V}$(x, y, z) a smooth vector field, the line integral of $\underline{V}$ along C is equal to

$$\int_C \underline{V} \cdot \underline{dr} = \int_C v_1 dx + v_2 dy + v_3 dz$$

$$= \int_a^b \underline{V}(x(t), y(t), z(t)) \cdot \underline{P}'(t)dt = \int_C \underline{V} \cdot \underline{T}\, ds$$

Green's theorem is an integral identity applying to line integrals in the plane. For Ω a closed bounded domain in $\mathbb{R}^2$ having a piecewise smooth boundary Γ with the usual orientation

$$\int_\Gamma v_1(x, y)dx + v_2(x, y)dy = \iint_\Omega (\partial_x v_2(x, y) - \partial_y v_1(x, y))\, dx\, dy$$

For $\underline{V}$(x, y, z) a continuous vector field in domain Ω in $\mathbb{R}^3$ the following properties are equivalent:

1. *$\underline{V}$ is path independent in Ω*

2. *$\underline{V}$ is conservative in Ω*
3. *$\underline{V}$ is a potential field in Ω*

If $\underline{V}$ is C^1 and Ω is simply connected then these conditions are equivalent to

4. *$\underline{V}$ is irrotational*

For W a simply connected domain in $\mathbb{R}^3$ and $\underline{V} \in C^1(\Omega)$

$$\text{curl}\underline{V} = 0 \textit{ in } \Omega \textit{ if and only if } \underline{V} = \textbf{grad}\, F \textit{ for } F \in \mathbb{C}^2(\Omega)$$

$$\text{div}\, \underline{V} = 0 \textit{ in } \Omega \textit{ if and only if } \underline{V} = \text{curl}\, G \textit{ for } G \in \mathbb{C}^2(\Omega)$$

A surface S in $\mathbb{R}^3$ may be described explicitly, implicitly or parametrically. The following are parametric descriptions of some common surfaces:

Sphere of radius R centered at the origin: $\varphi \in [0, \pi], \vartheta \in [0, 2\pi)$

$x = R \operatorname{Sin} \varphi \operatorname{Cos} \vartheta,\ y = R \operatorname{Sin} \varphi \operatorname{Sin} \vartheta,\ z = R \operatorname{Cos} \varphi,$

Torus of radius a with cross sectional radius of b: $\varphi, \vartheta \in [0, 2\pi)$

$x = (a + b \operatorname{Cos} \varphi)\operatorname{Cos} \vartheta,\ y = (a + b \operatorname{Cos} \varphi)\operatorname{Sin} \vartheta,\ z = b\operatorname{Sin} \varphi$

Cone with vertex angle of α: $\vartheta \in [0, 2\pi),\ s \in [0, h]$

$x = s \operatorname{Sin} \alpha \operatorname{Cos} \vartheta,\ y = s \operatorname{Sin} \alpha \operatorname{Sin} \vartheta,\ z = s \operatorname{Cos} \alpha$

Paraboloid: $\vartheta \in [0, 2\pi),\ s \in [0, h]$

$x = as \operatorname{Cos} \vartheta,\ y = as \operatorname{Sin} \vartheta,\ z = s$

Surface of revolution: $\vartheta \in [0, 2\pi),\ s \in [a, b]\ f \in \mathbb{C}[a, b]$

$x = s \operatorname{Cos} \vartheta,\ y = s \operatorname{Sin} \vartheta,\ z = f(s)$

If S is a smooth surface in $\mathbb{R}^3$ then if S is described parametrically in terms of one of the variables x, y or z then the surface integral of the continuous scalar field F(x, y, z) is given by one of the following

$$\iint_S F\, dA = \iint_{U_{xy}} F(x, y, f(x, y)) \sqrt{f_x^{\,2} + f_y^{\,2} + 1}\, dx\, dy$$

$$= \iint_{V_{xy}} F(x, g(x, z), z)\sqrt{g_x^{\,2} + g_z^{\,2} + 1}\, dx\, dz$$

$$= \iint_{W_{xy}} F(x(y, z), y, z)\sqrt{h_y^{\,2} + h_z^{\,2} + 1}\, dy\, dz$$

The surface integral of the continuous vector field $\underline{V}$ is defined to be

$$\iint_S \underline{V} \cdot \underline{N}\, dA = \iint_S v_1 dy\, dz + v_2 dx\, dz + v_3 dx\, dy$$

$$= \iint_S (v_1 \operatorname{Cos} \alpha + v_2 \operatorname{Cos} \beta + v_3 \operatorname{Cos} \varphi)\, dA$$

where $\underline{N} = \operatorname{Cos} \alpha\, \underline{i} + \operatorname{Cos} \beta\, \underline{j} + \operatorname{Cos} \varphi\, \underline{k}$ is the unit normal vector to S, with

$$\text{Sec}\,\alpha = \sqrt{h_y^2 + h_z^2 + 1}$$

$$\text{Sec}\,\beta = \sqrt{g_x^2 + g_z^2 + 1}$$

$$\text{Sec}\,\varphi = \sqrt{f_x^2 + f_y^2 + 1}$$

If S is given parametrically in the form

$$\{x = x(s, t), y = y(s, t), z = z(s, t): (s, t) \in R \textit{ in } \mathbb{R}^2\}$$

then

$$\iint_S \underline{V} \cdot \underline{N}\, dA = \iint_{R_{st}} \left(v_1 \frac{\partial(y, z)}{\partial(s, t)} + v_2 \frac{\partial(z, x)}{\partial(s, t)} + v_3 \frac{\partial(x, y)}{\partial(s, t)} \right) ds\, dt$$

For Ω a closed bounded domain in $\mathbb{R}^3$ with piecewise smooth boundary S we have the following identities:

if $\underline{V} \in \mathbb{C}^1(\Omega)$ $\iint_S \underline{V} \cdot \underline{N}\, dA = \iiint_\Omega \text{div}\underline{V}\, dx\, dy\, dz$ *(Gauss' Theorem)*

if $\underline{F} \in \mathbb{C}1(\Omega)$ $\iint_S F \cdot \underline{N}\, dA = \iiint_\Omega \textbf{grad}\, F\, dx\, dy\, dz$

if $\varphi, \psi \in \mathbb{C}_1(\Omega)$ $\iiint_\Omega \nabla^2\varphi\, dx\, dy\, dz = \iint_S \nabla\varphi \cdot \underline{N}\, dA$

and

$$\iiint_\Omega \psi\nabla^2\varphi\, dx\, dy\, dz = \iint_S \nabla\varphi \cdot \underline{N}\, dA - \iiint_\Omega \nabla\varphi \cdot \nabla\psi\, dx\, dy\, dz$$

$$\iiint_\Omega (\psi\nabla^2\varphi - \varphi\nabla^2\psi)\, dx\, dy\, dz = \iint_S (\psi\nabla\varphi \cdot \underline{N} - \varphi\nabla\psi \cdot \underline{N})\, dA$$

For S a smooth bounded but not closed surface with oriented boundary Γ

if $\underline{V} \in \mathbb{C}^1(\Omega)$ $\int_\Gamma \underline{V} \cdot \underline{T}\, ds = \iint_S \text{curl}\, \underline{V} \cdot \underline{N}\, dA$ *(Stokes' Theorem)*

Combining these integral identities with the equivalent properties of vector fields stated earlier, we can show that for a simply connected domain Ω in $\mathbb{R}^3$ the following are equivalent conditions on a vector field $V \in C^1(\Omega)$:

1. *curl* $V = 0$ *in* Ω
2. $V =$ ***grad*** F *in* Ω *for some* $F \in C^2(\Omega)$
3. $\int_C V \cdot T\, ds = 0$ *for every simple closed curve C in* Ω

and the following are equivalent conditions on a vector field $V \in C^1(\Omega)$:

1. *div* $\underline{V} = 0$ *in* Ω
2. $\underline{V} =$ *curl* $\underline{G}$ *in* Ω *for some* $G \in C^2(\Omega)$
3. $\iint_S \underline{V} \cdot \underline{N}\, dA = 0$ *for every closed surface S in* Ω.

Index